Building Contract Dictionary

Also of interest

The JCT Intermediate Form of Contract
Second Edition
David Chappell & Vincent Powell-Smith
0-632-03965-5

The JCT Design and Build Form
Second Edition
David Chappell & Vincent Powell-Smith
0-632-04899-9

The JCT Minor Works Form of Contract
Second Edition
David Chappell & Vincent Powell-Smith
0-632-03967-1

Powell-Smith & Sims' Building Contract Claims
Third Edition
David Chappell
0-632-03646-X

Parris's Standard Form of Building Contract
Third Edition
David Chappell
0-632-02195-0

Building Contract Dictionary

Third Edition

David Chappell
Derek Marshall
Vincent Powell-Smith
Simon Cavender

Foreword by His Honour Judge Peter Bowsher QC

Blackwell
Science

Blackwell Science Ltd, a Blackwell Publishing company
Editorial offices:
Blackwell Science Ltd, 9600 Garsington Road, Oxford OX4 2DQ, UK
  Tel: +44 (0) 1865 776868
Blackwell Publishing Inc., 350 Main Street, Malden, MA 02148-5020, USA
  Tel: +1 781 388 8250
Blackwell Science Asia Pty, 550 Swanston Street, Carlton, Victoria 3053, Australia
  Tel: +61 (0)3 8359 1011

First edition published by The Architectural Press 1985
Second edition published by Legal Studies & Services (Publishing) Ltd 1990
Third edition published by Blackwell Science 2001
Reprinted 2004

Library of Congress Cataloging-in-Publication Data
Building contract dictionary/David Chappell ... [et al.].–3rd ed.
    p. cm.
  Rev. ed. of: Building contract dictionary/Vincent Powell-Smith and
  David Chappell. 1985.
  Includes bibliographic references.
  ISBN 0-632-03964-7
  1. Contruction contracts–Great Britain–Dictionaries. I. Chappell, David.
  II. Powell-Smith, Vincent. Building contract dictionary.

KD1641.A68 P69 2001
343.41'078624'03–dc21   2001035006

ISBN 0-632-03964-7

A catalogue record for this title is available from the British Library

Set in Times and produced by Gray Publishing, Tunbridge Wells, Kent
Printed and bound in India using acid-free paper
by Thomson Press (I) Ltd, India

For further information on Blackwell Publishing, visit our website:
www.blackwellpublishing.com

PREFACE

It is just over ten years since the second edition of this book was published. It was conceived as a handy reference for architects, quantity surveyors and other construction professionals as well as for contractors and their staff. Since the first edition in 1985, it has been gratifying to learn that a number of construction lawyers have also found the book useful. It is hoped it will also provide a useful reference for civil engineers, even though it does not address engineering contracts specifically, It is, as the title suggests, a *dictionary* and nothing more. Our treatment is not exhaustive and we do not claim that the definitions are authoritative. The book is a *vade mecum* and not a legal textbook — there are a good many of those and we have included a selected list for further reading.

A broad view has been taken of the words and phrases to be included so that, although they are not all purely contractual, they are all likely to be encountered in connection with building contracts. With a few exceptions, we decided against including definitions of Latin terms, because good legal dictionaries are readily available and we wish to keep this book to a manageable size. The selection of words, phrases and concepts for inclusion is our own, but we have valued the suggestions of many practitioners.

Partly as a result of the suggestions we have received and partly due to the many changes in law, legislation and building contracts over the last ten years, the book has substantially changed and, in terms of numbers of entries and coverage, enlarged. Some of the lengthy tables in previous editions have been removed leaving just a few tables and other illustrations which seem to be really useful. All the original entries have been reconsidered and updated in the light of case law and legislation, in particular the Housing Grants, Construction and Regeneration Act 1996, the Contracts (Rights of Third Parties) Act 1999, the Late Payment of Commercial Debts (Interest) Act 1998, the Human Rights Act 1998, the Construction (Design and Management) Regulations 1994, the Scheme for Construction Contracts (England and Wales) Regulations 1998 and the Civil Procedure Rules. We have referred to a wide range of contracts including JCT 98, IFC 98, MW 98, WCD 98, PCC 98, MC 98, ACA 3, GC/Works/1 (1998), NEC, NSC/C, DOM/1, DOM/2.

Any book of this kind will omit words that should be included and vice versa. We will be glad to receive, care of the publishers, any suggestions for inclusions or deletions for incorporation in a future edition.

Where terms within a definition have their own entry, this has been indicated as 'qv'. Related terms and their definitions have been listed at the end of the main entry as 'See also'.

It need hardly be said that we are indebted to the authors of the leading standard textbooks, and there are a number of other people to whom we owe special thanks. First to His Honour Judge Peter Bowsher QC for writing a foreword. Next to the late Professor Vincent Powell-Smith who had the idea for this book and who played a major part in laying down its solid foundation. We are also grateful to Blackwell

Science for undertaking the third edition and to Jane Oldfield and the staff of the RIBA Information Unit for digging out useful facts. We owe a special debt of thanks to Anthony Speaight QC who kindly read through the text and made many helpful suggestions for its improvement. We alone, however, take responsibility for the text in its finished state.

We are grateful to RIBA Publications for permission to reproduce some of their standard forms. The front page of the Housing Grants, Construction and Regeneration Act 1996 is reproduced under the terms of Crown Copyright Policy Guidance issued by HMSO.

The text is corrected, so far as we are aware, until 31 December 2000, but a few later developments have been noted at proof stage.

David Chappell, *Tadcaster*
Derek Marshall, *Tadcaster*
Simon Cavender, *London*

FOREWORD

by His Honour Judge Peter Bowsher QC

When starting to read law as an undergraduate, the first law book I bought was *Wharton's Law Lexicon*, a book I still have and use. On appointment as an Official Referee, I bought the Penguin dictionaries of *Building*, *Civil Engineering*, *Architecture* and *Electronics*. If I had known that this book was then in existence in its first edition, I would have bought that also. It would have been an enormous help to me.

I congratulate the authors and warmly recommend their work.

Peter Bowsher
St Dunstan's House
Fetter Lane
London

ABBREVIATIONS

ACA 3	Association of Consultant Architects Form of Building Agreement 1998
ARB	Architects Registration Board
ARCUK	Architects Registration Council
BPF	British Property Federation
CD 81	JCT Standard Form of Building Contract With Contractor's Design 1981
CDM Regulations	Construction (Design and Management) Regulations 1994
CE/99	RIBA Conditions of Engagement for the Appointment of an Architect
CIC	Construction Industry Council
CIMAR	Construction Industry Model Arbitration Rules
CIS	Construction Industry Scheme
CPR	Civil Procedure Rules
EDI	Electronic data interchange
DOM/1 and DOM/2	Standard Form of Sub-Contract for Domestic Sub-Contractors
FIDIC	Federation Internationale des Ingenieurs-Conscils
GC/Works/1 (1998)	General Conditions of Government Contracts for Building and Civil Engineering Works 1998
GMP	Guaranteed maximum price
ICE	Institution of Civil Engineers
IFC 84	JCT Intermediate Form of Building Contract 1984
IFC 98	JCT Intermediate Form of Building Contract 1998
IN/SC	IFC 84 and 98 Domestic Sub-Contract
JCT	Joint Contracts Tribunal
JCT 63	JCT Standard Form of Building Contract 1963
JCT 80	JCT Standard Form of Building Contract 1980
JCT 98	JCT Standard Form of Building Contract 1998
JCT 87	JCT Management Contract 1987
LLP	Limited liability partnership
MC 98	JCT Management Contract 1998
MW 80	JCT Agreement for Minor Building Works 1980
MW 98	JCT Agreement for Minor Building Works 1998
NAM/A	IFC 84 and 98 Named Sub-Contractor Articles of Agreement
NAM/SC	IFC 84 and 98 Named Sub-Contractor Conditions
NAM/T	IFC 84 and 98 Named Sub-Contractor Tender
NEC	Engineering and Construction Contract (formerly the New Engineering Contract)
NHBC	National House Building Council

NSC/A	JCT Standard Form of Nominated Sub-Contract Articles of Agreement
NSC/C	JCT Standard Form of Nominated Sub-Contract Conditions
NSC/N	JCT Standard Form for Nomination Instruction for a Sub-Contractor
NSC/T	JCT Standard Form of Nominated Sub-Contract Tender
NSC/W	JCT Standard Form of Employer/Nominated Sub-Contractor Warranty
PC	Prime cost
PCC 92	JCT Standard Form of Prime Cost Contract 1992
PCC 98	JCT Standard Form of Prime Cost Contract 1998
qv, qvv	*quod vide* – indicates a term that has a definition of its own
PFI	Private Finance Initiative
PM	Project manager
RIBA	Royal Institute of British Architects
SFA/92 and SFA/99	RIBA Standard Form for the Appointment of an Architect
SMM7	Standard Method of Measurement of Building Works, 7th Edition
SW/99	RIBA Small Works Agreement
TCC	Technology and Construction Court
WCD 98	JCT Standard Form of Building Contract With Contractor's Design 1998
Works Contract/2	Works Contract Conditions

Abandonment of work A phrase used in the arbitration provisions of GC/Works/1 (1998) (clause 60 (2) (a)). Completion or abandonment of the work marks the point at which any reference to arbitration may be opened. Abandonment of the works must entail complete stoppage of all the works and the clear intention not to continue at some future date. It implies removal of all the contractor's men and sub-contractors from the site and may be construed as an intention to repudiate the contract.
See also: **Repudiation.**

Abatement The term 'abatement of action' refers to the interruption of legal proceedings following an application, usually by the defendant, stating reasons why the proceedings should not continue. The most common instance in the construction industry is probably the application of the limitation period (see: **Limitation of actions**), but it could be an objection to the form or place of the claimant's (qv) claim.

Abatement in relation to nuisance (qv) refers to the right of the person who suffers injury or damage by reason of the nuisance to act personally to remove the cause. Care must be taken not to interfere with another party's rights and, in any case, abatement of nuisance is not looked upon with favour by the courts, unless there is an emergency, because other remedies are available by application to the courts. Local authorities may serve abatement notices in respect of statutory nuisances.

In the construction industry, the word is most commonly used to refer to the process of reducing a price or value, e.g. when a valuation is reduced to take account of the fact that some work is not properly executed. Abatement of price is often confused with deduction of money or set-off (qv) or counterclaim (qv) from the price. There is a very clear difference in law between abatement of price and set-off against a price and although the end result in money terms may be exactly the same, there are circumstances in which one will be allowable and not the other[1].

Abeyance Technically, where a right is not presently vested in anyone, and in this sense of no importance in building contracts. Generally, when something is said to be 'in abeyance' what is meant is that it is in a state of being suspended or temporarily put aside.

Abrogate To repeal or annul, and hence *abrogation* which refers to the annulling or repealing of a law by legislation.

Absolute Full, complete and unconditional. It is possible to have an absolute duty of care. Absolute liability (sometimes known as 'strict liability' (qv)) is liability

---

[1] *Mellows Archital Ltd* v. *Bell Projects Ltd* (1997) 87 BLR 26.

irrespective of the degree of care taken. No proof of negligence or default is required. It is sufficient only that a particular incident has occurred. This type of liability may be imposed by statute (qv).
See also: **Liability; Strict liability.**

Absolute assignment The assignment (qv) or transfer of an entire debt (qv) or other legal right, as opposed to merely part of it, and without any conditions attached.

Abstract of Particulars The phrase used in GC/Works/1 (1998) to refer to the supplement which contains important terms and details which, in other forms of contract, are usually set out in an Appendix (qv). It lists modifications to the printed conditions, gives the date for completion, the amount of liquidated damages and the length of the maintenance period (qv). It also names the 'Employer' and the 'Project Manager (PM)' (qv). Two addenda set out dates after acceptance for the provision of certain information which is relevant in the case of a disruption claim and the length of time for any sub-contract nominations.

Abut In physical contact with. There must be actual contact between part of the premises and the road or other feature which will produce some measurable frontage.

ACA Form of Building Agreement The ACA Form of Building Agreement was first published by the Association of Consultant Architects in October 1982. The second edition was published in September 1984. The latest revision was carried out in 1998 to take account of the Housing Grants, Construction and Regeneration Act 1996 (qv) although it should be noted that there is no express provision in the contract to take account of the contractor's suspension rights for failure to pay under s. 112 of the Act. The contractor is left to apply the legislation directly.

By providing alternatives in a number of key clauses, the standard terms allow a variety of contractual arrangements. The employer, in conjunction with the architect, will decide which of the alternatives is to apply.

The contractor's basic obligation (clause 1.1) is to 'execute and complete the works in strict accordance with the contract documents', i.e. (contract drawings (qv), the time schedule (qv), either a schedule of rates or bills of quantities/schedule of activities (qvv), and, optionally, a specification (qv)). The contractor must 'comply with and adhere strictly to the Architect's instructions' issued under the agreement and is entitled to payment for compliance unless the matter is already covered by the contract sum or results from his default.

Clause 2, covering contract documentation, gives two alternatives. Alternative 1 is traditional and requires the architect to issue further information. Alternative 2 applies where the contractor undertakes to supply further information. It must be submitted to the architect for comment.

The time schedule sets out important stages of the job and provides a list for the insertion of key dates. Where he has undertaken to provide additional drawings, etc. the contractor warrants under clause 3.1 that:

— The works will comply with any performance specification or requirement contained in the contract documents (qv).
— Any part of the works to be designed by him will be fit for its required purpose.

If he is responsible for the design in whole or in part, clause 6.6 requires him to take out professional indemnity insurances in respect of his negligence or that of his sub-contractors, suppliers, etc.

Clause 1.1.1 states that, once possession of the site (or appropriate part) is given by the employer, the contractor shall then immediately begin the works and proceed 'regularly and diligently' (qv) and in accordance with the time schedule so that the works are completed 'fit and ready for taking-over by the employer' by the due or extended date. There is provision for general damages as an alternative to liquidated damages for delay. In both cases the architect's clause 1.1.2 certificate of delay must be issued before deduction.

Extensions of time are dealt with under clause 11.5 which provides alternative criteria. Alternative 1 limits the grounds to 'any act, instruction, default or omission of the employer or of the architect on his behalf' whether authorised by the agreement or not. Alternative 2 is more traditional and lists such things as *force majeure* (qv) and insurance contingencies.

Clause 11.7 provides for a mandatory review of extensions of time granted by the architect. Clause 11.8 gives the architect power to order acceleration or postponement. The architect's decisions are reviewable on arbitration (if the arbitration option applies) or by the adjudicator. The time schedule must be revised (clause 11.9) if the contract period is extended or an acceleration or a postponement instruction is issued.

The contractor is responsible for his sub-contractors and suppliers (clause 9.9), but the architect's consent must be obtained to subletting (clause 9.2). Provision is made in clauses 9.4 and 9.5 for sub-contractors to be named either in the contract documents or by way of an architect's instruction regarding provisional sums.

Clause 5 is important. It provides for the contractor to ensure proper management of the works, appoint a site manager and employ only appropriately skilled and qualified people on the works. This duty is backed up by the sanction that the architect (clause 8.1 (b)) may require the dismissal from the works of any incompetent person.

Architect's instructions are dealt with by clause 8. Certain instructions can be issued at any time up to completion of all the contractor's obligations. Procedures for valuation are covered in clause 17 which requires the contractor to submit written estimates of the time and the amount of any loss or expense. Work not forming part of the contract may be carried out by the employer's own contractors subject to certain provisos.

The claims clause (clause 7) is very broad, dealing with any act, omission, default or negligence of the employer or the architect which disrupts the regular progress of the whole or part of the works. Loss or expense resulting from architect's instructions is excepted, being dealt with under clause 17.

The scheme of certificates and payments requires the contractor to submit interim applications, with supporting documents, on the last working day of each month up to and including the month in which taking-over occurs and thereafter, as and when further amounts become due either to the contractor or to the employer (clause 16.1). The architect is to issue his certificate within 10 working days of the contractor's application. Payment of 95% of the amount stated as due must be made by the employer within a further 10 working days. There is an alternative (B) for stage payments. Failure to pay is a ground for termination under clause 20.2 (a). Final payment is governed by clause 19. The contractor must submit his final account with vouchers within 60 working days after the end of the maintenance period (qv) and the architect must issue his final certificate (qv) within 60 working days after the contractor has completed all his contractual obligations. No certificate relieves the contractor of any liability under the contract.

A special feature is the optional provision for disputes to be settled by a named conciliator. There is also adjudication (qv), arbitration (qv) and litigation (qv). If litigation is adopted, the contract purports to confer on the courts 'full power to open up, review and revise' the architect's opinions etc., but in light of *Beaufort Developments (NI) Ltd* v. *Gilbert-Ash NI Ltd* (1998)[2] it is not now necessary.

A special edition of this contract (BPF edition of the ACA Form of Building Agreement) is also available, and is adapted for use with the British Property Federation system of contracting (see: **BPF System**). It contains some minor differences, such as fewer alternative clauses and the use of the term 'client's representative' (qv) rather than 'architect'.

Some of the innovatory features of ACA 3 have been incorporated in GC/Works/1 (1998) (qv).

Acceleration of work
Under the general law, the architect has no power to instruct the contractor to accelerate work. The contractor's obligation is to complete the work within the time specified, or where no particular contract period is specified, within a reasonable time (qv). The contractor cannot be compelled to complete earlier than the agreed date unless there is an express contract term authorising the architect to require acceleration.

ACA 3, clause 11.8 empowers the architect to issue an instruction to bring forward dates shown on the time schedule (qv) for the taking-over (qv) of any part of the works, but this power may not be exercised unreasonably and an appropriate adjustment must be made to the contract sum.

In other cases, if the employer wishes the work to be completed earlier (or more usually to be completed on time despite unavoidable delays) a special agreement must be negotiated, and will generally involve extra payment.

---

[2](1998) 88 BLR 1.

Architects sometimes believe that provisions such as JCT 98, clause 25.3.4.2, and IFC 98, clause 2.3 give them the power to instruct acceleration measures because such clauses state that the contractor must 'do all that may reasonably be required' to the satisfaction of the architect to proceed with the works. A clause like this is often erroneously referred to by architect and contractor alike as the 'acceleration clause'. That, however, is not its true function nor even part of its function. It does not empower the ordering of acceleration. It is there to ensure that the contractor proceeds with the work diligently (qv), taking notice of the architect's wishes but not to an extent involving the use of additional resources.

If the architect does issue an instruction to accelerate and the contractor obeys, the legal position is probably that the contractor is in breach of contract and he is not entitled to payment. Much depends upon the authority, whether ostensible or implied, of the architect if it is contended that the instruction is given as agent for the employer. It is possible that the contractor is entitled to reasonable payment on the basis of an implied contract or *quantum meruit* (qv). If the employer has authorised the instruction, the contractor is likely to be able to make a successful claim.

There is a cumbersome acceleration clause in the Management Contract MC 98, which may be contrasted with the manifestly better provision in GC/Works/1. MC 98, clause 3.6 only applies if so stated in the Appendix. The procedure is elaborate. The architect issues a preliminary instruction to the contractor giving details of the employer's acceleration requirements, and the contractor similarly instructs any works contractor who is affected. Both management contractor and any works contractor affected are entitled to make reasonable objection to the preliminary instruction, which must then either be withdrawn or varied to meet the objection. The works contractor must also give notice to the management contractor of the revised time he requires for completion, and either the lump sum he requires or else a statement saying that he wishes the financial consequences to be dealt with under the works contract ascertainment provisions. Once the architect has dealt with any objections and received all the necessary information from each affected works contractor, he issues a formal acceleration instruction.

Under GC/Works/1, clause 38, if the employer wishes to have the works (or a section) completed before the date(s) for completion, it can direct the contractor to submit priced proposals for achieving the accelerated date, together with any consequential amendments to the programme, or to submit an explanation of why the contractor cannot achieve accelerated completion. The employer's direction specifies the period within which this must be done. If the employer accepts the contractor's proposals, he must notify the contractor in writing setting out (a) the accelerated completion date (b) the amendments to the programme (c) the revised contract sum (d) a revised stage payment chart and (e) any other agreed relevant amendment. This is the formal acceleration instruction. The contractor may, of his own volition, submit proposals to the authority for early completion of the works or a section of them; if he does so and his proposals are accepted, a formal acceleration instruction will then be issued.

A contractor may often base a claim for loss and/or expense (qv) on 'acceleration', sometimes referred to as 'constructive acceleration'. The claim usually proceeds on the basis that if the contractor is entitled to an extension of time but the architect refuses to extend the contract period and exhorts the contractor to finish on the completion date, the contractor is entitled to accelerate the work in order to avoid having to pay liquidated damages (qv). There is no contract provision which deals with this situation and the claim is essentially one of damages for the architect's breach of his obligation to properly carry out his duty to extend the contract period. The contractor's problem is causation (qv). His remedy for the architect's breach is adjudication or arbitration. In practice, the contractor may not wish to risk incurring large sums in liquidated damages which he cannot recover later. However, the chances of success are not good and the contractor would have to make out a compelling case probably showing that the architect's refusal was final and the likely liquidated damages would cause the contractor to become insolvent.
See also: **Agency; Postponement.**

Acceptance The act of agreeing to an offer (qv) which constitutes a binding contract. Acceptance may be made in writing, orally or by conduct. Acceptance by conduct would occur if the offeree acted in such a way as to observe the terms of the offer and clearly show that he intended to be bound by it. Acceptance must be unqualified or there is no contract. A qualified acceptance may amount to a counter-offer (qv). Thus, if contractor A offered to build a house for employer B for the sum of £20 000, and B 'accepted' subject to a reduction in price for the omission of the garage, B is said to have made a counter-offer. The original offer is terminated and B cannot later decide to accept it.

Where the offeror has stipulated a way of acceptance, the offer can generally only be accepted in that way. So where postal acceptance is required, an oral acceptance will not usually suffice. It may be possible to accept an offer by an alternative method, depending upon the precise construction (qv) of the offer or on the conduct of the offeror. It is generally advisable, where possible, to comply with any stipulations.

Where an offer is accepted by post it takes effect upon posting, not receipt by the offeror[3]. This applies even where the offeror has posted a withdrawal of his offer which has not been received by the offeree. These so-called postal rules do not apply to modern 'instantaneous' forms of communication, such as telephone or telex[4]. There are no decisions relating to dictated telegrams, faxes or e-mails[5] etc.

If a tender is received by the employer and a letter of acceptance sent, it may be that a binding contract has already come into existence. Where the tender identifies a form of contract, unless there are terms which remain to be agreed

---

[3]*Henthorn* v. *Fraser* [1892] 2 Ch 27, 33.
[4]*Entores Ltd* v. *Miles Far East Corporation* [1955] 2 QB 327.
[5]See also: **Electronic data interchange (EDI).**

or the parties have agreed that no contract exists prior to the signing of the formal documents, it is likely that there will be a binding contract between the parties[6]. This point is often overlooked by architects who, for example, may wrongly refuse to issue certificates until the formal contract documents are signed.

See also: **Letters of intent; Subject to contract.**

Accepted programme A precise term used in clause 11.2 (14) of the NEC (qv) to refer to the programme, if any, identified in the contract data (qv). Alternatively, it is the latest programme accepted by the project manager. Submission, acceptance and revision of a programme are dealt with by clauses 31 and 32.

See also: **Engineering and Construction Contract (NEC); Programme.**

Accepted risks The term used in GC/Works/1 (1998), clauses 1 (1), 19 and 36 to describe the risks which may affect the works but which are outside the contractor's control. Presumably they are termed 'accepted' to indicate that they are accepted by the employer. In any event, that is the effect. In other words, they are the risks accepted by the employer. Clause 1 (1) defines 'accepted risks' as pressure waves caused by the speed of aircraft or other aerial devices; ionising radiations or contamination by radioactivity from any nuclear fuel or from nuclear waste from the combustion of nuclear fuel, radioactive, toxic, explosive or other hazardous properties of any explosive nuclear assembly including any nuclear component and war, invasion, act of foreign enemies (whether or not war has been declared), civil war (qv), rebellion, insurrection (qv) or military or usurped power.

Under clause 19 the contractor must make good or compensate the employer for any loss or damage which arises out of, or is connected with, the execution of the works. If a claim is made or proceedings are brought against the employer in respect of loss or damage, the contractor must reimburse reasonable costs, but the employer must reimburse the contractor's reasonable costs or expenses to the extent that the loss or damage is caused by the employer or its agents' default, accepted risk, unforeseeable ground conditions or other circumstances outside the contractor or his sub-contractor's control. The contractor is entitled to an extension of time under clause 36 (2) (d) for any delay caused by the occurrence of an accepted risk.

Access to neighbouring land The Access to Neighbouring Land Act 1992 was intended to deal with the difficult problem which arises when it is necessary to enter upon a neighbour's land in order to carry out work. Neighbours could be held to ransom where the work was essential to deal with weather ingress or structural problems. The Act applies only to England and Wales and it deals with 'basic preservation works'. The term is broad and it includes, but is not necessarily restricted to, such things as maintenance or repair of a building,

---

[6]*G. Percy Trentham Ltd* v. *Archital Luxfer Ltd* [1993] 1 Lloyds Rep 25 per Steyn LJ at 29–30.

clearance or repair of a drain or cable, treatment or cutting back of any growing thing and the filling in or clearance of a ditch.

An application must be made to the court which must be satisfied that the work is reasonably necessary for preservation and that it cannot be carried out without substantial difficulty unless entry onto the adjoining land is possible. The court cannot make an order if the adjoining owner would suffer interference with use or in enjoyment of the land or if he would suffer hardship. The court may include whatever terms and conditions it deems appropriate to protect the adjoining owner's property or privacy. These terms may include the payment of money to the adjoining owner by the person desiring to carry out the work.

Access to works The contractor has an implied right of access to the works insofar as the access is controlled by the employer, otherwise it would be impossible for him to carry them out.

Under clause 25.4.12 of JCT 98, failure by the employer to give ingress or egress to or from the site is a ground for extension of time. It may also give rise to a money claim under clause 26.2.6. There are several provisos attached:
— The access must be across adjoining or connected land, buildings, way or passage.
— Such land, etc. must be in the possession *and* control of the employer.
— The means of access must have been stated on the drawings or in the bills of quantities (qv).
— The contractor must have given such notice, if any, that he is required to give.

It is not a breach of contract where access is impeded by third parties over whom the employer has no control, e.g. pickets[7]. Similarly, no extension can be awarded or money claim allowed if the employer fails to obtain permission for the contractor to cross a third party's property, though that might well amount to a breach of contract by the employer if he has expressly undertaken to obtain such access. There is also a strange provision in the clauses for extension of time and money if the employer has failed to give such access as the *architect* and the contractor have agreed between them. This seems to be a surprising extension of the architect's power to bind the employer.

ACA 3 makes no specific provision for extension of time or money on the ground that the employer has failed to provide access, but failure to provide agreed access would give rise to such claims on the ground of the employer's default: clause 11.5, both alternatives; clause 7.

The position under GC/Works/1 (1998) is that the employer's failure to give access would give rise to an extension of time under clause 36 (2) (b), but only delay in being given possession of the site would give rise to a prolongation or disruption claim under clause 46 (1) (b). Clause 26 (Site admittance) is not relevant. It merely refers to the power of the PM (project manager) (qv) to refuse admission to such persons as the PM shall think fit.

---

[7]*LRE Engineering Services Ltd* v. *Otto Simon Carves Ltd* (1981) 24 BLR 127.

Accident An unlooked-for mishap or an untoward event neither designed nor expected[8]. Its actual meaning in a contract or elsewhere is a question of interpretation[9]. In general, accident is no defence to an action in tort (qv) and in some cases the happening of an accident may itself give rise to a *prima facie* case of liability. This is known as *res ipsa loquitur* ('the thing speaks for itself') which was explained in *Scott* v. *London & St Katherine's Docks Co* (1865):

> 'Where the thing is shown to be under the management of the defendant or his servants, and the accident, is such as in the ordinary course of things does not happen if those who have the management use proper care, it affords reasonable evidence in the absence of explanation by the defendants, that the accident arose from want of care'[10].

For example, objects do not usually fall from scaffolding unless there is negligence, so if a visitor to site is injured by a bucket falling on his head from scaffolding, the maxim will apply.

See also: **Inevitable accident.**

Accommodation works Works such as bridges, fences, gates, etc. which are carried out and maintained by statutory undertakers (qv), e.g. the Department of the Environment, Transport and the Regions, Railtrack, etc. for the accommodation or convenience of the owners or occupiers of adjoining land. For example, there is a statutory obligation on Railtrack (as successor to the former railway companies) to fence off land used for the railway from adjoining land.

Accord and satisfaction 'The purchase of a release from an obligation whether arising under contract or tort by means of any valuable consideration, not being the actual performance of the obligation itself. The accord is the agreement by which the obligation is discharged. The satisfaction is the consideration which makes the agreement operative'[11]. Accord and satisfaction bars any right of action. If a contractor agrees to accept part payment and to release the employer from payment of the balance, this will be valid if the agreement is supported by fresh consideration (qv) or if the agreement is executed as a deed (qv). There must be true accord, under which the creditor *voluntarily* agrees to accept a lesser sum in satisfaction[12]. The essential point is that the creditor must voluntarily accept something different from that to which he is entitled[13]. Although writing is not legally necessary, it is prudent to arrange that the agreement should be recorded formally in a letter or other document, e.g. if a legal action is being compromised a suitable formula might be 'I accept the sum of £x in full and final settlement of all or any claims . . . and I will forthwith instruct my solicitors to serve notice of discontinuance'. Ultimately, whether there has been an effective agreement is a question of fact.

---

[8] *Fenton* v. *Thorley* [1903] AC 443.
[9] *J. & J. Makin Ltd* v. *London & North Eastern Railway Co* [1943] 1 All ER 645.
[10] (1865) 3 H & C 596 per Erle CJ at 601.
[11] *British Russian Gazette & Trade Outlook Ltd* v. *Associated Newspapers Ltd* [1933] 2 KB 616 at 643.
[12] *D & C Builders Ltd* v. *Rees* [1966] 2 All ER 837.
[13] *Pinnels Case* (1602) 5 Co Rep 117a.

Accrued rights or remedies of either party This is a phrase used in JCT 98, clause 28.4, MC 98, clause 7.11, and IFC 98, clause 7.11 with reference to the rights and duties of the parties following the contractor's determination of his employment under the contract. It does not refer merely to cases where the right or remedy is a claim for breach of contract but also to other rights and remedies, e.g. the architect's right to issue an instruction requiring rectification of defective work[14]. Since under the contract the employer acquires a right to have the defective work remedied at the time it was carried out, this is an 'accrued right' for the purposes of the clause.

See also: **Rights and remedies.**

Acknowledgement of service The formal step by which a defendant responds to the service of a claim form (qv) enclosing (or if served later, upon receipt of[15]) particulars of claim (qv) and records whether he accepts the court's jurisdiction (qv)[16] and his intention to dispute some or all of the claim[17]. The procedure is governed by Part 10 of the Civil Procedure Rules (CPR) (qv). An acknowledgement should be filed at court within 14 days after the service of the claim form or particulars of claim, whichever is later[18]. Alternatively, a defendant may serve a defence (qv)[19]. Where part the claim is admitted and part disputed, a defendant may merely serve an admission and either an acknowledgement or a defence.

Act of God An archaic legal phrase meaning a sudden and inevitable occurrence caused by natural forces. The test is whether or not human foresight and prudence can reasonably recognise its possibility so as to guard against it[20]. Lightning, earthquake (at least in the UK) and very extraordinary weather conditions come within the concept. An Act of God does not in itself excuse contractual performance, but it may do so on the true interpretation of the terms of the contract. Some insurance policies and contracts for the carriage of goods provide that there is no liability for losses caused by Act of God. There appear to be no reported cases involving Act of God in the context of the construction industry, although some contractors may refer to it as an excuse for non-performance or a ground for terminating the contract. What they usually mean is the similar but wider concept of *force majeure* (qv).

See also: **Frustration; Vis major.**

Act of Parliament A statute (qv). It is primary legislation as distinct from statutory instruments and regulations which are secondary legislation. It is the

---

[14]*Lintest Builders Ltd* v. *Roberts* (1980) 10 BLR 120.
[15]CPR Rule 9.1(2).
[16]CPR Rule 10.1(3)(b). To contest the court's jurisdiction, a defendant must follow the procedure set out in CPR Part 11.
[17]Where the defendant admits the claim, he should file an admission complying with CPR Part 14 – CPR Rule 9.2(a).
[18]CPR Rule 10.3(1).
[19]CPR Rule 9.2(b) – the defence must comply with CPR Part 15.
[20]*Greenock Corporation* v. *Caledonian Railway Co* [1917] AC 556.

formal expression of the will of Parliament and sets out the law in written form, e.g. the Housing Grants, Construction and Regeneration Act 1996.

Proposed legislation is introduced in the form of a bill which must pass through all the requisite stages in both Houses of Parliament and then receive the Royal Assent. The majority of modern Acts of Parliament are public general statutes which are of general application. Local Acts (qv) are private statutes of local application.

An Act of Parliament is divided into several parts:
— The *short title* by which the Act is known.
— The *long title* which sets out the purpose of the Act in general terms.
— The *enacting formula* which runs 'Be it enacted by the Queen's most Excellent Majesty, by and with the advice and the consent of the Lords Spiritual and Temporal, and Commons, in this present Parliament assembled, and by the authority of the same, as follows:'
— The *numbered sections* which contain the substance of the Act. Each is divided into sub-sections, paragraphs and subparagraphs as appropriate.
— *The marginal notes* to each section.
— Various *schedules* which contain matters of detail, repeals, etc.

The modern practice is for Acts to state broad general principles leaving matters of detail to be covered by regulations made by a minister by secondary legislation in the form of a statutory instrument (qv).

Figure 1 shows the first page of an Act of Parliament.

Action A civil legal proceeding by one party against another. The purpose may be to gain a remedy, enforce a right, etc. Actions may be *in persona* (against an individual — the defendant) or *in rem* (against an item of property). Criminal proceedings are termed 'prosecutions'.
See also: **Defendant; Plaintiff; Pleadings; Statement of case.**

Activity Schedule A term referred to in the NEC (qv), main option A: 'Priced contract with activity schedule' and main option C: 'Target contract with activity schedule'. The activity schedule in each option is to be identified in the contract data (qv) in each option. Although not defined, it seems that the activity schedule is identical to the schedules of activities (qv) in the BPF System (qv) and the activity schedule referred to in JCT 98 and IFC 98. The activity schedule is subject to change, at least under main option A (clause 54.2), but it is not clear how the lump sum prices for each activity are recalculated.
See also: **Priced Activity Schedule.**

Ad hoc For this purpose. The Latin term used to refer to an appointment for a particular purpose and usually in contrast to an appointment *ex officio* (by virtue of office).

Ad idem Literally, 'at the same point', but also 'agreed' or 'of the same mind'. Negotiating parties are said to be *ad idem* when they have reached agreement on all the terms of contract.

Ad idem

Housing Grants, Construction and Regeneration Act 1996

short title

1996 CHAPTER 53

An Act to make provision for grants and other assistance for housing purposes and about action in relation to unfit housing; to amend the law relating to construction contracts and architects; to provide grants and other assistance for regeneration and development and in connection with clearance areas; to amend the provisions relating to home energy efficiency schemes; to make provision in connection with the dissolution of urban development corporations, housing action trusts and the Commission for the New Towns; and for connected purposes.

long title

[24th July 1996]

date of royal assent

B E IT ENACTED by the Queen's most Excellent Majesty, by and with the advice and consent of the Lords Spiritual and Temporal, and Commons, in this present Parliament assembled, and by the authority of the same, as follows:—

enacting formula

PART I

GRANTS, &C. FOR RENEWAL OF PRIVATE SECTOR HOUSING

CHAPTER I

THE MAIN GRANTS

Introductory

section

1.—(1) Grants are available from local housing authorities in accordance with this Chapter towards the cost of works required for—

side note

Grants for improvements and repairs, &c.

sub-section

(a) the improvement or repair of dwellings, houses in multiple occupation or the common parts of buildings containing one or more flats,

(b) the provision of dwellings or houses in multiple occupation by the conversion of a house or other building, and

paragraph

Figure 1   First page of an Act of Parliament (reproduced by permission of HMSO).

Addendum bills A term used to describe bills of quantities (qv) produced to modify the bills originally prepared. Common reasons for preparing addendum bills are:

— To make a reduction on the lowest tender figure if it exceeds the employer's budget. In this case they are usually termed 'bills of reductions'.

— When standard house types are designed and standard bills of quantities are prepared, addendum bills are often necessary for use on individual contracts to quantify minor variations from the standard to accommodate such items as steps and staggers in terraces or otherwise identical dwellings. A point is reached when it becomes more convenient to take off a completely fresh set of quantities and the process of amendment starts again.

Addendum bills of the first type are not popular with any of the parties to the construction process. They can be confusing and lead to errors unless both original bills and addendum bills are fully cross-referenced. For example, the original bills may include an item for pointing in a particular type of mastic. The addendum bills may show that the mastic has been omitted and a different, superior mastic added back. The addendum bills are, of course, referenced to the originals but the originals are often not referenced to the addendum, because when they were prepared there were no addendum bills. It is possible, therefore, that the contractor may overlook the change unless he checks through both documents. Some alterations will be clear from the drawings, which should reflect the situation shown in the original bills plus addendum bills. Unfortunately an item such as mastic will often simply be termed 'mastic' on the drawing, without any indication of the type. The contractor would be required to correct his mistake at his own cost, but he would be understandably angry about it. When faced with addendum bills, contractors should take care to go through their working copy of the original bills, noting in the margin where the addendum bills take effect.

If possible, addendum bills should be avoided unless they are very short. Their advantages – cheapness and speed – could be negatived if they lead the contractor to make a major blunder.

Addition See: **Extra work.**

Additional variation percentage A fixed percentage addition to valuations of additional work to allow for the cost of any disruption caused by architect's instructions for variations. The percentage is specified by the contractor in his tender. This is a somewhat unusual method of dealing with cost-based claims and no standard form contract makes provision for it. It is not to be confused with the standard provision in, for example, JCT 98 clause 13.5.3.3 that in any valuation allowance must be made for any addition to or reduction of preliminary items.

See also: **Liquidated prolongation costs.**

Adjacent Lying near to but not necessarily adjoining[21]. It is a phrase sometimes found in building contracts in relation to access (see: **Access to works**) to the site and is contrasted with 'adjoining' which suggests a degree of contiguity. JCT 98, clause 25.4.12, for example, recognises as a ground for extension of time the employer's failure to give access to the site over land which is in his possession and control and which is 'adjoining or connected with the site'. That sub-clause does not extend to an agreement to give access over *adjacent* land, though failure by the employer to do so where he has agreed access with the contractor might well amount to a breach of contract at common law and give rise to a common law claim by the contractor.

Adjoining (property) Few building sites stand in isolation and so the rights of owners of adjoining property must always be considered. There is no general right of access over adjoining property, even for the purpose of carrying out essential repairs. Care must therefore be taken to ensure that the works are set out so that no trespass (qv) to neighbouring property occurs. Maintenance can be a problem, particularly in regard to older property where the setting out of building works may have been somewhat informal and subsequently property has been divided into parcels without much thought to future repairs.

The Access to Neighbouring Land Act 1992 (qv) was intended to deal with such problems.

See also: **Party wall; Support, right of.**

Adjudication In English common law, it refers to the decision of a court, especially in regard to bankruptcy. In Scots law, it is concerned with the attachment of land, usually in relation to a debt. In the special context of building contracts it means to decide an issue judicially. Standard form contracts such as ACA 2, CD 81, GC/Works/1, edition 3 and the respective sub-contracts have always had, sometimes limited, provision for adjudication of various kinds.

Following the coming into force of the Housing Grants, Construction and Regeneration Act 1996 (qv), otherwise known as the 'Construction Act', on 1 May 1998 and the equivalent Construction Contracts (Northern Ireland) Order 1997 (qv) on 1 June 1999, every construction contract as defined in the Act must contain specific provisions so that a party to such a contract has the right to refer any dispute (qv) arising under the contract to adjudication. The provisions are contained in s. 108 of the Act (article 7 of the Order). They require that a construction contract must:

— Enable a party to give notice at any time (this has been held to include the period after determination)[22].
— Have a timetable with the object of appointing the adjudicator and having the dispute referred within 7 days of the notice.

---

[21] *Wellington Corporation* v. *Lower Hutt Corporation* [1904] AC 773.
[22] *A & D Maintenance and Construction Ltd* v. *Pagehurst Construction Services Ltd* (1999) CILL 1518.

— Impose a limit of 28 days from the date of referral for the adjudicator to reach a decision or a longer period by agreement by the parties after referral.

— Allow the adjudicator to extend the 28 days by up to 14 days if the referring party agrees.

— Require the adjudicator to act impartially (that does not necessarily require the adjudicator to be independent).

— Enable the adjudicator to take the initiative in finding out the facts and the law.

The contract must also include provisions:

— That the adjudicator's decision is binding until either the dispute is decided by arbitration or legal proceedings (as the contract may provide) or the parties agree that it is final. The courts have shown themselves ready to enforce an adjudicator's decision, provided that he had the jurisdiction to decide the dispute, even when it can easily be demonstrated that the decision is plainly wrong.

— That neither the adjudicator nor his employees or agents are liable for anything done or omitted in acting as adjudicator unless the act or omission was done in bad faith.

All the standard form construction contracts include adjudication provisions which comply with the Act, e.g. JCT 98 clause 41A, IFC 98 clause 9A and MW 98 supplemental condition D. Among other things, the contracts give the adjudicator wide powers to use his own knowledge and expertise, open up and revise certificates and decisions, visit the site and take technical or legal advice. If the contract does not comply with the Act, the adjudication procedures in the Scheme for Construction Contracts (England and Wales) Regulations 1998 apply or the Scheme for Construction Contracts (Scotland) Regulations 1998 or the Scheme for Construction Contracts in Northern Ireland Regulations (Northern Ireland) 1998 as appropriate. There are also a number of other procedures which comply with the Act and which can be easily incorporated by the draftsmen of bespoke contracts. The Construction Industry Council (CIC) procedure and the Technology and Construction Court (TeCSA) rules are among the best known. In practice, problems occur, because the referring party may have a considerable time to put together the referral document while the respondent has seven days at most to reply. This is usually referred to as an 'ambush'. The seven days will include every day except bank holidays (see: **Day**) so that in the case of a complicated issue work through the weekend is the norm. Other difficulties arise, because a party may refer an exceedingly complex matter to adjudication and may evidence it with the help of many files of documents. The dispute may well be better dealt with in arbitration, but a party has the right to have it adjudicated. The principle behind adjudication is that it should be a quick method of resolving disputes with decisions which have temporary binding effect and it should be inexpensive. For that reason the Act says nothing about costs and the adjudicator can usually award only his own fees and expenses against the losing party. Otherwise, each side pays its own costs. Although it has been held that an adjudicator has the power to award the costs of the winning party against the loser, the better view now supported by the courts is that the adjudicator has no such power unless the parties expressly or by necessary implication give it to him[23].

Administrative receiver A receiver or manager of the whole (or substantially the whole) of a company's property appointed by or on behalf of the debenture holders[24]. A person dealing with an administrative receiver in good faith and for value is not concerned to inquire whether the receiver is acting within his powers[25]. He is deemed to be the company's agent unless the company goes into liquidation. He is personally liable on any contract entered into by him in the carrying out of his functions (unless the contract provides otherwise) and is entitled to an indemnity out of the company's assets in respect of that liability[26].

Admissibility of evidence The purpose of evidence (qv) is to establish facts in court or before a tribunal. In England and Wales the law of evidence is mainly exclusionary, i.e. it deals largely with what evidence may or may not be introduced. Admissibility deals with the items of evidence which may be brought before the court. The main basic rule is that the evidence must be *relevant* to the matter under enquiry.
— Hearsay (qv) evidence is now admissible in civil proceedings[27]. The statutory regime requires a party wishing to rely upon hearsay evidence to serve notices upon the other parties; however, failure to comply does not render the evidence inadmissible, rather it may go to the weight attached by the court to that evidence[28].
— Extrinsic evidence (qv) is generally inadmissible.
— Opinion evidence (see: **Expert witness**) is limited to experts.

Advance payment The system of advance payment was introduced into some JCT standard contracts (i.e. the private editions of JCT 98 and IFC 98) following the Latham Report (qv). If the employer and the contractor agree that an advance payment should be made by the employer to the contractor, the amount agreed and the date for payment must be inserted in the appendix to the contract together with a schedule showing the times and amounts of repayments. A form of bond is available if required. It is difficult to envisage a situation in which a bond would not be required for an advance payment of this kind. The provisions of JCT 98 and IFC 98 are identical. They require any reimbursement due on the advance payment to be deducted in the certificate.

Advances on account A term used in GC/Works/1 (1998), clause 48, to refer to the payments which the contractor is entitled to receive during the progress of

---

[23]*Northern Developments (Cumbria) Ltd* v. *J. & J. Nichol* [2000] BLR 158.
[24]Insolvency Act 1986, s. 29(2)(a).
[25]1986 Act, s. 42(3).
[26]1986 Act, s. 44(l).
[27]Section 1, Civil Evidence Act 1995.
[28]Section 2(4), Civil Evidence Act 1995.

the execution of the works at monthly intervals. Now, of course, the Housing Grants, Construction and Regeneration Act 1996 (qv) s. 109 requires all construction contracts (qv) to make provision for periodic payments where the duration of the work is specified or agreed to be less than 45 days. The provision is similar to those clauses in other contracts providing for payment through interim certificates (qv).

Adverse possession Occupation of land inconsistent with the rights of the true owner, commonly called 'squatter's rights'. Title to land may be acquired by adverse possession under the Limitation Act 1980. If a landowner allows a third party to remain in possession of his land for 12 years (30 years in the case of Crown Land) without payment of rent or other acknowledgement of title the squatter may acquire a possessory title and the original owner's title is excluded. A mere demand that the land be vacated is not sufficient to interrupt the period[29].

Acquiring a possessory title is not easy. Mere occupation of the land is insufficient. 'Acts must be done which are inconsistent with the (owner's) enjoyment of the soil for the purpose for which he intended to use it'[30]. There is much relevant case law. Periodical cultivation of a piece of unmarked land was held to be insufficient to establish a possessory title in *Wallis's Cayton Bay Holiday Camp Ltd* v. *Shell-Mex & BP Ltd* [1975] where Lord Denning MR summarised the position aptly:

> 'Possession by itself is not enough to give a title. It must be adverse possession. The true owner must have discontinued possession or have been dispossessed and another must have taken it adversely to him. There must be something in the nature of an ouster of the true owner by the wrongful possessor . . . Where the true owner of land intends to use it for a particular purpose in the future, and so leaves it unoccupied, he does not lose his title simply because some other person enters on to it and uses it for some temporary purpose, like stacking materials, or for some seasonal purpose, like growing vegetables'[31].

In contrast, in *Rudgwick Clay Works Ltd* v. *Baker* (1984)[32], the incorporation of a piece of land into the curtilage of a house showed an intention to possess the land permanently and was capable of amounting to adverse possession. The incorporation was inconsistent with the use of the land for future mining operations. The question as to whether adverse possession has been established is one of fact.

Boundaries (qv) are frequently varied by adverse possession, e.g. when a fence is re-erected by a householder, and it is in this connection that problems are caused in building contract situations.

See also: **Adjoining (property); Boundaries; Possession; Site; Title.**

Adverse weather conditions The changing nature of the weather has always been the enemy of building work which generally takes place exposed to the

---

[29]*Mount Carmel Investments Ltd* v. *Smee and Another* [1988] EGCS 99.
[30]*Leigh* v. *Jack* (1879) 5 Ex D 264.
[31][1975] 3 All ER 575 at 580.
[32]Unreported.

elements. At common law, bad weather as such does not excuse the contractor if he is delayed as a result[33].

Extraordinary weather 'such as could not reasonably be anticipated' may amount to an Act of God (qv) or *force majeure* (qv).

The realities of the situation are recognised by most forms of contract which allow for bad weather to varying degrees and provide for an extension of time (qv) to be awarded under certain circumstances.

JCT 98, clause 25.4.2 and IFC 98, clause 2.4.2 list 'exceptionally adverse weather conditions' as a relevant event (qv) entitling the contractor to claim an extension of time. JCT 63, clause 23 (b) referred to 'exceptionally inclement weather'. The change in wording makes clear that the wording is now intended to cover all exceptional weather which has an adverse effect on the construction work; for example, a hot summer, which would scarcely be classed as 'inclement weather'. Excessive heat and drought can be just as damaging to progress as snow or frost.

Adverse weather conditions would embrace any weather conditions which were contrary to the ideal in any particular circumstance, and the contractor must be taken to have contemplated the possibility of such weather as part of his contractual risk[34]. The qualifying word 'exceptionally' is, therefore, of the utmost importance. In order to show that weather conditions were exceptionally adverse, the contractor may have to provide meteorological records for a lengthy period – 10 or 20 years – to show that the weather was 'exceptional' for the area for the time of year. It is the kind of weather which may be expected at the particular site which is important at the particular time when the delay occurs. 'Exceptional' does not refer to the period during which the works are delayed[35].

Thus, in most areas of England and Wales snow is not exceptional in January, but it is in July. In some areas, however, and at some altitudes, snow would not necessarily be exceptional in early summer. Even if the weather conditions are exceptional, they may not necessarily be 'adverse' because the weather must interfere with the works at the particular stage when the exceptionally adverse weather occurs. This depends on the stage of the construction work at the particular time. If some internal works can continue, for example, the contractor would generally have no valid claim. The contractor is expected to allow in his tender and his programme (qv) for anticipated weather conditions in the area, having regard to historical data, the time of year and the location of the site. This allowance is or should be reflected in the tender price. Often the situation is not clear-cut and, for example, some work may continue on internal fittings at the same time as external work is delayed due to exceptionally adverse weather conditions. In such cases, the architect must enquire carefully into the contractor's master programme (qv) before reaching a decision.

---

[33] *Maryon* v. *Carter* (1830) 4 C & P 295.
[34] *Jackson* v. *Eastbourne Local Board* (1886) HBC 4th edn vol 2 p. 81.
[35] *Walter Lawrence & Son Ltd* v. *Commercial Union Properties (UK) Ltd* (1984) 4 Con LR 37.

GC/Works/1 (1998) does not allow weather conditions as a circumstance entitling the contractor to claim an extension of time but the project manager (qv) could order suspension of the work or any part of the work to avoid the risk of damage from the weather under clause 40 (2) (g), in which case the contractor might be entitled to make a claim for extension of time.

ACA 3 makes no specific references to the weather. However, clause 11.5 (alternative 2) allows *force majeure* (qv) as a basis for a claim for extension of the time and wholly exceptional and unanticipated weather conditions, e.g. extraordinary rainfall, extraordinary snow, etc. could qualify under this head. This is not, however, as wide as under JCT 98 or GC/Works/1.
See also: **Extension of time.**

A fortiori argument *A fortiori* means so much more; or, with stronger reason. It is commonly heard in judicial utterances when a particular case is being considered. Reference is made to a rule which applies to another case and it is thought that the case under consideration shows a stronger reason for application of the same rule and, therefore, the rule should apply to the case under consideration as well.

Such an argument is open to a variety of logical criticisms, notably that there may well be reasons why one rule should apply to the first case and a different rule to the case under consideration.

Affidavit A sworn written statement of evidence sometimes used in civil actions. Affidavit evidence may be given:
— By agreement.
— If the judge or arbitrator so decides.
— Formerly in relation to applications for summary judgment (qv) in the High Court.

The content of the affidavit may be strictly factual or simply the opinion of the person swearing to it. The architect who is required to give affidavit evidence will give his solicitor a statement of the points he wishes to make. The solicitor will prepare the actual document, then the architect (referred to as 'the deponent') swears (or affirms) that it is true and signs it before an authorised person. Authorised persons include a Justice of the Peace, a solicitor (other than the one who has drawn up the affidavit) or a court official. Documents attached to, and referred to in, an affidavit are called exhibits.
See also: **Evidence; Oaths and affirmations.**

Affirmation of contract Where there is a breach of contract of a kind such as to amount to a repudiation (qv) which would entitle the innocent party to terminate (see: **Determination**) his obligations, the innocent party may choose to affirm the contract and treat it as still being in force. The breach itself does not bring the innocent party's obligations to an end automatically; he must first decide how to treat the breach. Only if he accepts the breach do his obligations end. If he refuses to accept the breach, obligations under the contract continue in force. In such circumstances the innocent party will still have a right to damages (qv), and in an appropriate case, e.g. a contract for the sale of land, he may obtain an injunction ordering specific performance against the other party[36].

A not dissimilar situation arises where there is an actionable misrepresentation (qv) and the innocent party may likewise elect to affirm the contract. He then loses his right to rescind the contract.

Lapse of time may be evidence that the contract has been affirmed, but in general it may be said that clear words or actions are required, although standing by idly and remaining silent may also be sufficient where it was inconsistent with accepting the repudiation.

See also: **Rescission.**

Agency An agent is a person exercising contractual powers on behalf of someone else, the important point being that the principal is bound by the acts of his agent. The architect is the employer's agent under the ordinary building contract, even though he has a duty to act fairly between the parties[37].

The agency relationship can be created by express appointment or by implication. It may also arise where someone, without prior authority, contracts on someone else's behalf and the latter ratifies or adopts the contract. Agency may also sometimes be implied from a particular relationship between the parties where one has apparently held out the other as his agent. This situation commonly arises where employees holding administrative functions contract on behalf of their employers.

The key concept is that of the agent's authority. An agent has *actual authority* according to the terms of his appointment, but he has *apparent authority* according to the type of functions he performs. Agency may also be implied from a course of dealing between the parties. It is therefore important to determine what acts fall within an agent's usual or apparent authority. For example, the manager of a builders' merchant's depot may act for the owner in all matters connected with the business. Those dealing with him are not bound by any limitations placed upon his authority by his employer unless they have notice of those limitations. An agent's primary duty is to see that he acts in his principal's interests and he must not abuse his position. He is in a fiduciary (qv) relationship to his principal. Thus, if an agent makes an unauthorised profit for himself in the course of his agency he can be compelled to hand over any profit wrongfully made. He also forfeits any agreed remuneration. Similarly an agent is under a strict duty to account for all property coming into his hands on the principal's behalf. In carrying out his duties the agent must use ordinary skill and diligence and, except in certain circumstances, he cannot delegate the performance of his duties to another[38] – *delegatus non potest delegare* (qv). Delegation may be expressly or impliedly authorised by the principal.

---

[36] *Hasham* v. *Zenab* [1960] AC 316.
[37] *Sutcliffe* v. *Thackrah* [1974] 1 All ER 319; *Pacific Associates Inc* v. *Baxter* (1988) 16 Con LR 90.
[38] *De Bussche* v. *Alt* (1878) 8 Ch D 286 at 310–311.

In general, an agent is not personally liable on a contract made on behalf of his principal, except where he fails to disclose the principal's existence or it is intended that he should be personally liable. However, if in fact the agent had no authority to contract, the aggrieved party may bring action against him for a breach of implied warranty of authority (qv). Usually, the agent drops out of the transaction once he has brought about a contract between his principal and the third party.

The agency relationship can be brought to an end by mutual consent or by performance. The principal may revoke the agent's authority and, in some cases, the relationship comes to an end automatically, e.g. on the death of the agent.

In the context of building contracts, the employer is only liable to the contractor for acts of his architect which are within the scope of his authority[39] and this principle is of importance since all the standard form contracts define closely the architect's powers. However, in many cases – particularly where the architect is an employee of the building owner – there will be instances where the exercise of his professional duties is sufficiently linked to the employer's attitude and conduct that he becomes the employer's agent so as to make the employer liable for his default[40]. In *Croudace Construction Ltd* v. *London Borough of Lambeth* (1986)[41], the local authority's chief architect named in a JCT contract was held to be the employer's agent and his failure timeously to ascertain or instruct the quantity surveyor to ascertain a contractor's money claim was held to be a breach of contract for which the council was liable in damages.

Agreement Although an agreement between two parties, in the sense of a meeting of minds, has no legal significance in itself, agreement is necessary for there to be a valid contract. Possibly for this reason, the word is often used to mean a contract. JCT 98 refers to 'Articles of Agreement' (qv) at the beginning of the contract, but from then on refers to 'Contract' or 'Conditions' (qv). The ACA form, however, uses 'Agreement' rather than 'contract' throughout the contract, e.g. 'This Agreement is made the...' at the very beginning of the document. The provisions for arbitration in building contracts are referred to as arbitration agreements to indicate that they have an existence which is quite separate to the building contract itself.

Agreement for Minor Building Works (MW 98) The JCT Agreement for Minor Building Works was first published in June 1968, revised in January 1980 and revised again in December 1998 to comply with the Housing Grants, Construction and Regeneration Act 1996 (qv) and certain recommendations of the Latham Report (qv). It is designed for use where minor building works are to be carried out for an agreed lump sum and where an architect or contract administrator has been appointed on behalf of the employer. It is for use where

---

[39]*Stockport Metropolitan Borough Council* v. *O'Reilly* [1978] 1 Lloyds Rep 595.
[40]See the first instance decision in *Rees & Kirby Ltd* v. *Swansea City Council* (1983) 25 BLR 129.
[41](1986) 6 Con LR 70.

a lump sum offer has been obtained based on drawings and/or specifications and/or schedules but without detailed measurements (in Scotland, bills of quantities are used). It is suggested that the Form is generally suitable for projects up to a value of £90,000 at 1998 prices. Contract value is not, however, the deciding factor, which is probably the complexity of the job.

The Form should not be used where any of the following are required:

— Nominated sub-contractors or suppliers (qv).
— Bills of quantities (qv) except in Scotland.
— Fluctuations (qv) in the value of labour or materials. Certainly, substantial amendments would need to be made to the Form as printed if any of these items were desired.

The Form consists of only eight main clauses, as follows:

1. *Intentions of the Parties*
1.1 Contractor's obligations
1.2 Architect's/Contract Administrator's duties
1.3 Reappointment of Planning Supervisor or Principal Contractor – notification to Contractor
1.4 Alternative B in the 5th Recital – notification by Contractor – regulation 7(5) of the CDM Regulations
1.5 Giving or service of notices or other documents
1.6 Reckoning period of days
1.7 Applicable law
1.8 Contracts (Rights of Third Parties) Act 1999 – contracting out
2. *Commencement and Completion*
2.1 Commencement and completion
2.2 Extension of contract period
2.3 Damages for non-completion
2.4 Practical completion
2.5 Defects liability
3. *Control of the Works*
3.1 Assignment
3.2 Sub-contracting
3.3 Contractor's representative
3.4 Exclusion from the Works
3.5 Architect's/Contract Administrator's instructions
3.6 Variations
3.7 Provisional sums
4. *Payment*
4.1 Correction of inconsistencies
4.2 Progress payments and retention
4.3 Penultimate certificate
4.4 Notices of amounts to be paid and deductions
4.5 Final certificate
4.6 Contribution, levy and tax changes
4.7 Fixed price
4.8 Right of suspension by contractor

5.   *Statutory Obligations*
5.1   Statutory obligations, notices, fees and charges
5.2   Value Added Tax
5.3   Statutory tax deduction scheme
5.4   Prevention of corruption
5.5   Employer's obligation – Planning Supervisor – Principal Contractor
5.6   Duty of Principal Contractor
5.7   Successor appointed to the Contractor as the Principal Contractor
5.8   Health and Safety file
6.   *Injury, Damage and Insurance*
6.1   Injury to or death of persons
6.2   Injury or damage to property
6.3A Insurance of the Works by Contractor – fire, etc.
6.3B Insurance of the Works and any existing structures by Employer – fire, etc.
6.4   Evidence of insurance
7.   *Determination*
7.1   Notices
7.2   Determination by Employer
7.3   Determination by Contractor
8.   *Settlement of disputes*
8.1   Adjudication
8.2   Arbitration
8.3   Legal proceedings

There are also the usual Articles of Agreement (qv) and Recitals (qv), the first of which defines the contract documents (qv). In addition, there are supplemental conditions which contain clauses dealing with A: contribution, levy and tax changes; B: value added tax; C: statutory tax deduction scheme; D: adjudication; and E: arbitration.

The express provisions are very much in common form. The contractor's basic obligation (clause 1.1) is to carry out and complete the works in accordance with the contract documents. He is to do all this with all due diligence (qv) 'and in a good and workmanlike manner'. There are no provisions for money claims (qv) for disruption or prolongation (but see below), and under the form such claims will have to be dealt with at common law. The extensions of time (qv) clause (2.2) only applies while the works are in progress and the architect has no power to grant an extension of time after the works have been completed.

An interesting feature of the form is clauses dealing with the final certificate (especially clause 4.5). Unlike JCT 98 (qv) the certificate is not stated to be conclusive evidence of performance to any extent. The final certificate referred to in clause 4.5 is merely the final certificate of payment and is issued on the basis of documentation submitted by the contractor. Similarly, in the defects liability certificate issued under clause 2.5, there is no requirement for the architect to state that the works have been completed to his satisfaction and after the end of the defects liability period questions as to liability must be dealt with at common law.

JCT Minor Works is a very short form of contract and no attempt has been made to cover all the situations envisaged by JCT 98 or IFC 98. In particular, it should be noted:

— Although there is provision for a quantity surveyor to be appointed, there is no indication of his duties. In most cases, there will be no quantity surveyor associated with the contract. If it is thought necessary to appoint one, he would no doubt act in an advisory capacity to the architect in valuing work done and variations.

— There is no provision for bills of quantities (qv) except in Scotland. Valuation of variations is to be carried out by using priced specification (qv), priced schedules or a schedule of rates (qv) provided by the contractor. Alternatively, the price may be agreed before the variation is carried out.

— There is no provision for the use of nominated sub-contractors or suppliers.

— The contract is on a fixed price basis with no provision for fluctuation in the price of labour or materials. Provision is made (clause 4.5 and Supplementary Memorandum) for contribution, levy and tax fluctuations, if appropriate.

— There is no provision for dealing with contractors' claims for loss and/ or expense although there is limited power for the architect to include payment of loss and/or expense associated with a variation arising from an architect's instruction, but it appears that the contractor is not required to make any specific application.

— There is no provision for the use of a clerk of works. This is a strange omission because it is more than likely that a clerk of works would be employed, part-time, on the larger of the 'minor' works. However, the point is easily rectified by a suitable insertion.

— The extension of time clause (2.2) is very broad and the contractor may claim an extension provided only that it is apparent that the works will not be completed by the completion date, that he notifies the architect (not necessarily in writing although it is clearly advisable), and the reasons for the delay are beyond his control. Such things as bad weather, strikes, late instructions, etc. are all covered by this clause, but delay due to sub-contractors or suppliers is expressly excluded. On jobs where there may be considerable expense caused to the employer as a result of late completion, this clause would be inadequate to safeguard his interests. The contract must be used with care by employer and contractor. Unforeseen problems invariably arise during construction and this form presupposes a considerable measure of goodwill on both sides. In particular, the clause appears to be inadequate to deal with all employer acts other than architect's instructions[42].

— The insurance provisions are fairly brief. There is no equivalent to the employer's non-negligent insurance to be found in JCT 98 and IFC 98.

---

[42] *Wells* v. *Army & Navy Co-operative Society* (1902) 86 LT 764.

There is no provision for the employer to take out insurance for new works and the required insurance level is 'specified perils' (qv) in all cases – there is no provision for 'all risks' (qv) insurance.

Agreement to negotiate English law does not recognise 'a contract to negotiate a contract'. In the context of the construction industry this is illustrated by *Courtney & Fairbairn Ltd* v. *Tolaini Brothers (Hotels) Ltd* (1974)[43] where an agreement 'to negotiate fair and reasonable contract sums' was held not to amount to a binding contract. There was no agreement on the price or any method by which the price was to be calculated. Since the law does not recognise a contract to make a contract, it cannot recognise a contract to negotiate a contract. Such an agreement fails for uncertainty (qv).

In fact, this proposition may not be as far-reaching as it appears because in some cases the courts may find means of filling gaps left in a contract[44]. The importance of the principle in building contracts is, however, that the parties should be agreed on all the essential terms of the contract[45]. The problem is largely important in relation to letters of intent (qv) and, in practical terms, it is essential to ensure that vital terms should not be left 'to be agreed' or 'subject to agreement' – phrases which are often seen in practice.

Depending upon the parties' intentions behind any such agreements to negotiate, it may be possible to attain similar objectives – 'lock-out' agreements prevent one party (usually the employer) from negotiating with other parties for a finite period[46]. Where Letters of Intent are used so as to allow a contractor to organise plant, materials and workforce, etc. it is good practice to state explicitly the intention behind the letter – has the contract been agreed? Are there specific terms which require agreement and, if so, what precise obligations is the employer prepared to give to the contractor (if any) in respect of any costs incurred as a part of the mobilisation?
See also: **Conditional contract; Subject to contract.**

Alien enemy A person whose state is at war with the UK, or a person, including a British subject, who is voluntarily resident or carrying on business in enemy or enemy-occupied territory.

Such persons are not permitted to bring actions in tort (qv) although they may defend an action against them. They may be allowed to leave the country or they may be interned. They cannot enter into a contract with a British subject and if a contract was made before the outbreak of war (qv), an alien enemy's rights are suspended except that he may defend an action in contract brought against him. Alien enemies may contract and enforce contracts if they are present in the UK by royal licence.

---

[43](1974) 2 BLR 100.
[44]*Foley* v. *Classique Coaches Ltd* [1934] 2 KB 1.
[45]*G. Percy Trentham Ltd* v. *Archital Luxfer Ltd* [1993] 1 Lloyds Rep 25.
[46]See, e.g. *Pitt* v. *PHH Asset Management Ltd* [1994] 1 WLR 327.

All risks The risks for which works insurance must be taken out under JCT 98, clauses 22A, 22B and 22C.2; IFC 98, clauses 6.3A, 6.3B and 6.3C.2. A new definition of risks was introduced in 1986 and largely replaces the previous clause 22 or 6.3 perils which are now called 'specified perils' (qv). The main additional risks in all risks are impact, subsidence, theft and vandalism. The full definition included in JCT 98 is:

'Insurance which provides cover against any physical loss or damage to work executed and Site Materials and against the reasonable cost of removal and disposal of debris and of any shoring and propping of the Works which results from such physical loss or damage but excluding the cost necessary to repair, replace or rectify

1   Property which is defective due to
 .1   wear and tear,
 .2   obsolescence,
 .3   deterioration, rust or mildew;
2   Any work executed or any site materials lost or damaged as a result of its own defect in design, plan, specification, material or workmanship or any other work executed which is lost or damaged in consequence thereof where such work relied for its support or stability on such work which was defective;
3   Loss or damage caused by or arising from
 .1   any consequence of war, invasion, act of foreign enemy, hostilities (whether war be declared or not), civil war, rebellion, revolution, insurrection, military or usurped power, confiscation, commandeering, nationalisation or requisition or loss or destruction of or damage to any property by or under the order of any Government *de jure* or *de facto* or public, municipal or local authority;
 .2   disappearance or shortage if such disappearance or shortage is only revealed when an inventory is made or is not traceable to an identifiable event;
 .3   an Excepted Risk (qv);
     and if the contract is carried out in Northern Ireland
 .4   civil commotion;
 .5   any unlawful, wanton or malicious act committed maliciously by a person or persons acting on behalf of or in connection with an unlawful association; 'unlawful association' shall mean any organisation which is engaged in terrorism and includes an organisation which at any relevant time is a proscribed organisation within the meaning of the Northern Ireland (Emergency Provisions) Act 1973; 'terrorism' means the use of violence for political ends and includes the use of violence for the purpose of putting the public or any section of the public in fear'.

Alteration or amendment of contract The forms of contract in common use in the construction industry have been carefully drafted to take account of most of the situations which regularly arise during the course of building works. The forms are regularly updated in line with decisions of the courts. The employer may wish to incorporate some special provisions in a particular contract to suit his own requirements. It is perfectly feasible to alter or amend a standard form provided:

—   The contractor is made aware of the alterations or amendments at the time of tender or at least before the contract is executed.

— If proposed after the contract is executed, both parties must expressly agree the proposed amendments.
— The amendments are carried out carefully so that no inconsistencies result.
— The amendments do not contravene legislation.

It is always advisable to obtain the assistance of a person specialising in building contracts and construction law if anything but minor amendments are needed, as any amendments may have wide-ranging ramifications.

Most forms provide for certain deletions to be carried out (for example, the insurance provisions in clause 22 of JCT 98) and the printed instructions must be followed minutely. There are pitfalls, however, if more radical alterations are required. The principal danger concerns the current JCT forms which are negotiated with all sides of industry and, therefore, are not caught by the provisions of the Unfair Contract Terms Act 1977. Extensive tampering with the terms of the JCT contracts may well cause them to be considered as the employer's 'written standard terms of business' under s. 3 of the Act and/or to be construed *contra proferentem* (qv).

Two other common problems are worth mention. The employer sometimes wishes to stipulate that the building must be completed in sections on particular dates. In order to do this effectively, great care must be taken to make the appropriate alterations throughout the contract, otherwise the employer may find himself, for example, unable to deduct liquidated damages (qv) for late completion of some or all of the sections[47]. JCT Forms, with the exception of MW 98, have Sectional Completion Supplements to overcome these problems and these should always be used where sectional completion is desired.

If the employer wishes to amend clause 25 of JCT 98, he will lose his entitlement to 'freeze' fluctuations after completion date (qv) unless he also strikes out the appropriate clause in the fluctuation provisions (clauses 38.4.8.1, 39.5.8.1 or 40.7.2.1).

Any amendments must be made on the printed form itself and signed or initialled by both parties. It is not sufficient merely to refer to amendments in the bills of quantities (qv) or specification (qv) because most forms contain a clause giving priority to the provisions of the printed form over any of the other contract documents (qv). An alternative is to delete the priority clause and allow the usual principles of interpretation of contracts to prevail.
See also: **Priority of documents.**

Ambiguity Something which is of unclear or of uncertain meaning; a word, phrase or description which may have more than one meaning.

ACA 3, clause 1.5 refers to ambiguities in the contract documents. This clause is somewhat broader than the similar clause 2.3 in JCT 98 and clause 1.4 in IFC 98 which refer to discrepancies (qv). If a clause in, say, the specification (qv) can be read so as to have two very different meanings, it is possible to

---

[47]*Trollope & Colls Ltd* v. *North West Metropolitan Regional Hospital Board* (1973) 9 BLR 60; *Bramall & Ogden Ltd* v. *Sheffield City Council* (1983) 1 Con LR 30.

argue, under the JCT Form, that one of the meanings would give rise to no discrepancy and, therefore, that is the meaning to be used. Under the ACA Form, however, it would appear to be sufficient, to bring the clause into operation, that an ambiguity exists. It might conceivably be to the contractor's advantage to plead ambiguity, although he would have to show that it could not have been found or foreseen at the date of the agreement.

Ancient monument A historical or archaeological building or site scheduled by the Secretary of State for the Environment under s. 1 of the Ancient Monuments and Archaeological Areas Act 1979, as amended. In the case of monuments in England, this duty is in fact carried out by the Ancient Monuments Branch of the Department of Environment after consultation with the Historic Buildings and Monuments Commission. Under s. 2 of the 1979 Act, it is an offence to carry out construction work to the scheduled monument without consent. The 1979 Act also introduced the concept of 'areas of archaeological importance' or archaeological areas (qv).

See: JCT 98, clause 34.3, ACA 3, clause 14 and GC/Works/1, clause 32 (3), as to what is to happen if 'fossils, antiquities (qv) and other objects-of interest or value' are found on site.

Anticipatory breach of contract When one party to a contract states that he will not carry out his obligations before the time for carrying out the obligations has arrived. Such breaches of contract (qv) range in severity from the minor to a total refusal to carry out any obligations under the contract. Depending upon the nature of the breach, it might well amount to a repudiation (qv) of the contract. The other party may accept the breach or may wait until the actual time for performance. If he accepts, he may sue immediately for breach of contract or for repudiation. Alternatively, he may wait until the actual time for performance and then sue. This latter course may be dangerous as circumstances may change to favour the defaulting party.

Antiquities Ancient relics of various kinds. In building works, they could be parts of ancient structures or artifacts, coins or works of art.

Most standard forms of contract have provision for ownership on discovery and for safeguarding such items until they can be examined and removed from site. JCT 98, clause 34, ACA 3, clause 14; GC/Works/1 (1998), clause 32 (3), NEC, clause 73.1. In practice, many small items such as coins are easily 'lost' unless the likelihood of discovery is appreciated and constant supervision of excavation is maintained. The discovery of larger antiquities, such as ancient pavements, etc. is often greeted with dismay by employer and contractor alike because of the probable delay to the works.

For a fuller consideration of discoveries upon the site see also: **Ancient monument; Archaeological areas; Fossils; Title; Treasure trove.**

Appeal An application and proceedings before a court or tribunal, higher than the one which decided an issue, for reconsideration of that decision.

At common law there is no right of appeal from a superior court (see: **Courts**), but rights of appeal have been created by various Acts of Parliament. Appeals generally relate to questions of law or procedure and only rarely to questions of fact.

In litigation (qv), Part 52 of the Civil Procedure Rules (CPR) (qv) has recently altered the appeals structure in England and Wales. Any party now wishing to appeal a decision of any judge generally requires permission (formerly known as leave) of that judge or the court to which the appeal is being made. In the County Court, appeals from decisions of district judges go to circuit judges. Decisions of circuit judges, whether interim or in fast-track cases now generally go to the High Court. There is a second appeal, again subject to permission, to the Court of Appeal. In the High Court, decisions of masters, registrars or district judges still go to a High Court judge. Judgments from multi-track cases or specialist proceedings, whether in the County or High Court, go to the Court of Appeal. There is a further appeal from the Court of Appeal to the House of Lords where either the Court of Appeal or the House of Lords grant leave to appeal. In practice only important points of law go on appeal to the House of Lords. It is unclear with the use of the Court of Appeal as a 'second appeal' whether there will be a decline in matters going to the House of Lords.

In arbitration (qv), the Arbitration Act 1996 sets out the grounds upon which a party may challenge or appeal an award (qv) – challenges in relation to substantive jurisdiction (s. 67) or serious irregularity (s. 68) or appeal on a point of law (s. 69). The challenges under ss. 67 and 68 are not appeals as such, but are analogous. An appeal on a point of law can be made to the High Court, but permission is required (s. 69 (2)) and the Act imposes other restrictions (s. 70). See also: **Courts.**

Appearance In litigation in the High Court, this is the defendant's formal act indicating his intention to defend the case. This he does, personally or through his solicitor, by returning to the court office a form of acknowledgment of service.

The term is also used of the parties to an action being present in court when the proceedings are heard, either personally or by counsel or a solicitor.

Appendix An addition to a book or document, usually subsidiary to the main work.

The Appendix is an integral part of JCT 98 and in clause 2.1 it is expressly stated to be part of the contract documents. It is to be filled in, in accordance with the information given in the documents accompanying the invitation to tender, before the contract documents are executed. IFC 98 contains a similar Appendix. ACA 3 has a similar appendage entitled the 'Time Schedule'(qv). It is expressly stated to be one of the contract documents in part C of the Recitals (qv). NEC has 'contract data' (qv). If entries in the Appendix are filled in so as to be inconsistent with the provisions of the contract terms themselves then, at least under JCT terms, those entries will be construed *contra proferentem* (qv) and the printed contract terms will prevail[48].

---

[48] *Bramall & Ogden Ltd* v. *Sheffield City Council* (1983) 1 Con LR 30.

Appropriation of payments Setting apart money for a specific purpose out of a larger sum. It usually arises when there are different debts between the same debtor and creditor or when payments are made on account of work done by a contractor in relation to particular items of work, e.g. variations. This cannot be done if there is only one contract and the variations have been ordered under it. The question can only arise if extra work was ordered outside the terms of the contract and if the employer has paid money generally on account.

For example, a contractor is due to be paid £5000 on contract A, £150 on contract B and £50 on contract C with the same employer. The employer may send a single cheque for £5150. The employer should state how he has made up the payment, e.g. £5000 for contract A, £100 for contract B and £50 for contract C, leaving £50 owing for contract B. If the person making the payment fails to appropriate it, it is open to the person receiving the payment to do so. In some cases this may be advantageous, for example where one of the debts has become statute-barred (qv).

Approval and satisfaction Most contracts, either in the printed conditions or in annexed documents such as bills of quantities (qv) or specification (qv), make provision for approval to be obtained to materials, workmanship or operations. The extent of the approvals required varies from contract to contract. It is sometimes expressed as being 'to the satisfaction of...'. The provision is extremely important, with implications which are not always obvious.

In building contracts there are three possible sources of approval:
— The employer.
— The architect.
— A statutory authority, e.g. through building control.

Unless expressly excluded, the expression of satisfaction by employer or architect is binding on the parties to the contract. Approval by a statutory authority is not final and binding contractually because it represents only an additional safeguard for the employer. The architect, for example, may require a higher standard than the building control officer.

Where the architect's approval is specified in addition to the requirement that the work is to be in accordance with the contract, his approval will not override the latter requirement. Thus, if the architect approves of some materials which are not strictly in conformity with the contract, the employer can require the contractor to substitute different materials at a later stage, because the courts have held the two requirements to be cumulative[49]. Even though the contract may not expressly state it, the courts will expect the architect's satisfaction to be reasonable (qv). For example, if the specification required one priming coat, one undercoat and one coat of gloss paint to be applied to internal doors, to the architect's satisfaction, it would not be considered reasonable if two coats of gloss paint were required to obtain a finish which met with his approval. However, the architect is entitled to withhold his approval until the best possible finish is achieved, given the limited specification.

---

[49] *National Coal Board* v. *William Neill & Son* [1984] 1 All ER 555.

Approval and satisfaction

Neither the architect nor the employer is entitled to withhold approval without a genuine reason. For example, the architect's refusal to accept the contractor's making good at the end of the defects liability period (qv) simply to avoid the release of retention money is not a genuine reason. The architect must be acting within his authority when he requires work or materials to be to his satisfaction. As far as the contractor is concerned, that authority can only be discovered by examining the contract documents. If there is no requirement for the architect's approval then, strictly, his approval need not be sought. However, the contractor will still have express and implied obligations under the contract to carry out the work correctly. Moreover, the architect must not certify for payment work which is defective. In practice, many aspects of the contract imply the architect's approval.

JCT 98 and IFC 98 have clauses which make the final certificate (qv) conclusive about the architect's satisfaction where he has expressly reserved something for his satisfaction[50]. Neither ACA 3 nor GC/Works/1 has a similar provision. The provision is of enormous importance for the architect. Many specifications are littered with provisos that work or materials must be to the architect's satisfaction or 'to the architect's approval'. Often a 'catch-all' clause will be attempted such as: 'Unless otherwise stated, all materials and workmanship must be to the architect's satisfaction'. The result may be that the final certificate becomes conclusive about the architect's satisfaction with all materials and workmanship.

In the context of the JCT 98 and IFC 98 Forms, these clauses have the opposite effect to what the architect probably intended. Instead of being limited in effect, the final certificate becomes conclusive evidence that all materials and workmanship are to the architect's satisfaction. There is no obligation on the architect to express his approval of the work as it progresses, indeed he would be most unwise to do so. His approval cannot be implied through silence. Approval may be implied through the issue of certificates (qv), but usually there is a clause restricting such implication, e.g. JCT 98, clause 30.10. The Court of Appeal has muddied the waters by apparently ignoring this clause to give some weight to certificates issued at practical completion (qv) and making good of defects (see: **Defects liability period**)[51]. In practice, the architect can hardly escape from giving certain approvals as the work proceeds, otherwise he may rightly be regarded as extremely uncooperative, and probably in breach of his duties under the contract. JCT 98 clause 8.2.2 requires that if work is specified as to be to the architect's reasonable satisfaction, he must express any dissatisfaction within a reasonable time of the work being executed.

Depending upon the particular terms of the contract, the architect's approval or satisfaction will be subject to review on adjudication (qv), arbitration (qv) or litigation (qv). However, it should be noted that, in making provision for the appointment of a new architect following the death or ceasing to act for any reason of the original architect, JCT contracts stipulate that the new architect

---

[50]For a fuller discussion of the situation before the final certificate clauses were amended in JCT contracts, see: **Final certificate**.

[51]*Crown Estates Commissioners* v. *John Mowlem & Co Ltd* (1994) 70 BLR 1.

may not disregard or overrule any certificate or instruction given by the original architect. It is important that the contractor's interests be safeguarded where the employer has sole control over the situation. The provision cannot mean that certificates or instructions of a previous architect can never be changed. The provision is there to make clear that if the new architect, with the approval of the employer, believes it necessary to change a previous decision, it will rank as a variation.

Approved documents Documents issued under s. 6 of the Building Act 1984 giving 'practical guidance with respect to the requirements of any provision of building regulations' (qv). Their legal effect is stated in the Act. If proceedings are brought against a contractor by a local authority (qv) for contravention of the Building Regulations and he has complied with the requirements of an 'approved document', his compliance will tend to remove liability. Conversely, he is not liable automatically if he fails to comply, but the onus is then on the contractor to show that he has met the relevant functional requirements of the regulations in some other way.
See also: **Building control.**

Approximate quantities Quantities which are not accurately measured, but merely roughly or approximately measured. JCT 98 has a special edition in both local authority and private versions 'With Approximate Quantities'. Effectively, it is a remeasurement contract. Even JCT 98 'With Quantities', which is intended to have an accurately measured set of bills of quantities (qv), acknowledges that some quantities may be measured approximately. Hence, JCT 98, clause 25.4.14 makes the execution of work, for which an approximate quantity is not a reasonably accurate forecast of the quantity of work required, a relevant event and, therefore, ground for an extension of time if delay is caused as a result. Likewise, clause 26.2.8 provides the same ground as a 'matter' entitling the contractor to apply for loss and/or expense.

In an otherwise standard bill of quantities, approximate quantities may be given with regard to items of work such as substructure or drainage where the extent of the work that will have to be done simply cannot be properly or even reasonably accurately measured.

It is not uncommon for the quantity surveyor to include items in bills of quantities for the excavation of rock or running sand or for the necessity to excavate below the water table. The quantity is only given as an estimate. As the work proceeds, it is remeasured at the rates which the contractor has inserted against the item in the bills of quantities. Approximate quantities (or sometimes provisional sums (qv)) are also taken for such things as cutting holes through walls and floors for plumbing and other services. They are often taken from a schedule supplied by the specialist concerned and are commonly referred to as 'builder's work'.

The term 'provisional quantities' (qv) is often used erroneously to describe approximate quantities.

Arbitration The settlement of disputes by referring the matters at issue to the decision of an independent tribunal consisting of one or more arbitrators (qv). It is an essential feature of arbitration that the parties agree to be bound by the decision of the third party, which is called an award (qv). In Scotland, the arbitrator is styled 'the arbiter' and his award is called a 'decree arbitral' and different statutory provisions apply.

Arbitration is regulated by the Arbitration Act 1996 which applies to England, Wales and Northern Ireland. It requires agreement. Under s. 5, the arbitration agreement must be in writing if it is to fall within the Act, though the agreement (see: **Arbitration agreement**) can be entered into before or after the dispute has arisen.

All the standard forms of building contract contain an arbitration agreement: JCT 98, article 7A and clause 41B; ACA 3, clause 25B; GC/Works/1, clause 60 are typical. Some offer the arbitration agreement as the default position and if the parties wish to resolve disputes by legal proceedings they have to deliberately choose them. Such provisions make it a term of the contract that disputes between them shall be settled by arbitration. The effect is that neither party can refer the dispute to litigation unless the other party agrees. This is because under s. 9 of the Act either party can require the court to order a stay of proceedings (qv) while the matter is decided in arbitration. The court now has no discretion and can only refuse a stay if it is satisfied that the arbitration agreement is 'null and void, inoperative, or is incapable of being performed'.

The arbitrator may be appointed by agreement or else by an agreed third party, e.g. the President of the RIBA, or other professional body.

Arbitration is an excellent method of settling construction industry disputes, although in the majority of cases it is no cheaper than litigation (qv) and may be marginally more expensive, since the parties are responsible for the arbitrator's fees and expenses, the cost of a room for the hearing and ancillary costs.

The usual standard form contracts confer very wide powers on the arbitrator to 'open up, review and revise any decision, opinion, instruction, certificate' etc. of the architect. It used to be thought that no corresponding power was available to the court. But this view has now been comprehensively dismissed and the courts have intrinsic power to do what the arbitrator may do only by power conferred under the contract[52].

Essentially, arbitration is a voluntary process and so the powers of the arbitrator are limited, especially as regards joining third parties, compelling the attendance of witnesses etc. although s. 43 of the Act does enable any party with the agreement of the arbitrator to apply to the High Court to compel witnesses to attend.

Arbitration procedure is flexible and may be adapted by agreement to suit the needs of the parties, but in practice, in most major arbitrations, normal

---

[52]The House of Lords in *Beaufort Developments (NI) Ltd* v. *Gilbert-Ash NI Ltd* (1998) 88 BLR 1 which overruled the decision in *Northern Regional Health Authority* v. *Derek Crouch Construction Co Ltd* (1984) 26 BLR 1.

court procedures are followed. The normal stages in a hearing are:
— Preliminary meeting at which the parties agree with the arbitrator to determine procedure, timetable, etc.
— Service of statement of case, defence (qvv), etc. These define the matters in dispute.
— Disclosure of documents (qv) followed by each party inspecting the other's documents.
— Exchange of factual witness statements and, in appropriate cases, reports of expert witnesses.
— The hearing when each party or his advocate presents his case, calling witnesses. Although the normal rules of evidence (qv) are generally followed, there is some flexibility.
— The arbitrator makes his award (qv) which is final and binding on the parties.

In 1988 the JCT published Arbitration Rules which were incorporated in all the JCT arbitration agreements. The 1998 contracts replaced them with the Construction Industry Model Arbitration Rules (CIMAR) (qv). The Rules provide three alternative procedures for the conduct of arbitrations and give the parties themselves an opportunity to choose which of the three procedures apply. There are also strict time scales.

The courts retain wide powers to control arbitrations, and under the Arbitration Act 1996 there is in effect a system of appeals (qv) against an arbitrator's award for errors of law. The court can order the arbitrator to give reasons for his decision but, as a result of case law development, it is difficult to obtain leave to appeal against an award except on substantial matters of law which it is in the public interest should be resolved. In practice, the courts are reluctant to interfere unless it can be shown, e.g., that the arbitrator has made a serious error, irregularity, etc.

Arbitration agreement/clause Section 6 of the Arbitration Act 1996 defines an 'arbitration agreement' as 'an agreement to submit to arbitration present or future disputes'. The arbitration agreement must be in writing if the arbitration is to be governed by the Act.

The majority of standard form building contracts contain a provision committing the parties to submit future disputes to arbitration, and these are sometimes called 'agreements to refer'. JCT 98, article 7A and clause 41B; IFC 98, article 9A and clause 9B; MW 80, article 7A and supplemental condition E; ACA 3, clause 25B; and GC/Works/1 (1998), clause 60 are typical arbitration agreements. The essential point is that there must be a contractual obligation to arbitrate. Although expressed as a separate clause in the contract, it is actually considered to be a separate agreement between the parties which survives determination or liquidation of the parties.

It is possible to incorporate an arbitration agreement if it is referred to in very clear words[53]. Failure to do so may not be fatal[54].

---

[53] *Aughton Ltd* v. *M. F. Kent Ltd* (1992) 57 BLR 1.
[54] *Roche Products Ltd* v. *Freeman Process Systems* (1996) 80 BLR 102.

If a contract contains no arbitration agreement there is nothing to prevent the parties coming to an *ad hoc* agreement after the dispute has arisen, but this is rare in practice.

Arbitrator An impartial referee selected or agreed upon by the parties to a dispute to hear and determine the matter in dispute between them. In one sense, an arbitrator resembles a judge, but unlike a judge he derives his jurisdiction from the consent of the parties. The procedure which he follows is a matter to be determined from the express or implied terms of the arbitration agreement (qv) and various powers are conferred on arbitrators by Act of Parliament, specifically the Arbitration Act 1996.

The arbitrator must be impartial – he owes duties equally to both parties – and he must act in a judicial manner. 'He stands squarely between the two parties having no special affiliation to either'[55].

In building contracts, it is usual for there to be an arbitration agreement providing for the parties to agree on an arbitrator but, failing agreement, the standard form contracts provide for an arbitrator to be appointed by the President or Vice-President of some appropriate professional body, e.g. the Royal Institute of British Architects or the Chartered Institute of Arbitrators.

The arbitrator may be chosen for his professional expertise or technical knowledge, but certain important basic rules must be observed:
—   The arbitrator must not have an interest in the dispute or a subsisting relationship with either party which might affect his impartiality.
—   He must have a general technical knowledge of the technicalities of the matter in dispute.
—   He must be able to act judicially.

The Chartered Institute of Arbitrators, 24 Angel Gate, City Road, London EC1V 2RS, is the professional organisation concerned with arbitration. It runs training courses for prospective arbitrators and training includes a period of pupillage. The selection of an arbitrator who is listed on one of the Institute's panels of arbitrators may be some guarantee of his professional competence as an arbitrator and, in fact, the majority of appointing bodies nominate as arbitrators only those who are members of one of the Institute's panels.

An arbitrator is essentially the servant of the parties and his fees are paid by them. There is no recommended scale of fees.

Arbitrator's award See: **Award.**

Archaeological areas 'Areas of archaeological importance' may be designated by the Secretary of State for the Environment and certain local authorities under s. 33 of the Ancient Monuments and Archaeological Areas Act 1979. Once an area has been designated, it is an offence to carry out any operations in it which disturb the ground without serving 'an operations notice' on the borough (district) council. This brings various controls into play. Few such areas have been designated except in historic cities, e.g. York and Chester.

---

[55]Mustill & Boyd *Commercial Arbitration*, 2nd edn, 1989, p. 223, Butterworths.

Architect's Appointment Introduced in July 1982 by the Royal Institute of British Architects (RIBA) to assist in the agreement of fees, services and responsibilities between the architect and his client. It was the successor to the RIBA Conditions of Engagement (qv) and followed the report of the Monopolies and Mergers Commission on architects' and surveyors' services, which recommended the abolition of the mandatory fee scale. There was also a Small Works Edition (effective from September 1982) for jobs where the total construction cost did not exceed £80,000 (1989 prices). The document was arranged in four parts:

One – Preliminary and Basic Services provided by the architect.

Two – Other Services normally charged on a time or lump sum basis.

Three – Conditions of Appointment, normally applying.

Four – Recommended methods of calculating fees for services and expenses.

A Memorandum of agreement tied up the appointment on a firm legal basis.

The document was superseded by the Standard Form of Agreement for the Appointment of an Architect (SFA/92) in 1992 (actually a suite of documents) and the revised editions in 1999 (SFA/99 (qv), CE/99, SW/99). The Architect's Appointment is still occasionally used by architects although it does not comply with the Housing Grants, Construction and Regeneration Act 1996 (qv).

Architects Registration Board (ARB) The successor body to the Architects Registration Council (qv). It was established on 1 April 1997. The principal Act is now the Architects Act 1997 which consolidates the Architects (Registration) Acts 1931–69. Persons wishing to use the title 'Architect' for business purposes must be registered with ARB. The Board is keen to enforce the restriction.

Registration may be achieved by:
— gaining a qualification after passing an examination which is recognised by ARB; and
— completing at least two years' practical experience supervised by an architect, one of the years being undertaken after completion of a recognised five year course of study and gaining the qualification; and
— passing a written and/or an oral examination in professional practice recognised by ARB.

Architects can be removed from the register as follows:
— If an architect makes application in writing stating the grounds.
— If an architect fails to pay the annual retention fee at the appropriate time after a written request to do so.
— If at the time of registration the architect was subject to a disqualifying decision in another European Economic Area (EEA) state of which ARB was unaware.
— If an architect fails to notify ARB of a change of address after being requested to do so.
— If an architect is guilty of unacceptable professional conduct or serious professional incompetence or has been convicted of a criminal offence which may have relevance to fitness to practise as an architect.

The duties of ARB are stated as follows:
— To maintain a register of architects and to publish it.
— To prescribe the admission criteria.
— To protect consumers from misconduct or incompetence by architects.
— To require appropriate evidence from firms wishing to practise under the title 'Architect'.
— To draw up a code of conduct.
— To prosecute unregistered persons who practise under the title 'Architect'.

An unregistered person who practises or carries on business under any name containing the word 'architect' is guilty of a criminal offence. Such a person is liable to a heavy fine. A person who is a member of the RIBA but not registered may not be styled 'Chartered Architect' and may not even use the affixes FRIBA, ARIBA or RIBA, because they contain the prohibited word[56]. Only the title and not the function is protected. An unregistered person can style themselves 'Architectural Consultant' or 'Architectural Designer' if they so wish. It may be that such titles will be put onto the proscribed list in due course.

The members of ARB consist of:
Seven members elected in accordance with an electoral scheme made by the Board with the approval of the Privy Council after consultation with bodies which are representative of architects. All registered persons may take part in such elections.
— Eight members who are appointed by the Privy Council after consultation with the Secretary of State and others. These members represent the interests of the users of architectural services and the general public and no registered person may be appointed.

Architects Registration Council (ARCUK) Set up by the Architects (Registration) Act 1931 to control the architectural profession in England, Wales, Scotland and Northern Ireland. The Act was followed by the Architect's Registration Acts of 1939 and 1969. ARCUK ceased to exist in 1997, being replaced by the Architects Registration Board (qv).

Arrangement, deed or scheme of Someone who is unable to pay his debts may agree with his creditors to discharge his liabilities by composition or part payment. This can be done privately or by application to the High Court or County Court. If the deed of arrangement is executed privately, the provisions of the Deeds of Arrangements Act 1914 must be complied with. A deed of arrangement is a contract and its effect depends on its own terms.

A Scheme of Arrangement is an insolvent debtor's proposal for dealing with his debts by applying his assets or income in proportionate payment of them. Part VIII of the Insolvency Act 1986 deals with such voluntary arrangements. The scheme must be approved by the creditors or the majority of them. The court has power in bankruptcy (qv) proceedings to approve a voluntary scheme in

---

[56]*Jones (on behalf of the Architects Registration Board of the United Kingdom)* v. *Ronald Baden Hellard* (1998) 14 Const LJ 299.

lieu of adjudging the debtor bankrupt. The term is often used for schemes proposed by limited companies in like circumstances. A statutory procedure is laid down and a company Scheme of Arrangement requires the approval of the court. It may compromise claims, alter the rights of shareholders or resolve other difficulties.

Under Part VIII of the Insolvency Act 1986, in specified circumstances the court may declare a moratorium for an insolvent debtor who intends to make a proposal to his creditors for a composition in satisfaction of his debts or a scheme of arrangement of his affairs. This voluntary arrangement requires the approval of the creditors and if so approved is implemented under supervision; it is not subject to the 1914 Act.

JCT 98, clause 27.3.1 lists the contractor 'making a composition or arrangement with his creditors' as a ground for determination of the contractor's employment under the contract by the employer. IFC 98, clause 7.3.1 does the same. Both contracts entitle the contractor to determine under clauses 28.3.1 and 7.10.1 respectively if the employer makes a similar composition or arrangement. ACA 3, clause 20.3 lists as a ground for termination by either party if the other 'shall make or offer to make an arrangement or composition with its creditors', but notice is required. GC/Works/1 (1998), clause 56 (6) (C) includes voluntary arrangements under Part VIII of the 1986 Act in its definition of insolvency as a ground for determination.

Articles of Agreement 'Articles' generally means clauses. The Articles of Agreement are the formal opening parts and Recitals (qv) of the JCT and ACA forms of contract.

Artificial person An entity, other than a human being, which is recognised in law as a legal person capable of acquiring rights and duties. A corporate body, such as a local authority, a limited company, or the bishop of a diocese. In general, a corporate body can only exist if it has been formed under the authority of the state and today the only methods of incorporation are a charter from the Crown and an Act of Parliament.
See also: **Corporation; Limited company; Local authority.**

Artists and tradesmen A phrase found in JCT 63, clauses 23 (h) and 24 (l) (d). The full phrase reads: 'artists, tradesmen or others'. It has been held that the words 'or others' were not to be construed *ejusdem generis* (qv) and could refer to statutory authorities engaged by the employer under contract and not carrying out their statutory duties[57]. Thus, in practice, the phrase refers to anyone engaged by the employer under a separate contract, sometimes called 'employer's licensees'.

The phrase has been completely removed from the JCT 80 form and clauses 29, 25.4.8 and 26.2.4 substituted a much clearer wording which is wider in scope than the JCT 63 provisions.

---

[57] *Henry Boot Construction Ltd* v. *Central Lancashire Town Development Corporation* (1980) 15 BLR 1.

As soon as possible A stricter obligation than 'forthwith' (qv) or 'in a reasonable time'. If an act is to be done as soon as possible all circumstances must be taken into account[58]. Therefore, supply of goods as soon as possible means the supply within the time that would be enough to carry out the supply assuming that the supplier had everything necessary and taking account of other actions to which he was already committed[59].
See also: **Directly; Immediately.**

Ascertain To find out for certain. Compare the use of this word with *estimate* (qv).
It is used in the JCT 98, IFC 98, and ACA 3 forms in relation to financial claims. The contracts intend that the calculation of money due to the contractor is to be an extremely accurate process rather than a rough assessment or the expression of an opinion or a fortuitous guess. It has been held:

> 'Furthermore "to ascertain" means "to find out for certain" and it does not therefore connote as much use of judgment or the formation of opinion had "assess" or "evaluate" been used. It thus appears to preclude making general assessments as have at times to be done in quantifying damages recoverable for breach of contract.'[60]

The JCT 98 and ACA 3 forms, clauses 26.1, 34.3.1 and 15.3, 17.5 respectively, make clear that it is the architect's duty to ascertain the amount of a claim. He can, if he wishes, delegate the ascertainment to the quantity surveyor. He is well advised to do so because the quantity surveyor is specially qualified to carry out this work. However, in *R. B. Burden Ltd* v. *Swansea Corporation* (1957)[61], under an earlier version of what is now JCT 98, it was indicated that the architect need not accept the quantity surveyor's quantification. Under JCT 98, if the architect delegates the function of ascertainment to the quantity surveyor it is thought that the architect is bound by it. Note that the architect has no power to delegate the initial decision to the quantity surveyor; that is, whether or not there is a valid claim. A curious variation on that theme is to be found in JCT 98, clause 13.4.1.2, alternative A, where the contractor is entitled to submit a priced statement (qv). If he does so, he is entitled also to submit an estimate of the extension of time and loss and/or expense required. Clause A7.1 places the responsibility for accepting or not accepting the contractor's estimate on the quantity surveyor 'after consultation with the Architect'. Consultation (qv) does not imply agreement and it is within the experience of one of the authors that the quantity surveyor might accept the estimate against the architect's opinion. That would have the effect of fixing a new date for completion without the architect operating the extension of time clause (clause 25) and of irrevocably agreeing the amount to be certified as loss and/or expense.
The limited nature of the quantity surveyor's general powers under JCT contracts was confirmed in *County & District Properties Ltd* v. *John Laing*

---

[58] *Verlest* v. *Motor Union Insurance Co Ltd* [1925] 2 KB 137.
[59] *Hydraulic Engineering Co Ltd* v. *McHaffie* (1879) 4 QBD 670.
[60] *Alfred McAlpine Homes North Ltd* v. *Property and Land Contractors Ltd* (1995) 76 BLR 65 per Judge Lloyd at 88.
[61] [1957] 3 All ER 243.

Construction Ltd (1982):

> 'His authority and function under the contract are confined to measuring and quantifying. The contract gives him authority, at least in certain instances, to decide quantum. It does not in any instance give him authority to determine liability, or liability to make any payment or allowance.'[62]

Assent Agreement or compliance. It is also used to describe the formal act of a deceased person's executor to give effect to a gift made to a legatee.

Assignment and sub-letting Assignment is the legal transfer of a right or benefit from one to another. As a general rule, a party cannot assign the burden of a contract[63]. So, for example, a contractor 'A' might assign the rights to receive retention monies from the employer 'B' to 'C'[64]. The transfer of any benefit may generally be effected at law or in equity:

— Statutory assignment under s. 136 of the Law of Property Act 1925 requires any assignment to be in writing, to be absolute and written notice must be given to the other party.

— An equitable assignment of a legal interest need not be in writing, although the equitable assignment of an equitable interest must be in writing[65].

Where there has been an effective assignment of rights, unless the assignor can show financial loss, he will normally only have a right to nominal damages (qv)[66].

It is possible for the parties to restrict the ability to assign any rights[67]. Where a purported assignment takes place, contrary to the parties' agreement, the assignment will be ineffective as between 'B' and 'C', but the assignment will be binding as between 'A' and 'C' such that 'A' may sue for 'C' 's losses[68]. Such a prohibition may not, however, prevent the assignment of the 'fruits of performance' or cause(s) of action arising out of performance[69].

JCT 98, clause 19.1 prohibits either party assigning the contract without the written consent of the other party. ACA 3, clause 9.1 also forbids assignment save the contractor may assign any monies due or which become due to him, and the employer may assign his rights after taking-over (qv). GC/Works/1, clause 61 prohibits the contractor from assigning or transferring the contract or any part, share or interest under it without the written consent of the employer.

Causes of action, in other words, the right(s) to bring a claim for a debt or damages can be transferred so long as the assignee 'C' has a sufficient interest

---

[62] (1982) 23 BLR 1, per Webster J at 14.
[63] For example, *Nokes* v. *Doncaster Amalgamated Collieries Ltd* [1940] AC 1014.
[64] See, for example, *Re Tout and Finch Ltd* [1954] 1 All ER 127.
[65] Section 53 (1) (c) of the Law of Property Act 1925.
[66] *Alfred McAlpine Construction Ltd* v. *Panatown Ltd* [2000] 3 WLR 946 (HL).
[67] *Linden Gardens Trust Ltd* v. *Lenesta Sludge Disposals Ltd; St Martins Property Corporation Ltd and St Martins Property Investments Ltd* v. *Sir Robert McAlpine & Sons Ltd* [1994] AC 85.
[68] *Linden Gardens Trust Ltd* v. *Lenesta Sludge Disposals Ltd; St Martins Property Corporation Ltd and St Martins Property Investments Ltd* v. *Sir Robert McAlpine & Sons Ltd* [1994] AC 85.
[69] *Linden Gardens Trust Ltd* v. *Lenesta Sludge Disposals Ltd; St Martins Property Corporation Ltd and St Martins Property Investments Ltd* v. *Sir Robert McAlpine & Sons Ltd* [1994] AC 85.

recognised by law in the benefit of the assignment[70]; for example X could not assign a right for damages for personal injuries to Y as Y has no legitimate interest in the benefit.

If a contractor no longer wishes to perform his part of the contract, he may have the contract performed vicariously, unless the contract required personal performance; e.g. a sculpture being completed by a particular artist. In other words, a contractor may sub-contract or sub-let part of the works. In the absence of such a personal requirement, an employer could not refuse vicarious performance[71]. In any event, such vicarious performance will not release the contractor from liability for non-performance[72]. Most standard form contracts allow for the sub-letting of works with the employer's consent which cannot be unreasonably withheld (see JCT 98, clause 19.2.2; IFC 98, clause 3.2 and ACA 3, clause 9.2). The exception is GC/Works/1, clause 62 (1) which prohibits sub-letting unless the employer accepted a sub-letting proposal prior to the award of the contract or the PM (project manager) (qv) gives his prior written consent.

An employer may assign the benefit of a building contract, i.e. the right to have the contractor carry out the works, unless the rights in question are of a personal nature[73]. Equally, an employer may assign the benefit of any warranties given by the contractor, where the warranty is capable in law of assignment. The assignee can be in no better position than the employer in relation to the question of damages. Where the warranty provides for the return of a contractor to remedy defects, this is enforceable and damages would flow in the usual way for breach. If the warranty merely vouched that the building was defect-free, the assignee could only sue for losses actually suffered by the assignor or which would have been suffered had the employer retained the building and the benefit of the warranty[74].

The only way for one party to transfer the benefit and burden of a contract to another party and be released himself from any further obligations is by way of novation (qv). In essence, all three parties agree to discharge the first contract between 'A' and 'B' and to replace it with a new contract between 'A' and 'C' on identical terms.

The differences between novation (qv), assignment and sub-letting have been admirably set out as follows:

'(a) *Novation* This is the process by which a contract between A and B is transformed into a contract between A and C. It can only be achieved by agreement between all three of them, A, B and C. Unless there is such an agreement, and therefore a novation, neither A nor B can rid himself of any obligation which he owes to the other under the contract. This is commonly expressed in the proposition that the burden of the contract cannot be assigned, unilaterally. If A is entitled to look to B for payment under the contract, he cannot be compelled to look to C instead, unless

---

[70] *Trendtex Trading Corporation* v. *Crédit Suisse* [1980] 3 All ER 721.
[71] *British Waggon Co Ltd* v. *Lea* (1880) 5 QBD 149 (DC).
[72] *British Waggon Co Ltd* v. *Lea* (1880) 5 QBD 149 (DC).
[73] *Tolhurst* v. *Associated Portland Cement Manufacturers (1900) Ltd* [1903] AC 414.
[74] *Darlington Borough Council* v. *Wiltshier Northern Ltd* [1995] 1 WLR 68 (CA).

there is a novation. Otherwise B remains liable, even if he has assigned his rights under the contract to C...

(b) *Assignment* This consists in the transfer from B to C of the benefit of one or more obligations that A owes to B. These may be obligations to pay money, or to perform other contractual promises, or to pay damages for a breach of contract, subject of course to the common law prohibition on the assignment of a bare cause of action. But the nature and content of the obligation, as I have said, may not be changed by an assignment. It is this concept which lies, in my view, behind the doctrine that personal contracts are not assignable...Thus if A agrees to serve B as chauffeur, gardener or valet, his obligation cannot by an assignment make him liable to serve C, who may have different tastes in cars, or plants, or the care of his clothes ...

(c) *Sub-contracting* I turn now to the topic of sub-contracting, or what has been called in this and other cases vicarious performance. In many types of contract it is immaterial whether a party performs his obligations personally, or by somebody else. Thus a contract to sell soya beans, by shipping them from a United States port and tendering the bill of lading to the buyer, can be and frequently is performed by the seller tendering a bill of lading for soya beans that somebody else has shipped.'[75]

Many building contracts include clauses restricting the right to assign. If a party purports to assign in such circumstances without consent, the assignment is of no legal effect[76]. JCT 98, clause 19.1 forbids either party to assign 'the contract' without the other's written consent. ACA 3, clause 9.1 also forbids assignment with the proviso that the contractor may assign any monies due or to become due to him under the contract and the employer may assign any of his rights after taking-over of the works. This is a perfectly sensible provision. The employer, for example, may wish to assign the building to someone else after it is completed. However, he remains liable to the contractor for any further payments as they become due. GC/Works/1 (1998), clause 61 provides that the contractor is not to assign or transfer the contract, or any part, share or interest under it without the written consent of the employer.

Following traditional practice in the building industry, most contracts allow the contractor to sub-let part of the works with the consent of the employer. Unlike the provision governing assignment, it is common for the contract to warn that consent to sub-letting must not be unreasonably withheld (see: JCT 98, clause 19.2.2, IFC 98, clause 3.2 and ACA 3, clause 9.2).

GC/Works/1, clause 62 (1) prohibits sub-letting unless the employer has accepted a sub-letting proposal prior to the award of the contract or the prior written consent of the project manager has been obtained.

See also: **Novation; Privity of contract; Contracts (Rights of Third Parties) Act 1999; Sub-contract; Sub-contractor.**

Attachment of debts
Another name for garnishee proceedings. The procedure is employed in High Court actions where a judgment for the payment of money

---

[75] *St Martins Property Corporation Ltd and St Martins Property Investments Ltd* v. *Sir Robert McAlpine & Sons Ltd* and *Linden Gardens Trust Ltd* v. *Lenesta Sludge Disposals Ltd, McLaughlin & Harvey PLC and Ashwell Construction Company Ltd* (1992) 57 BLR 57 per Staughton LJ at 76, CA.
[76] *Helstan Securities Ltd* v. *Hertfordshire County Council* (1978) 20 BLR 70.

has been obtained against a debtor to whom money is owing by another person. The judgment creditor can then obtain an order that sums owing by the third party should be attached to satisfy the judgment debt. This has the effect of preventing the third party from paying his creditor until the court has considered the matter.

See also: **Garnishee order.**

Attendance Sub-contract NSC/C, clause 3.15.1 states: 'General attendance shall be provided by the Contractor free of charge to the Sub-Contractor and shall be deemed to include only use of the Contractor's temporary roads, pavings and paths, standing scaffolding, standing power operated hoisting plant, the provision of temporary lighting and water supplies, clearing away rubbish, provision of space for the sub-contractor's own offices and for the storage of his plant and materials and the use of messrooms, sanitary accommodation and welfare facilities'. If the nominated sub-contractor requires any items of special attendance, he is to set them out in tender NSC/T. A further definition of the terms 'general attendance' and 'special attendance' can be found in the **Standard Method of Measurement** (qv).

Attestation The practice of having contracts or other documents signed or sealed in the presence of a witness who also signs and adds his address and description as evidence that the document was properly signed or sealed. One witness is generally sufficient. A dictionary definition of 'attest' is 'to witness any act or event'. Different forms of attestation clause are used in the case of contracts executed as deeds and those which are merely executed by hand. Except in Northern Ireland, there is no longer a requirement for a seal (qv) to be attached to a document when it is executed as a deed and the mere fact of a seal is no longer evidence that a document is a deed unless other criteria are met. A document can be executed as a deed, without sealing, by a company if it is signed by two directors or a director and the company secretary and by an individual who must sign in the presence of a witness who attests the signature. A company may still affix its common seal if desired. In each case, it is essential that the document makes clear on its face that it is executed as a deed[77]. **Figure 2** shows specimen attestation clauses.

Available/availability Capable of being used. Particularly available at the place where a thing can be used[78]. JCT 98 clause 5.5 refers to the contractor keeping a copy of the contract drawings, the unpriced bill of quantities, the descriptive schedules or other like documents, the master programme and the drawings and details upon the works 'so as to be available to the Architect'. The obligation would clearly not be satisfied nor would the materials be 'available' if it was at the contractor's head office. JCT 98 clause 25.4.9 provides that the Government's exercise of any statutory power which restricts the availability of labour is a relevant event if it directly affects the execution of the works.

---

[77]Law of Property (Miscellaneous Provisions) Act 1989 and the Companies Act 1989.
[78]*Roberts* v. *Dorman Long & Co Ltd* [1953] 2 All ER 428.

(i) For a simple contract (under hand)

AS WITNESS THE HANDS OF THE PARTIES HERETO

Signed on behalf of _____ (Employer)

in the presence of _____

Signed on behalf of _____ (Contractor)

in the presence of _____

(ii) For a specialty contract (a deed)

EXECUTED AS A DEED BY THE EMPLOYER namely _____

by affixing hereto its common seal

in the presence of _____

LS

or* acting by two directors or one director and the company secretary whose signatures are set out below

namely _____ (Director)

Signed _____

and _____ (Director/Company Secretary)

Signed _____

AND EXECUTED AS A DEED BY THE CONTRACTOR namely _____

by affixing hereto its common seal

int the presence of _____

LS

or* acting by two directors or one director and the company secretary whose signatures are set out below

namely _____ (Director)

Signed _____

and _____ (Director/Company Secretary)

Signed _____

* If a company registered under the Companies Acts

Figure 2   Sample attestation clauses.

Avoidance Setting aside or making void (qv), especially a contract, e.g. when one party withdraws from a voidable (qv) contract. Where a bond (qv) contains a condition providing that it is void on the happening of a certain event it is said to be 'conditioned for avoidance'.

Award The decision of an arbitrator (qv). The arbitrator's award must be:
— Final (the arbitrator may make interim awards).
— Certain in its meaning.
— Consistent in all its parts.
It must:
— Deal with all matters referred to arbitration.
— Comply with any special directions in the submission.

Provision is made in s. 47 of the Arbitration Act 1996 for *interim awards* to be made at any time, e.g. in respect of matters of principle or for part of the sum claimed.

The award is usually in writing and the date of the award can be decided and stated by the arbitrator or tribunal (s. 54 (1)). If not so stated, the date is the date on which the award is signed by the arbitrator. If there are several arbitrators, it is the date on which the last arbitrator signs (s. 54 (2)). The award usually contains reasons, although this was not always the case, but s. 70 of the Arbitration Act 1996 empowers the High Court to order an arbitrator to state the reasons for his award if there is any appeal on a question of law under the Act. Section 66 of the Act provides for an award to be enforced in the same way as a judgment or order of the court.

Bailee A person to whom the possession (qv) of goods is entrusted by the owner for a particular purpose, with no intention of transferring the ownership (qv). A common example in the construction industry is that of the hirer of plant. The bailee (hirer) receives both possession of the plant and the right to use it, in return for a price to be paid to the bailor (owner). A bailee has qualified ownership of the goods.
See also: **Hire.**

Bailment The legal relationship which exists where goods are lent to or deposited with another person on the condition that they will be re-delivered to him or to his order in due course. Bailment may be gratuitous, e.g. a simple loan, or as a pledge or pawn. It may also be for reward, e.g. hire. A common example of bailment is where goods are left with someone for repair and in such a case the bailee of uncollected goods is given a power of sale under the Torts (Interference with Goods) Act 1977.

Bankruptcy The procedure by which the state takes over the assets of an individual who is unable to meet his debts (see also: **Insolvency**). The purpose is two-fold:
- — To ensure equal distribution of assets among creditors, subject to an order of preference (see: **Liquidation**).
- — To protect the debtor from the pressing demands of his creditors and to enable him to start again.

The procedure is laid down in Part IX of the Insolvency Act 1986 and if the court makes a bankruptcy order, the official receiver (qv) will be receiver and manager of the bankrupt's estate until the creditors have appointed a trustee in bankruptcy (qv) in whom the bankrupt's property vests. Very complex rules are laid down for the administration of the bankrupt's estate. In most cases, a bankrupt is automatically discharged from bankruptcy three years from the date of the bankruptcy order. Undischarged bankrupts are precluded from holding certain offices and carrying out certain functions as well as being subject to onerous obligations during the continuance of the bankruptcy.

Banwell Report A report produced by a government committee under the chairmanship of Sir Harold Banwell. The proper title of the report is *The Placing and Management of Contracts for Building and Civil Engineering Work* (HMSO 1964). Far-reaching recommendations were made for tendering and contract procedures. Following the report, open tendering (where tenders are invited from any contractor who cares to apply) was discouraged and the membership of the National Joint Consultative Committee for Building was broadened to embrace a wider spectrum of the industry. The Codes of Procedure for Single- and Two-stage Selective Tendering (qv) resulted.

Barrister A lawyer who is a member of one of the Four Inns of Court and who has been called to the Bar by his Inn. He has rights of audience in all courts in England and Wales, but only limited contact with clients. A growing number of barristers specialise in construction law.

Base date An expression referred to in JCT contracts from which certain events are related. For example, IFC 98 in clause 5.5 states: 'To the extent that after the Base Date the supply of goods and services to the Employer becomes exempt from value added tax ...'. Although it is included in the definitions in both JCT 98 and IFC 98, the definition is nothing more than a reference to the date in the Appendix (qv). Therefore, the base date can be any date agreed by the parties. The expression was substituted for the term 'date of tender' in 1987 on the basis that the date of tender might change if tenderers were given an extension of time for any reason. It was considered more practical to stipulate a date which would not vary for the purpose of setting a datum for such things as fluctuations. In practice, the base date is usually set as what would previously have been the date of tender.

Basic method/Alternative method Referred to the systems of nominating sub-contractors under JCT 80, clause 35. These methods have been replaced by the 1991 method which is now incorporated into JCT 98. The systems of nomination were intricate. Very briefly, the difference between the methods was that, under the 'basic method,' the employer used NSC/1 and NSC/2 to obtain tenders and the employer/nominated sub-contractor agreement before nomination on NSC/3 and the signing of sub-contract NSC/4. In the 'alternative method' some other method of picking the sub-contractor was used, NSC/2a was used for the employer/nominated sub-contractor agreement (but its use was optional) and sub-contract NSC/4a was signed. There was an optional standard document (NSC/1a) for obtaining tenders from potential sub-contractors and an optional nomination document (NSC/3a).
See also: **Nominated sub-contractors.**

Basic prices See: **Schedule of rates.**

Beneficial occupation A phrase sometimes used by contractors who contend that if the employer takes possession of a project before practical completion (qv), he is thereby precluded from recovering liquidated damages (qv), because he has the benefit of occupying the premises and it is the failure to achieve occupation which liquidated damages are intended to compensate. The contention appears to have no basis in law. Indeed there is case law to suggest that liquidated damages may be recovered even though the employer has taken possession, provided that practical completion has not been certified[79].

---

[79] *BFI Group of Companies* v. *DCB Integration Systems Ltd* (1987) CILL 348.

Best endeavours A phrase used in the JCT 98 and IFC 98 contracts, clauses 25.3.4.1 and 2.3 respectively. It must be read in the context of the contract in order to determine its meaning. In these contracts it is the duty of the contractor to use his best endeavours to prevent delay. The carrying out of the duty is a pre-condition to the awarding by the architect of an extension of time.

Best endeavours, in this context, means that the contractor must constantly do everything reasonably practicable to prevent delay, short of incurring substantial additional expenditure. It has been described as doing everything prudent and reasonable to achieve an objective[80] and again as what a prudent, determined and reasonable party who was acting in his own interests would do to achieve the objective[81]. In the majority of cases, best endeavours means simply that the contractor must continue to work regularly and diligently (qv) and nothing more. Put another way, provided the contractor is working regularly and diligently and he has not contributed to the delay by his own fault, he can be said to have used his best endeavours. The point is often disputed. If, for example, the contractor could reduce delay by switching a gang of bricklayers from one portion of the work to another and does not do so, it could reasonably be said that he is not using his best endeavours. Similarly, if the contractor foresees delay, he must reprogramme if it is practicable to do so[82].

Bias A tendency or inclination to decide an issue influenced by external considerations and without regard to its merits. It is essential that the architect avoids actual or apparent bias in exercising his functions under the building contract, especially as regards certifying or giving or withholding approval or consent. The architect as certifier must act fairly, reasonably and independently as between employer and contractor. Failure to act fairly or acting as a result of improper pressures or influence will result in the decision being of no effect. For example, in *Hickman & Co* v. *Roberts* (1913)[83] the architect was instructed by the employer not to issue a certificate until the contractor's account for extras was received, and the architect advised the contractor accordingly. The House of Lords held that the need for the architect's certificate was dispensed with and the contractor was entitled to sue without a certificate.

In arbitration proceedings, the arbitrator must show no bias. The Arbitration Act 1996 does not refer to 'bias', but under s. 24 a party may apply to the court for removal of an arbitrator on the grounds that circumstances exist that give rise to justifiable doubts as to his impartiality. This probably covers situations where unfairness might be suspected or foreseen, as where a person with close links with one of the parties accepts an appointment as arbitrator[84]. Bias against one party will also disqualify a person from appointment as an arbitrator. The question always is whether there is a predisposition to decide

---

[80]*Victor Stanley Hawkins* v. *Pender Bros Pty Ltd* (1994) 10 BCL 111.
[81]*IBM UK* v. *Rockware Glass* (1980) FSR 335.
[82]*John Mowlem & Co* v. *Eagle Star and Others* (1995) CILL 1047.
[83][1913] AC 229.
[84]*Veritas Shipping Corporation* v. *Anglo-American Cement Ltd* [1966] 1 Lloyds Rep 76.

for or against one party without proper regard to the merits of the dispute. If that question is answered affirmatively, then the courts can intervene.

Apparent animosity to one party or his witnesses amounts to bias[85].

The question of judicial bias has recently been considered by both the House of Lords[86] and the Court of Appeal[87]. The general principle that the Court of Appeal applied is that where a judge had a more than *de minimis* (qv) direct personal interest, he was automatically disqualified from hearing or continuing to hear any application or trial. The judge's knowledge of any such interest was irrelevant. A party with full knowledge of any such interest might waive (qv) any right to object. It is submitted that the courts are likely to adopt a similar approach in relation to the statutory test set out in s. 24 of the Arbitration Act 1996.

See also: **Arbitration; Arbitrator.**

Bid A contractor's price for carrying out work, submitted in competition with others. Another name for the contractor's tender (qv). The buyer at an auction sale makes a bid, i.e. offer, which the auctioneer is free to accept or reject.

Bill of sale A document under which a person transfers his property in personal chattels (qv) to someone else without transferring possession (qv) of them. In general terms it is a document creating a security and a bill of sale is a document of title (qv). The general position of the parties is similar to that of parties to a legal mortgage of land.

It is the substance of the transaction rather than its form which is decisive, e.g. where an owner of goods sells them to someone else and agrees to take back the goods on hire-purchase and the real object of the transaction is to provide security, no title in the goods will pass to the purchaser[88].

The rules governing bills of sale are complex. The Bills of Sale Acts 1878 and 1882 apply to most bills of sale. Bills of sale must be registered in the Central Office of the Supreme Court in London within seven days of their execution. If not registered the security is void (qv). The Acts apply only where the bill of sale is made by an individual, but the Companies Act 1985 requires company charges to be registered.

Bill of variations Sometimes known as the 'final account' or 'computation of the adjusted contract sum'. It is prepared by the quantity surveyor and should be completed within the period named in the contract and before the issue of the final certificate (qv). However, the final account is not a pre-condition to the issue of the final certificate[89].

The contractor must present the quantity surveyor with all the documents necessary for him to produce a detailed list of all the variations from the

---

[85]*Catalina* v. *Norma* (1938) 82 SJ 698.

[86]*R* v. *Bow Street Metropolitan Stipendiary Magistrate ex parte Pinochet Ugarte (No. 2)* [2000] 1 AC 119.

[87]*Locabail (UK) Ltd* v. *Bayfield Properties Ltd* [2000] QB 451.

[88]*North Central Wagon Finance Co Ltd* v. *Brailsford* [1962] 1 All ER 502.

[89]*Penwith District Council* v. *P. Developments Ltd*, 21 May 1999, unreported.

original bills of quantities (qv). These documents may take the form of invoices, sub-contractors' accounts, measurements, etc. The quantity surveyor normally prices all the items which have already been agreed or which he can price using the method set out in the particular contract. Any remaining items are sometimes settled by negotiation, but in most contract forms, e.g. JCT 98, the quantity surveyor is required to make the decision. It is good practice to send the finished bill to the contractor prior to the final certificate whether or not the contract requires it.

JCT 98 refers to the procedure in clause 30.6.1. ACA 3 refers to 'computing the Final Contract Sum' in clause 19.1, GC/Works/1 (1998) refers to the final account in clause 49.

See also: **Final account.**

Bills of quantities A detailed list of all the work and materials required to produce a building. Its main purpose is to allow rates to be fixed for every item of work and materials and thus to arrive at a total price; the resultant rates become the basis for valuing variations. The bill is commonly divided into two sections:

— The preliminaries (separated into fixed and time related charges).
— The measured work.

The preliminaries contain factual information of a general nature to help the contractor arrive at a price for the work. Among items to be included are the following:

— Name and address of the employer.
— Name and address of the architect.
— Name and address of the quantity surveyor.
— Description of the site including access and working space.
— Visiting the site.
— Trial holes.
— Inspection of drawings.
— Possession.
— General description of the works.
— Form and type of contract.
— Plant, tools and vehicles.
— Safety, health and welfare.
— Notices and fees.
— Setting out.
— General foreman, person-in-charge.
— Maintenance of roads.
— Safeguarding the works.
— Police regulations.
— Obligations and restrictions imposed by the employer.
— Water for the works.
— Lighting and power for the works.
— Injury to persons and property and damage to the works.
— Insurance of employer's liabilities.

— Clearing away.
— Temporary roads, etc.
— Temporary sheds, offices, messrooms, sanitary accommodation, etc.
— Temporary offices for use of architect, quantity surveyor, clerk of works.
— Temporary telephone facilities.
— Temporary screens, fencing, hoarding, etc.
— General scaffolding.
— Works by the local authority or statutory undertaking.
— Nominated sub-contract works.
— Nominated suppliers.
— Protecting the works.
— Drying the works.
— Removal of rubbish.

A full description of the items to be included in the preliminaries can be found in the Standard Method of Measurement, 7th edition, pp. 17–22.

The quantities, together with the relevant description of the item, make up the major portion of the bills and they are usually arranged under trades, approximately in the order in which they will be carried out. Each trade has a preamble which provides a general description of the work for the contractor, followed by a series of detailed descriptions of every part of the particular trade. Alongside each description the quantity is expressed in lineal, square or cubic measurement (e.g. lin m or m^1, sqm or m^2, cu m or m^3) or enumerated (e.g. No. 5). Weights are given in kg. Space is provided for the contractor to insert a rate opposite each item and to arrive at a total for each item by multiplying the quantity by the rate. The prices can be totalled by the contractor at the foot of each page and the totals carried to a collection at the end of each trade. The collections are gathered together on a sheet at the end of the document and totalled to arrive at a price for the whole work.

The method of measurement adopted for the bills should be clearly stated therein. Where there are exceptions, they must be specifically stated in relation to particular items. Usually this is the Standard Method of Measurement of Building Works (SMM), compiled by the Royal Institution of Chartered Surveyors and the Construction Confederation, the seventh edition of which (SMM 7) became operative on 1 July 1988. It was revised in 1998. It is a set of rules based on the Common Arrangement of Work Sections which provides a standard measurement basis.

Use of the SMM is mandatory on JCT contracts; thus, if on such contracts it is desired to depart from the SMM for any reason, the departure must be clearly stated at the beginning of the bills.

The preparation of bills of quantities is normally undertaken by a quantity surveyor from the architect's production information: drawings, schedules and specifications (qv). The traditional process is complex and usually involves the following stages:

— Taking off, i.e. taking the measurements from architect's drawings.
— Abstracting, i.e. the gathering together of quantities for like items.

— Billing, i.e. writing up the final bills of quantities. A detailed specification (qv), which should be prepared by the architect, is commonly bound into the bills between the preliminaries and the quantities and is usually referred to as the Trade Preambles. Bills of quantities are included as one of the contract documents in JCT 98 (With Quantities editions), ACA 3 and GC/Works/1 With Quantities (1998) contracts. They are often referred to as the 'contract bills'.

The use of computerised systems has simplified this task.

It is sometimes useful to prepare what is known as 'Bills of approximate quantities' in order to determine rates for various items of work and materials when the precise quantity cannot be established before tenders are required to be submitted. By this method a rough idea of the total price can be obtained before work begins and the individual rates are applied to the actual quantities once these are fixed, which enables fast, accurate and indisputable remeasurement to proceed as the job progresses. Such a bill is often used for works of alteration or renovation or in the rarer instances where work must begin before the architect's production information can be completed.

See also: **Bill of variations; Discrepancies; Schedule of Activities; Schedule of Rates.**

Body of deed The operative parts of a deed (qv) which set out the terms of the agreement between the parties.

Bonds A bond is a contract under seal (qv) where a bank, insurance or parent company assumes obligations towards a beneficiary (usually the employer). The bond must be honoured according to its precise terms – have the event(s) occurred which trigger the obligation to pay[90]? In the absence of fraud[91], upon the happening of the relevant event(s) and/or valid demand (if required) payment must be made without proof or conditions[92].

There are several different types of bond, but in general their purpose is to guarantee payment of a fixed sum by way of compensation for non-performance of a contractual obligation. In this context, a *performance bond* is an undertaking given by an insurance company, bank or other surety to indemnify the beneficiary (usually the employer) against the contractor's failure to perform the contract. In the UK, performance bonds are usually 'on default' bonds, i.e. their operation is conditioned on the contractor defaulting on his obligation, but in international practice 'on demand' bonds are common. 'On demand' bonds are becoming more common in the UK also and they tend to be favoured by the financial institutions providing the bonds, because when they receive a demand made in conformity with the criteria set

---

[90]*Howe Richardson Scale Co Ltd* v. *Polimex-Cekop and National Westminster Bank Ltd* [1978] 1 Lloyds Rep 161, 165 (CA).
[91]See *Turkiye IS Bankasi AS* v. *Bank of China* [1996] 2 Lloyds Rep 611.
[92]*Edward Owen Engineering Ltd* v. *Barclays Bank International Ltd* [1978] QB 159 per Lord Denning MR at 171.

out in the bond, they can pay without the necessity of carrying out any kind of investigation.

There is a growing number of English cases relating to performance bonds, which are common in the building industry and are commonly required by local and other public authorities. A performance bond is a strong weapon in the employer's hands to ensure prompt completion and is, in effect, a thinly-disguised solvency guarantee. The effect of a bond, like that of any other contract, depends on its precise wording and this can be particularly important where it is vital to have funds available to complete the works after the original contractor has defaulted[93]. The standard bond has been held to be a 'default' bond and a guarantee. Moreover, the surety may be entitled to set off or counterclaim any amounts due, before making payments[94].

JCT 98 and IFC 98 have two bond forms bound into the document. One is the bond for use where the employer and the contractor agree to operate the advance payment (qv) provisions. The other bond is used where the employer has agreed that payment can be certified for off-site materials (qv) and he has provided a list. The provision of a bond is one of the mandatory criteria which must be satisfied if the value of materials which are not uniquely identified is to be certified. It is wise also to use the bond for uniquely identified materials. These two bonds are on terms agreed between the Joint Contracts Tribunal (qv) and the British Bankers' Association.

Other types of bond which are occasionally used are retention bonds: to be given by a contractor in return for early release of, or in place of, traditional retention; and bid bonds: used to guarantee a contractor's intentions in the case of complicated or large-scale projects.

Bonus clause A clause included in a contract with the object of encouraging the contractor to complete the works before the contractual completion date by offering additional money for early completion. The Engineering and Construction Contract (ECC) (qv) is the only current standard form to include a bonus for early completion. This is at a set amount per day and is payable to the contractor for each day by which completion of the works, as certified by the project manager, precedes the date for completion or extended completion.

The amount of money specified need not bear any relationship to the amount the employer pre-estimates he will gain through early completion. Therefore, there is no necessity for a bonus amount to be a genuine pre-estimate of gain. The employer may stipulate any sum he thinks fit. In this, it is quite unlike liquidated damages (qv).

Problems may arise if the amount of the bonus is not as great as any figure for liquidated damages in the contract. Moreover, default by the employer which prevents the contractor from earning the bonus may result in the contractor recovering its amount as damages for breach of contract[95] but, unless

---

[93] *Paddington Churches Housing Association* v. *Technical & General Guarantee Co Ltd* [1999] BLR 244.
[94] *Trafalgar House Construction (Regions) Ltd* v. *General Surety and Guarantee Co Ltd* (1995) 73 BLR 32.
[95] *Bywaters* v. *Curnick* (1906) HBC, 4th edn, vol. 2, p. 393.

the clause expressly provides to this effect, circumstances beyond the builder's control which delay completion will not entitle him to the bonus[96]. It is also open to question whether a bonus clause is a significant incentive to the contractor. It is perhaps better to specify a shorter contract period at the tendering stage so that the contractor may price accordingly. Many contractors argue that a bonus clause must always be present in any contract which includes a liquidated damages clause – presumably on a 'carrot and stick' principle. This argument is without any legal foundation.

GC/Works/1 (1998), clause 38(4) empowers the contractor to submit to the project manager written proposals which he thinks will 'enhance the buildability of the Works, reduce the cost of the Works or the cost of maintenance or increase the efficiency of the completed Works'. If accepted, any savings are shared equally between employer and contractor, completion and programme dates are adjusted and any necessary extension of time is given. This is not the same as a bonus clause, but it operates on the same principle, i.e. that the contractor will be rewarded if he achieves some advantage for the employer. Clause 38A provides for a bonus if the Abstract of Particulars (qv) so states.

Boundaries The demarcation lines between the ownerships of land. Boundaries should be defined in the title deeds although frequently they are obscure. Common reference is to walls, fences, hedges and watercourses. Ownership usually, but not invariably, extends to the centre line of highways (qv) and watercourses. If the boundary is not clear from the title, it may be possible to settle the matter on site in the presence of both owners. There are certain presumptions which may be useful (see **Figure 3**). If the parties cannot agree, the matter can be settled, expensively, in court.

Encroachment over or under a boundary will give rise to an action for trespass (qv), but may also give rise to variation of the boundary by adverse possession (qv).

BPF System In December 1983 the British Property Federation (BPF), which represents the interests of property owners, published a manual describing a new system for the organisation of contract management. The manual has not been revised and it is still current although some supporting documentation has been withdrawn, revised or should be treated with care in view of changes in legislation, particularly the Housing Grants, Construction and Regeneration Act 1996 (qv) and the Contracts (Rights of Third Parties) Act 1999 (qv). The BPF System divides the design and building process into five stages – concept, preparation of brief, development of design, tender documentation and construction. The manual specifies these stages and notes the duties of the parties.

The system re-defines traditional roles, introducing 'a client's representative' (qv) who manages the project on behalf of the client, and a 'design leader' (qv) who has overall responsibility for pre-tender design and for sanctioning any contractor's design. A significant feature of the new system is that, in general,

---

[96]*Leslie* v. *Metropolitan Asylums Board* (1901) 68 JP 86.

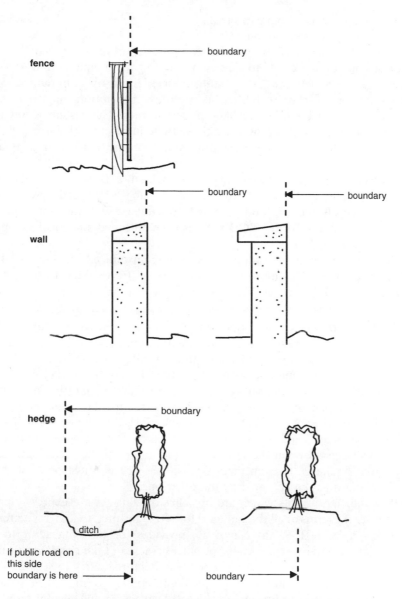

Figure 3 Boundaries: presumption of ownership.

payment is based on a 'Schedule of Activities' (qv) rather than on traditional bills of quantities (qv).

The BPF system aims to remedy the main problems which arise under the traditional system of contracting:

— Incomplete initial design, leading to extensive variations and disruption of the building programme.

— Higher costs than expected.
— Delays in completion.

The system emphasises the need for production of a detailed brief at the outset and for the client to determine his full requirements and at an early stage. It also sets out to establish clear lines of communication and demarcation of responsibilities as well as the notion of fixed fees for consultants and essentially a fixed price contract. The system is intended to be flexible and is not a rigid formula suitable in its entirety for every project. Properly operated, it may save time and cost and achieve high quality in building.

Subsequent to the publication of the manual, a BPF edition of the ACA Form of Building Agreement (qv) has been published, together with a Model Consultancy Agreement and various forms for use with the system. The JCT Design and Build Form (WCD 98) has been amended so as to be compatible with the system.

Major changes to the traditional procedure introduced are shown in **Table 1**. Copies of the BPF manual are obtainable from: The British Property Federation, 35 Catherine Place, London SW1E 6DY.

Breach of contract An unjustified failure to carry out obligations under the contract or a repudiation (qv) of contractual obligations. The breach may be total, i.e. refusing to perform the contract at all, in which case it is known as 'repudiation', or it may be partial. The breach may be of varying degrees of seriousness depending upon whether it is breach of a condition or a warranty (qvv). The typical common law remedy is to sue for damages and/or to treat the obligations under the contract as discharged. Specific performance (qv) is another remedy available in appropriate cases. The remedies applied by the court will depend upon the seriousness and nature of the breach. Breach does not itself discharge the contract; to do so the breach must be repudiatory in nature and must be accepted by the other party.

A number of events which are breaches of contract are expressly provided for under the terms of all the standard forms of contract, together with remedies. For example, JCT 98, at clause 26, provides for the contractor to obtain financial recompense for certain specified breaches of the employer. It must be noted, however, that it is always open to the injured party to seek damages at common law rather than through the contractual provisions if he so desires. See also: **Anticipatory breach of contract; Damages; Fundamental term.**

Bribery and corruption Promising, offering or giving money, secret commission, gifts, etc. to someone to influence his conduct. Secret dealings of this type as between, e.g. architect and contractor, would entitle the employer to terminate the architect's employment and to recover any commission paid[97]. The employer could also treat the building contract as at an end.

---

[97] *Reading* v. *Attorney-General* [1951] AC 507.

Table 1 Major changes in the BPF manual.

Adjudication	Disputes arising during the contract to be settled within five working days by an independent adjudicator. His decisions are subject to post-contract arbitration. The aim is to provide speedy settlement of disputes and prevent any clash of interests.
Architect	Provides pre-contract conceptual and architectural design and advises on architectural and design aspects of variations during construction. Has no authority to issue instructions to the contractor.
Bills of quantities	Not used in the recommended system although the contract makes provision for them as an option. Tenders are invited against drawings and specifications produced by the design team.
Consultants	Specialists – including architects and engineers – contracted by the client to produce design and cost services as well as those provided by the design leader.
Client's representative	The person or firm managing the project on the client's behalf. He may delegate his duties under the contract to any number of assistants.
Design leader	Has overall responsibilities for pre-tender design etc. and sanctions any contractor's design through the client's representative. The team leader.
Design liabilities	Option for contractor to undertake a proportion of design.
Fixed-fee contracts	Professionals to work for a fixed fee. All contracts on a fixed price basis unless they last for more than two years when 80% only of fluctuations payable based on ACA Index.
Schedule of activities	Prepared by the contractor, it replaces priced bills. It is a priced schedule of the contractor's activities and forms the basis of his tender. Used to manage the project, monitor progress and for payment. Payment made on the basis of each completed activity, with provision for pro rata payment of preliminaries.
Sub-contractors	'Named sub-contractors' and suppliers if client wishes. Otherwise choice left to main contractor with whom full responsibility remains.
Variations	Special procedure laid down, based on contractor's estimates of cost.

Some building contracts deal expressly with this matter, e.g. GC/Works/1 (1998), clause 24; JCT 98, clause 27.4, and entitle the employer to determine the contract or employment under the contract. Corrupt practices are a criminal offence under the Prevention of Corruption Acts 1889 to 1916 whether in connection with a contract or otherwise.

Brown clause A contract term providing for liquidated prolongation costs (qv) is sometimes called a Brown clause.
See also: **Liquidated prolongation costs**.

Budget price A price given by a contractor which is not intended to be precise, but which is intended to cover the cost of all the work of which the contractor has been informed. Therefore, it is often a figure somewhat greater than the likely cost. A 'budget estimate' is similar in that it is a rough estimate possibly produced by the quantity surveyor (qv) or the contractor and pitched slightly higher than the likely price. It is not usually intended to be capable of acceptance to form a binding contract (but see **Estimate**).

Builder The individual, partnership or firm carrying out building works. Most contracts now refer to the 'contractor' (qv).

Building Defined in s. 121 of the Building Act 1984 as 'any permanent or temporary building and, unless the context otherwise requires, it includes any other structure or erection of whatever kind or nature (whether permanent or temporary)'. The definition proceeds to embrace vehicles, vessels, hovercraft or other movable objects of any kind under the heading: 'structure or erection', provided that circumstances prescribed by the Secretary of State prevail. The exercise of the Secretary of State's power is qualified and the circumstances must justify treating the object as a building.

The definition is very broad and it should be noted that the definition in the Building Regulations (qv) made under the Act is comparatively tight: 'any permanent or temporary building, but not any other kind of structure or erection'.

Building control The system of controls over the construction and design of buildings, other than planning controls. In England and Wales the basic framework is contained in the Building Act 1984 which consolidates all earlier primary statutory material, and in the Building Regulations (qv) which set out legal and constructional rules in greater detail. Local Acts (qv) also contain building control provisions. In Northern Ireland, the equivalent legislation is the Building Regulations (Northern Ireland) Order 1979 as amended by the Planning and Building Regulations (Amendment) (Northern Ireland) Order 1990.

In Scotland, the system of control is based on the Building (Scotland) Acts 1959–1970, as amended, and in regulations made under them, i.e. the Building Standards (Scotland) Regulations 1990.

See also: **Approved documents; Building Regulations.**

Building Employers' Confederation See: **Construction Confederation.**

Building line An imaginary line drawn parallel to the highway at a specified distance from the back of the footpath (if any). The dimensions are specified by the local planning authority as part of their overall responsibility for development control. The significance of the line is that no building or part of any building (with certain minor exceptions) may be erected between the building line and the highway. The authority has considerable discretion in fixing the line, depending upon all the circumstances. The main purpose of a building line is to ensure privacy and sight lines. Thus the building line on housing estates may be generally five metres, while in a town centre may well be the back of the footpath. Individual consultation with the local planning officer is necessary to establish the line required in any particular situation.

Building owner Usually, but not invariably, the person or firm known in most forms of building contract as 'the employer'. It is the person or firm which owns the site or will own the structure on completion (qv).

The Party Wall Act 1996 (qv) defines 'building owner' in section 20 as 'an owner of land who is desirous of exercising rights under this Act', thus giving the term a technical significance for the purposes of legislation.

Building Regulations The Building Regulations 1991 form the basis of the system of building control (qv) in England and Wales. In Northern Ireland, it is the Building Regulations (Northern Ireland) 1994.

They are set out in the form of functional requirements and are supported by a wide range of 'approved documents' (but not in Northern Ireland) which give practical guidance in respect of their provisions. The Regulations are arranged in three parts:
— General.
— Application.
— Procedural and miscellaneous provisions.

It should be noted that, in the context of building contracts, the contractor must comply with the building regulations: *Street* v. *Sibbabridge Ltd* (1980) unreported. Most of the standard form contracts make this clear, e.g. JCT 98, clause 6.1.1 imposes on the contractor an express duty to comply with all statutory obligations; such a term would be implied in any event.

Buildmark The trade mark of the NHBC Combined 10 year Warranty and Protection Scheme. The level of cover has been increased but the increase applies only to dwellings registered with the NHBC on or after 1 April 1999.

Buildmark does not cover against general wear and tear, condensation, normal shrinkage, failure to maintain the property or minor faults appearing after the second year. It does cover against certain specified risks which may be expensive to remedy. Essentially, physical damage to the dwelling caused by a defect due to a failure to comply with NHBC standards is covered during the first two years. The builder is liable for correction of the problem. For the remaining eight years, Buildmark covers against the full cost (if over £500) of putting right serious defects including contaminated land.

Further information is obtainable from http://www.nhbc.co.uk/houseinfo/guide.html
See: **National House Building Council.**

Burden of a contract The obligation which rests upon one party to a contract, e.g. under a building contract the contractor's obligation to execute and

complete the works. A contracting party cannot assign a contract so as to relieve himself of its burdens without the consent of the other party[98].
See also: **Assignment; Novation.**

Byelaw A form of delegated legislation (qv) made by local authorities and certain other public bodies and confirmed by some central government departments. They are a kind of local law enforceable in the courts which have power to review them and determine whether or not they have been properly made. Building control (qv) was formerly exercised through local building byelaws (now replaced by Building Regulations (qv)).

---

[98]*Tolhurst* v. *Associated Portland Cement Manufacturers (1990) Ltd* [1902] 2 KB 660.

C

Calderbank offer An offer to settle made in arbitration and expressed to be made 'without prejudice as to all matters except costs'. It must not be mentioned to the arbitrator until he has issued an award on all matters except costs when the arbitrator may treat it as having the same effect as a payment into court. It is so called after the case of *Calderbank* v. *Calderbank* (1975)[99] in which the procedure was judicially approved. Important constituents of such an offer are:

— It must be left open for acceptance for 21 days.
— It must offer to pay the other party's reasonable costs incurred up to the point of acceptance if accepted within 21 days. Thereafter, acceptance being subject to the offeror's costs incurred after the expiry of the 21 days being paid.
— It must say what is the position as to interest. Normally interest is included.
— It must say whether any counterclaim is taken into account in the offer.

Calendar month See: **Month.**

Capacity to contract The general law is that any person can enter into a binding contract. To this general rule there are a number of exceptions or qualifications. They may be summarised under the following heads:

— Corporations.
— Minors.
— Insane persons.
— Drunkards.
— Aliens.
— Agents.
— Unincorporated associations.

Corporations All corporations are restricted in their actions by the rules by which they were formed. For example, a company registered under the Companies Acts is restricted by its Memorandum of Association, a local authority is restricted by various statutes (qv). They may make binding contracts if such contracts are within the powers conferred upon them. If they attempt to make contracts outside their powers, such contracts are *ultra vires* (qv) and void.

Minors Persons under the age of eighteen. As a general rule a minor may only enter into a binding contract:

— For necessaries.
— For his benefit.

'Necessaries' include such things as food and clothing, but the concept is by no means clear-cut because items falling into the category of 'necessaries' will

---

[99][1975] 3 All ER 333.

61

depend upon circumstances. Contracts for the minor's benefit include contracts of apprenticeship and education. As with 'necessaries', the court will take all the circumstances into account in deciding whether a contract is for the minor's benefit.

All other contracts entered into by a minor are invalid. Thus contracts for the supply of goods or for payment of money cannot be enforced. Contracts which are of a long-term nature, such as the acquiring of an interest in land or a firm will become binding upon the minor unless he repudiates them before or soon after reaching the age of eighteen.

Insane persons Contracts are generally voidable, i.e. the legal relations which would otherwise be established can be avoided at the instance of one of the parties, provided the person was so insane when he made the contract that he did not know what he was doing and the other party was aware of it. If the insane person recovers his sanity, he may be bound by a contract made during the period of his insanity unless he repudiates the contract within a reasonable time.

Drunkards Contracts with drunken persons generally fall under the same rules as contracts with insane persons, but it is possible that the courts may have a broader discretion to set aside contracts purportedly made under the influence of drink.

Aliens Generally, in peacetime, an alien has the same capacity to contract as a British national (but see: **Alien enemy**).

Agents Capacity to form a binding contract on behalf of a principal depends upon the terms of the agency (qv).

Unincorporated associations (qv) Such groups are generally not capable of entering a contract and so cannot sue or be sued. There are, however, statutory exceptions, e.g. trade unions. This does not mean that no contract can be created, rather the parties to the contract may differ from those intended. Where a member of such a group purports to contract on behalf of the association, to the extent that he has authority, whether express or implied, he will act as agent for the members of the association and so all can be sued. If there is no such authority, there may be personal liability for the contracting member, but it will not bind the remaining members.

Care, duty of The legal obligation owed whereby the law requires care to be taken. The classic test is the so-called 'neighbour principle' set out by Lord Atkin in *Donoghue* v. *Stevenson*[100]. The imposition of such a duty fluctuates 'in accordance with changing social needs and standards' whereby 'new classes of persons legally bound or entitled to the exercise of care may from time to time emerge'[101]. The question posed by the court is whether it is fair, just and reasonable to impose liability[102]. This is a prerequisite of any claim in

---

[100][1932] AC 562.
[101]*Candler* v. *Crane, Christmas & Co* [1951] 1 All ER 426 per Asquith LJ at 441.
[102]*Henderson* v. *Merrett Syndicates Ltd* [1995] 2 AC 145 per Lord Goff at 180–181.

negligence. It is clear that a party may owe concurrent duties in contract and tort[103].

This area of law has been predominantly developed by case law and, as such, has changed over time – the House of Lords' decision in *Murphy* v. *Brentwood District Council*[104] marked a dramatic change from the high-water mark of the (now discredited) decision in *Anns* v. *Merton London Borough Council*[105]. It remains to be seen whether the decision in *Henderson* v. *Merrett Syndicates Ltd*[106] marks a move away from *Murphy*.

Examples of where a duty of care has been imposed:
— Manufacturers, etc. towards the ultimate consumer.
— Employers to employees.
— Architects towards third parties (qv).
— Builder to subsequent occupiers.

See also: **Negligence.**

Care, standard of In actions for negligence (qv) it is necessary to establish that the defendant has failed to discharge the duty of care expected of him. This standard of care is that of the 'reasonable man', who is a hypothetical creature of ordinary prudence and intelligence.

> 'Negligence is the omission to do something which a reasonable man guided upon those considerations which ordinarily regulate the conduct of human affairs, would do, or doing something which a prudent and reasonable man would not do.'[107]

However, if someone holds himself out as being capable of attaining a certain standard of skill, e.g. an architect, a contractor or an engineer, he must show the skill which is generally possessed by people in his trade or profession. So, when discharging the duties which he has contracted to do, the contractor or professional man is to be judged by the generally accepted standards prevalent at the time he carried out his work.

> 'Where you get a situation which involves the use of some special skill or competence, then the test as to whether there has been negligence or not is not the test of the man on the top of the Clapham omnibus, because he has not got this special skill. The test is the standard of the ordinary skilled man exercising and professing to have that special skill; it is well established law that it is sufficient if he exercises the ordinary skill of an ordinary competent man exercising that particular art.'[108]

This test has been approved time and time again. The terms of the contract may impose a higher standard, but generally the contractor must exercise in relation to his work the standard of care which is to be expected of a reasonably competent building contractor[109].

---

[103] *Henderson* v. *Merrett Syndicates Ltd* [1995] 2 AC 145.
[104] [1991] AC 398.
[105] (1978) 5 BLR 1.
[106] [1995] 2 AC 145.
[107] *Blythe* v. *Birmingham Waterworks Co* (1856) 11 Ex 781 per Alderson B at 784.
[108] *Bolam* v. *Friern Hospital Management Committee* [1957] 2 All ER 118 per McNair J at 121.
[109] *Worlock* v. *SAWS & Rushmoor Borough Council* (1982) 22 BLR 66.

The basic test establishes the degree of knowledge or awareness which the professional man ought to have. If, in fact, he has a higher degree of knowledge or awareness and acts in a way which, in light of that actual knowledge, he ought reasonably to have foreseen would cause damage, he may be liable in negligence even though the ordinary skilled professional would not have that knowledge[110].

See also: **Foreseeability.**

Case stated A procedure under the Arbitration Act 1950 by which the arbitrator could make his award in the form of alternatives hinging upon the interpretation of a point of law. The point was put to the High Court for resolution as a 'case stated'. The procedure was abolished by the Arbitration Act 1979, but is still applicable in Scotland.

The Arbitration Act 1996 has provision under s. 45 for the determination of points of law on the application of a party to the proceedings. The court is not empowered to consider such an application unless certain criteria have been satisfied. The application must be made with the agreement of all parties to the arbitration, the arbitrator must have given permission and the court must be satisfied that the application was made without delay and a decision is likely to save substantial costs.

See also: **Arbitration; Points of law.**

Cash discount Commonly considered to be a discount for prompt payment by the main contractor. It is allowed by sub-contractors and suppliers, usually for payment within a specified period. It is unusual for the contractor to be liable to pass the discount to the employer and, conversely, the employer does not guarantee to the contractor that it will be paid. It is a matter between the contractor and the sub-contractor. The purpose is to assist the contractor in his forward financing of the work. Some contractors look upon it as additional profit. Usual cash discounts are 2.5% from sub-contractors and 5% from suppliers. Where a payment period is stipulated, it appears that the contractor has no right to the discount unless he makes payment within the stipulated period. However, it has been held that where the discount is not made dependent upon payment within a specific period, the contractor is entitled to deduct such discount whenever payment is made[111]. If a provisional sum, on which the contractor expected to make money from cash discounts, is omitted, the contractor has no claim to the lost discount.

Causa causans The immediate cause. It is the last link in the chain of causation (qv) and must be recognised as different from the *causa sine qua non*, which is some earlier link but for which the *causa causans* would not have operated. In relation to monetary claims for direct loss and/or expense under building contracts, it means that the loss and/or expense must have been caused by the

---

[110]*Wimpey Construction UK Ltd* v. *D. V. Poole* (1984) 27 BLR 58.
[111]*Team Management Services plc* v. *Kier Management and Design Ltd* (1993) 63 BLR 76.

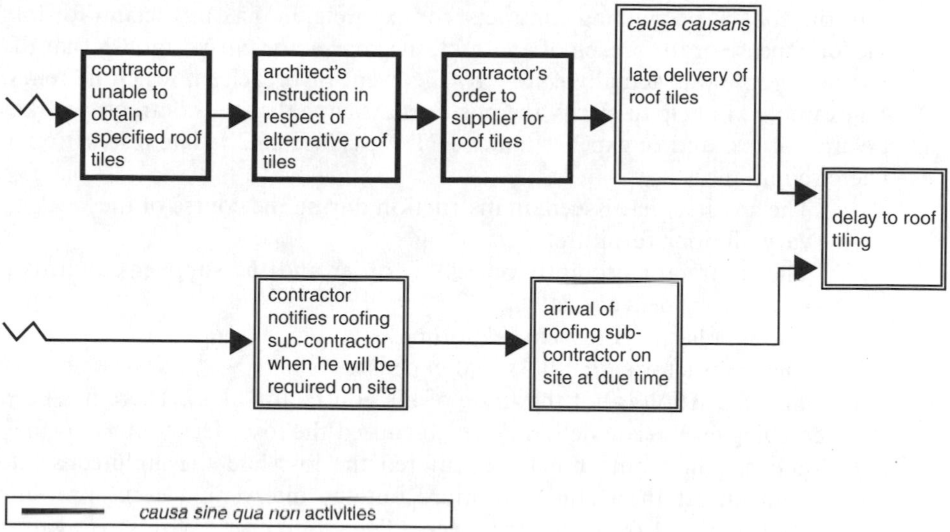

Figure 4   Chain of causation.

breach or act relied on and not merely be the occasion for it[112]. Many contractors have a very confused view of causation which leads them to submit claims which have no hope of success. For example, where there is a claim for direct loss and/or expense under JCT 98, clause 26.2.7 in respect of a variation, the loss and expense must flow from the variation order as a *causa causans*. If a variation order requires the contractor to obtain materials from a specified supplier who, in breach of his contract of sale with the contractor, delivers late or delivers defective materials, the *causa causans* is the supplier's breach of contract and not the variation order which is no more than a *causa sine qua non*. The simple precedence diagram in **Figure 4** should clarify the point. To take a quite different example, a Royal Mail delivery driver, involved in an accident, would blame his delivery instructions (a *causa sine qua non*) only at the risk of appearing ridiculous if the fault lay with his careless driving (the *causa causans*).

See also: **Causation; Foreseeability; Remoteness of damage.**

Causation The relationship between cause and effect. The concept is very important in the context of liability for negligence (qv). In many cases, the doing of a wrongful act starts off a series of events which lead to damage being suffered, and this is called by lawyers a 'chain of causation'. If liability is to be established, the original wrongful act must be connected, without interruption, to the loss or damage suffered or incurred by the injured party. Thus, if the effective cause of the damage was not the original event but some intervening event, the defendant will not be liable. (The legal term is *novus actus interveniens* — a new act coming in between.)

---

[112]*Weld-Blundell* v. *Stevens* [1920] AC 956.

In the context of building contracts, for example, in making a claim for loss and/or expense or for breach of contract, the contractor must establish that the loss or expense was actually *caused* by the event or breach on which he relies. An example will help to clarify the position. Assume that the contractor makes a claim for loss and/or expense under JCT 98, clause 26.2.1 (late instructions). The circumstances are:

— The architect has issued an instruction during the course of the work to vary all door furniture.
— The contractor promptly places his order and the supplier confirms a satisfactory delivery date.
— The supplier fails to deliver on time.
— The contractor suffers loss and expense.

The late instruction is not the *cause* of the contractor's loss. There has been an intervening event (late delivery) which caused the loss. The contractor may well argue that he would not have suffered the loss had the architect's late instruction not set the chain in motion, but the intervening event prevents recovery of damages from the employer. (The contractor's redress is against the supplier.) The intervening event might well be the contractor's own inefficiency. If, however, the architect's late instruction resulted in the contractor being unable to obtain a satisfactory delivery date and the supplier correctly delivered to such later date as was agreed, the late instruction would be the *cause* of the damage suffered by the contractor and a successful claim could result.

A graphic example of the concept of causation is found in *Lubenham Fidelities & Investment Co v. South Pembrokeshire District Council and Wigley Fox Partnership* (1986)[113] where negligent architects issued defective interim certificates and the contractors withdrew from site. The contractors lost their claim against the employer because they broke the chain of causation by persisting in suspension of the works despite the service by the employer of a preliminary notice of determination. They alone were responsible for the termination of the contracts. Although the negligence of Wigley Fox was the source of the events it was overtaken and overwhelmed by the contractors' serious breach of contract.

See also: **Causa causans; Foreseeability; Remoteness of damage.**

Caveat emptor Let the buyer beware. This is the basic common law rule in law of sale of goods, that the buyer purchases at his own risk and relies on his own judgment as to suitability or quality. Modern legislation has attenuated this principle particularly in the case of purchases by ordinary consumers, e.g. in most situations the Sale of Goods Act 1979 implies a condition that goods are of satisfactory quality (qv) and will be reasonably fit for their intended purpose. See also: The **Unfair Contract Terms Act 1977.**

CDM Regulations 1994 See: **Construction (Design and Management) Regulations 1994.**

---

[113](1986) 6 Con LR 85.

Certificates The expression of the architect's opinion in tangible form for the purposes specified in the contract[114]. All the standard forms of contract provide for the architect to issue certificates at various times. It is crucial that all certificates are issued promptly by the architect, otherwise the contractor may have a claim in damages since failure by the architect to issue a certificate required by the contract is a breach of contract for which the employer is liable.

Standard certification forms are available for use with some contracts and it is wise to use them. Where no form is available, a certificate must be specially prepared. A certificate may take the form of a letter, but to avoid any doubt, the letter should be headed 'Certificate of ...' and begin 'This is to certify ...'.

If the certificate is to be issued by an architect, it must be signed by an architect or by someone expressly empowered to sign on his behalf. In such circumstances the named architect will still be liable for any errors in the certificate.

Because a certificate is a contractual document, once issued it may not be altered or amended (except probably for obvious errors) unless this is empowered by the contract, e.g. ACA 3, clause 19.5. A certificate is not 'issued' merely because the architect signs it; it must be put into circulation, e.g. by being sent to the employer[115].

The effect of a certificate depends upon the actual wording of the contract[116]. In most standard form contracts an architect's certificate is a condition precedent (qv) to payment to the contractor, but if the architect refuses to issue the certificate, the contractor can sue without it[117].

Interference or obstruction with the issue of a certificate is, under most standard contracts, a ground on which the contractor may terminate his employment under the contract, e.g. JCT 98, clause 28.2.1.2.

See also: **Final certificate; Interim certificates.**

Chain of causation See: **Causation.**

Change of parties See: **Assignment and sub-letting; Novation.**

Charging order A judgment creditor can apply to the court for an order imposing a charge on a debtor's property as a means of enforcing his judgment. The court's discretion to charge a debtor's property in this way is derived from s. 1 of the Charging Orders Act 1979. The creditor is not entitled to the order as of right but the order will usually be made unless the debtor can persuade the court that in all the circumstances it should not be made. The charge may be enforced by an order for sale.

Section 75 of the Arbitration Act 1996 empowers the court to make orders charging property under s. 73 of the Solicitors Act 1974 or Article 71H of the Solicitors (Northern Ireland) Order 1976 in respect of arbitral proceedings as though the proceedings were in court.

---

[114] *Token Construction Co Ltd* v. *Charlton Estates Ltd* (1973) 2 BLR 3.
[115] *London Borough of Camden* v. *Thomas McInerney & Sons Ltd* (1986) 9 Con LR 99.
[116] *East Ham Borough Council* v. *Bernard Sunley & Sons Ltd* [1965] 3 All ER 619.
[117] *Page* v. *Llandaff Rural District Council* (1901) HBC, 4th edn, vol. 2, p. 316; *Croudace Construction Ltd* v. *London Borough of Lambeth* (1986) 6 Con LR 70.

Chattels Any property other than freehold land. *Chattels real* are leasehold interests in land in contrast to *chattels personal* which are all other things capable of being owned, e.g. goods and materials.
See also: **Personal property.**

Cheque, payment by Payment by cheque is only conditional payment. A creditor is not bound to accept a cheque in payment of his debt, but if he does so the debt will be discharged, provided the cheque is not dishonoured by the bank. Theoretically, under most of the standard form contracts, payment of amounts due on certificate ought to be made in legal tender (qv) because none of the standard forms makes provision for payment by cheque. There is nothing to prevent a special contract being drawn up to that effect.

In practice, payment by cheque (if the cheque is honoured) is sufficient, and it might be well argued that there is an established custom (qv) in the industry to that effect, and certainly if certificated payments have been accepted by cheque, and the cheques have been duly honoured, it is not thought that the courts would look kindly on a claim that a later payment by cheque amounted to a breach of contract (qv).

Choses in action; in possession Personal rights of property which are enforceable by legal action. Choses in action are intangible rights, such as a debt or the right to recover damages, in contrast with choses in possession (things in possession) which are items of personal property capable of physical possession. In general, they can be assigned (qv) and are transferred on death or bankruptcy (qv).
See also: **Personal property.**

Circuitry of action Claims by two parties which are effectively equal and opposite and cancel each other out[118].

Civil commotion A phrase used to describe a situation which is more serious than a riot (qv) but not as serious as civil war (qv)[119]. The essential element is one of turbulence or tumult, though it is not necessary to show that the acts were done at the instigation of an outside organisation. Civil commotion may amount to *force majeure*. The activities of protesters in public places may well amount to civil commotion on occasion.

JCT 98, clause 25.4.4 and IFC 98, clause 2.4.4 provide that civil commotion which delays the works is a ground for extension of time. Civil commotion which causes suspension of the works for a specified period is a ground on which the employer or the contractor may determine the contractor's employment (clause 28A.1.1.3) and it is also referred to in clause 22.2 as being one of the excluded risks if the contract is carried out in Northern Ireland.
See also: **Commotion; Disorder; Riot.**

---

[118]For example, see *Mifflin Construction Ltd* v. *Netto Food Stores Ltd* 26 October 1993 unreported; *Hydrocarbons Great Britain Ltd* v. *Cammell Laird Shipbuilders Ltd and Automotive Products plc* (1991) 53 BLR 84.
[119]*Levy* v. *Assicurazioni Generali* [1940] 3 All ER 427.

Civil Liability (Contribution) Act 1978 Section 1 (1) of this Act enables 'any person liable in respect of any damage suffered by another person (to) recover contribution from any other person liable in respect of the same damage (whether jointly with him or otherwise)'. For example, the building owner may sue the architect for negligence (qv). The architect may bring in the contractor and sub-contractors for contribution. This does not, however, apply against someone entitled to be indemnified by the tortfeasor, e.g. the employer under the JCT contracts.

A contribution can also be recovered by someone who has made a payment in *bona fide* settlement of a claim 'without regard to whether or not he himself is or ever was liable in respect of the damage'. In all cases, the amount of contribution is a matter for the court's discretion. The amount is to be 'just and equitable' having regard to the person's liability for the damage in question.

Problems have been identified, because there are occasions when a party may have to shoulder the burden of paying all the damages although liable for only a part of the loss. SFA/99 (qv) attempts to overcome this so far as architects are concerned by introducing what is known as a net contribution clause. The clause provides that the architect will be liable to pay only such part of any total damages as he would have to pay if all other parties also liable paid their own particular contribution.

See also: **Indemnity clauses.**

Civil Procedure Rules (CPR) These came into effect on 26 April 1999 and replaced the Rules of the Supreme Court (the 'White Book'). They are available in many formats: textbook, loose-leaf, on CD-ROM and on the internet. They result from the recommendations made by Lord Woolf following his review of civil justice[120]. The Rules attempt to address the criticisms that civil justice is too slow, too costly and too complex. Judicial case management lies at the heart of the reforms. The Rules apply to both High Court and County Courts. The rules are in parts and are written in simpler language than formerly. In addition there is a useful glossary. Some old expressions, such as 'writ' or 'pleadings' have been discarded in favour of more understandable terms, e.g. 'writ' is now 'claim form' (qv), 'pleadings' are now 'statements of case'. Each part of the rules is followed by the appropriate practice direction which sets out the administrative procedures. The overriding objective of the Rules is set out in Part 1, which is to enable the courts to deal justly with cases.

Civil war A continuous and large-scale state of hostilities, greater in scope than an insurrection (qv), between two or more sets of armed forces within a single state, often between the Government and an insurgent group. In most forms of contract the situation is covered by *force majeure* (qv), for example as grounds for an extension of time (qv).

See also: **Civil commotion; Commotion; Disorder; Riot.**

---

[120]Lord Woolf, *Access to Justice*, June 1996, Stationery Office.

Claim form This has replaced the former procedures by which to commence litigation (e.g. by writ (qv)). This change was introduced as part of the Civil Procedure Rules (CPR) (qv). All proceedings must now be commenced by the issuing of a claim form, generally in accordance with CPR Part 7[121]. The date of issuing of a claim form is the date from which the relevant time limits under the Limitation Act 1980 will be calculated[122].

A claim form is valid for serving on the defendant(s) (qv) for an initial period of four months[123], unless served out of the jurisdiction, in which case it is valid for six months[124]. The court has a discretion to extend the validity of the claim form[125].

Part 16 gives four essentials for a claim form. It must:
— contain a concise statement of the claim;
— state the remedy sought;
— if the claim is for money, contain a statement of value as set out in Rule 16.3; and
— contain any other matters which may be set out in a practice direction.
Importantly, the form must be verified by a statement of truth.

The claim form may contain within it or be accompanied by a separate document entitled particulars of claim (qv), although they can be served later. It is only upon receipt of the particulars of claim that a defendant is required to serve an admission, acknowledgement of service or defence[126].

A careful study of the appropriate Parts of the Rules is necessary for a full understanding of the claim form. Among other relevant matters are that if the claimant is a representative, the form must say what capacity it is. Similarly if the defendant (qv) is sued as a representative, the capacity must be stated. Rule 16.4 specifies the particulars of claim which must be included. If they are not so included, the claim form must state that those particulars will follow.

Significantly, the court may grant any remedy to which the claimant is entitled, even if the remedy is not specified on the form.

Claimant One who claims or who asserts a right. The term has always been used in arbitration (qv) and it tends to be used in adjudication (qv) documents rather than the cumbersome 'referring party'. It is now used for the party making a claim in litigation under the Civil Procedure Rules in place of the old style 'plaintiff'(qv).

Claims The dictionary defines 'claim' as 'an assertion of a right' and, under standard building contracts, the word conveys the concept of additional payment which the contractor seeks to assert outside the contractual machinery for valuing the work itself. The word is also used in respect of the contractor's applications for

---

[121]There is an alternative procedure under CPR Part 8, but this is only used in specialised contexts.
[122]See: **Limitation of actions.**
[123]CPR Rule 7.5 (2).
[124]CPR Rule 7.5 (3).
[125]CPR Rule 7.6.
[126]CPR Rule 9.2.

an award of extensions of time (qv). The main types of claim which may be made by a contractor are:

Contractual claims which are those made under specific provisions of the contract, e.g. one for 'direct loss and/or expense' under JCT 98, clause 26, or for 'expense' under GC/Works/1 (1998), clause 46. This type of claim is also occasionally described as being *ex contractu*, i.e. arising from the contract. Under JCT terms – and most other forms of contract – it is only claims of this type which the architect has authority under the contract to settle.

Common law claims which are those which arise apart from the express provisions of the contract. They include claims in tort (qv), claims for a *quantum meruit* (qv), claims for a *quantum valebat* (qv), claims for breach of express or implied terms of the contract or warranty (qv). All the current standard forms allow additional or alternative claims for breach of contract, based on the same facts. Usually they are based on implied terms relating to non-interference with the contractor's progress (see: **Hindrance or prevention**). They are sometimes called ex contractual or extra contractual claims.

Ex gratia claims are those without legal foundation and are usually made on moral or hardship grounds. Very rarely, there may be an advantage to meeting such a claim as a matter of grace, e.g. if the contractor is on the brink of insolvency (qv) and, as a result, the employer would face greater expense if the contractor could not carry on and completion contractors had to be employed.

In order to obtain payment under the provisions of the contract, any procedural requirements as to notices, etc. must be observed. Typically, the claims clause sets out (a) the grounds on which sums can be claimed (b) requirements as to notice (c) provision for payment, e.g. JCT 98, clause 26. All the current forms in use (except MW 98 which has no provision for the contractor to make a loss and/or expense claim) require notice in writing and impose restrictions on what is recoverable. Under JCT terms, the sums claimed must represent 'direct' loss and/or expense and must not be recoverable elsewhere under the contract. GC/Works/1 (1998), clauses 43 and 46, use the word 'expense' which must be 'beyond that otherwise provided for in or reasonably contemplated by the contract'. This is an objective test. There is no necessary link between money claims and extension of time, and the grant of an extension of time is not a precondition to a claim for direct loss and/or expense[127]. The confusion arises because some of the grounds for an extension of time are repeated as grounds for loss and/or expense. Most standard forms allow claims for both disruption and prolongation.

Contractors are often labelled 'claims conscious' on the basis that they are alive to their rights and make claims envisaged by the contract. The label takes no note of the validity or otherwise of such claims, and it is an unfair view of matters since the employer desires and has a right to expect an efficient contractor, and an efficient contractor will be efficient in all things – including his own claims. There are, of course, some contractors who make totally

---

[127]*H. Fairweather & Co Ltd* v. *London Borough of Wandsworth* (1987) 39 BLR 106.

unjustified, but time-consuming, claims either as a matter of routine on the basis that some will hit the target or because they have underpriced at tender stage. They are their own worst enemies and should not be labelled 'claims conscious', for they are nothing of the kind. They are simply inefficient.

Clause The numbered divisions or terms in a legal document or in a bill presented to Parliament are called clauses. All standard forms of contract have numbered terms for ease of reference. A new clause normally indicates a change in subject matter. Thus in IFC 98, clause 1.13 deals with 'Giving or service of notice or other documents' and the following clause, 1.14, deals with 'Reckoning periods of days'. The JCT standard forms refer to all clauses as 'conditions'. This is misleading because contract terms can be sub-divided broadly into 'conditions' (qv) and 'warranties' (qv) and the distinction is legally significant. Clearly, not every clause in the JCT conditions is a 'condition' in the legal sense.

Clerical errors Clerical errors in a contract or on a certificate will usually be treated as if the error had been corrected provided the error is perfectly obvious and the true words or clause numbers are equally obvious[128].
See: **Errors.**

Clerk of works An inspector employed on the works to ensure compliance with the contract provisions with regard to standards of materials and workmanship. The clerk of works is specifically mentioned in JCT 98 and GC/Works/1 (1998) forms of contract in clauses 12 and 4 (2) respectively. JCT 98 states that he is to be appointed by the employer and be under the direction of the architect. He may give 'directions' provided that they are in respect of matters for which the architect is expressly empowered by the contract to issue instructions, but they are of no effect unless the architect confirms them within two working days. The duty of the clerk of works is to act solely as an inspector. GC/Works/1 outlines duties in broadly similar terms without being specific on the matter of directions. ACA 3, and the JCT Agreement for Minor Building Works (qv) do not refer to a clerk of works but there is no reason why a clerk of works should not be employed if a suitable clause is inserted in the specification (qv).

SFA/99 (qv) and CE/99 refer to the employment of a clerk of works, where frequent or constant inspection is required. In practice, his duties will be somewhat broader than laid down in the contract as far as the architect is concerned. They will often include inspecting, reporting in detail, advising and generally being the eyes and ears of the architect on site. He must have a wealth of practical experience supplemented by sound technical knowledge.

The clerk of works is liable if he is negligent in the performance of his duties and this will reduce the architect's responsibility for inspection (see: **Inspection of the works**) in appropriate cases where the clerk of works is engaged by the employer. In *Kensington & Chelsea & Westminster Health Authority* v. *Wettern*

---

[128] *R. M. Douglas Construction Ltd* v. *C. E. D. Building Services Ltd* (1985) 3 Con LR 124.

Composites Ltd (1984)[129] the vicarious liability (qv) of the employer for the negligence of the clerk of works was considered.

Although the clerk of works is under the architect's direction and control, it was found on the facts that the clerk of works had been negligent, though to a lesser extent than the defendant architects. The judge described the relationship between clerk of works and architect as that 'of the Chief Petty Officer as compared with that of the Captain of the ship'.

The clerk of works was held 20% responsible and the employers were held to be contributorily negligent to the same extent, since they were vicariously liable for their employee's negligence. Damages were reduced accordingly. The negligent architects were responsible for the balance of 80% of the damages. It is very important that the duties of the clerk of works are clearly defined at the first site meeting to avoid difficult situations and misunderstandings arising during the contract.

The Institute of Clerk of Works of Great Britain Incorporated (ICW) was formed in 1882. It admits members, after examination, as Licentiate, Associate and Fellow. A useful publication is *Clerk of Works (Building)* produced by the ICW, which also publishes a selection of other publications.

Client One who employs a professional person. This word is used in the RIBA Conditions of Engagement (qv) documents SFA/99, CE/99 and SW/99 to describe the building owner or employer. These documents, published by RIBA Publications, set out the conditions to govern the relationship between architect and client.

Client's representative The term used, under the BPF System (qv) and its supporting form of contract, to describe the person or firm responsible for managing the project on behalf of and in the interests of the client (qv).

He may be an architect or other professional, or a project manager but, contractually, he performs the functions under the contract usually allotted to the architect. Under the BPF System, as far as his employer is concerned, he has a more extensive role, but under the BPF edition of the ACA contract (qv) his authority and powers are the same as those of the architect, although the client's representative is given a specific right to delegate his functions which the architect does not have.

Code of Procedure for Single-stage Selective Tendering 1996 A document
produced for the benefit of all who commission building work and which aims to introduce generally accepted standards into the traditional tendering procedure.

The Code is prepared by the National Joint Consultative Committee for Building in collaboration with:
— The Scottish Joint Consultative Committee.
— The Joint Consultative Committee for Building, Northern Ireland.

---

[129][1985] 1 All ER 346.

Reference is made to the duties of the parties under the CDM Regulations 1994 (qv).

The Code assumes that standard forms of building contract are to be used. If other forms of contract are used, some modification of detail may be necessary. There are clear benefits to all parties in the knowledge that a standard procedure will be followed in inviting and accepting tenders (qv).

Where tenders are to be invited for public works contracts in excess of a particular value, the procedure must follow the EU Directive 71/305/EEC as amended by 89/440/EEC and the provisions of the Code are so modified[130].

The Code recommends that the number of tenderers for a contract should be limited to a maximum of six. The number of tenderers is restricted because the cost of preparing abortive tenders will be reflected in prices generally throughout the building industry. In preparing a short list of tenderers, the following must be borne in mind:

— The firm's financial standing.
— Recent experience of building over similar contract periods.
— General experience and reputation of similar building types.
— Adequacy of management.
— Health and safety competence.
— Quality assurance position.
— Adequacy of capacity.

Each firm on the short list should be sent a preliminary enquiry to determine if it is willing to tender. The enquiry should contain:

— Job title.
— Description.
— Name of employer.
— Name of professional team.
— Name of planning supervisor.
— Location of site including plan.
— Approximate cost range.
— Number of tenderers.
— Principal nominated sub-contractors.
— Form of contract noting important additions or deletions.
— Procedure for correction of priced bills.
— Contract executed as a deed or under hand.
— Anticipated date for possession.
— Contract period.
— Anticipated date for despatch of tender documents.
— Length of tender period.
— Length of time tender must remain open for acceptance.
— Guarantee requirements.
— Special conditions.

Once a contractor has confirmed his intention to tender, he should do so. If circumstances arise which make it necessary for him to withdraw, he should

---

[130]Reference should be made to NJCC Procedure Note 19.

notify the architect before the tender documents are issued or, at the latest, within two days thereafter. If a contractor has expressed willingness to tender but is not chosen for the final short list, he must be informed immediately.

Note:

— Tender documents should be despatched on the stated date.
— Tenders must be submitted on the same basis.
— Alternative offers based on alternative contract periods may be admitted if requested on date of despatch of documents.
— Standard forms of contract should not be amended.
— A time of day should be stated for receipt of tender, and tenders received late should be returned unopened.
— The tender period should depend on the size and complexity of the job, but be not less than four working weeks (20 working days).

If any tenderer requires clarification of a point, he must notify the architect who should inform all tenderers of his decision. If a tenderer submits a qualified tender, he should be given the opportunity to withdraw the qualification without altering his tender figure, otherwise his tender should normally be rejected.

Under English law, a tender may be withdrawn at any time before acceptance (qv) which is why some tenders specify that the contractor has been paid a nominal sum (often £1) in consideration for keeping the tender open. Where there is consideration for keeping it open, the tender cannot be withdrawn. Under Scottish law, it cannot be withdrawn unless the words 'unless previously withdrawn' are inserted in the tender after the stated period of time that the tender is to remain open for acceptance.

After tenders are opened, all but the three lowest tenderers should be informed immediately. The lowest tenderer should be asked to submit his priced bills within four days. The other two are informed that they may be approached again. After the contract has been let, each tenderer should be supplied with a list of tender prices.

The quantity surveyor must keep the priced bills strictly confidential. If there is an error in pricing, the Code sets out alternative ways of dealing with the situation:

— The tenderer should be notified and given the opportunity to confirm or withdraw his offer (i.e. the total sum). If he withdraws, the next lowest tenderer is considered. If he confirms his offer, an endorsement should be added to the priced bills that all rates, except preliminary items, contingencies, prime cost and provisional sums, are to be deemed reduced or increased, as appropriate, by the same proportion as the corrected total exceeds or falls short of the original price.
— The tenderer should be given the opportunity of confirming his offer or correcting the errors. If he corrects and he is no longer lowest tenderer, the next tender should be examined. If he does not correct, an endorsement is required. Corrections must be initialled or confirmed in writing and the letter of acceptance must include reference. The lowest tender should be accepted, after correction or confirmation, in accordance with the alternative chosen.

Problems sometimes occur because the employer will see that a tender will still be the lowest even after correction. If the first alternative has been agreed upon and notified to all tenderers at the time of invitation to tender, the choice facing the tenderer should clearly be to confirm or withdraw. The employer may require a great deal of persuading to stand by the initial agreement in such circumstances. The answer to the problem is to discuss the use of the alternatives thoroughly with the employer before the tendering process. He must be made aware that the agreement to use the Code and one of the alternatives is binding on all parties. It is possible that an employer who stipulated alternative one and subsequently allowed price correction could be sued by, at least, the next lowest tenderer for the abortive cost of tendering.

The employer does not usually bind himself to accept any tender nor does he take responsibility for the costs of tendering. It may be that there are reasons why he will decide to accept a tender which is not the lowest. Although he is entitled to do so, it will not please the other tenderers. The Code is devised to remove such practices.

If the tender under consideration exceeds the estimated cost, negotiations should take place with the tenderer to reduce the price. The quantity surveyor then normally produces what is called 'reduction bills' or 'adjustment bills'. They are priced up and signed by both parties as part of the Contract Bills. See also: **Errors.**

Code of Procedure for Two-stage Selective Tendering 1996 Single-stage selective tendering is considered to be appropriate for most building contracts. Where it is thought desirable to involve the contractor in the design stage, two-stage tendering is usual. This document is produced for the benefit of all who commission building work and aims to introduce generally accepted standards into the procedure. The Code is prepared by the National Joint Consultative Committee for Building in collaboration with:
— The Scottish Joint Consultative Committee.
— The Joint Consultative Committee for Building, Northern Ireland.

Reference is made to the duties of the parties under the CDM Regulations 1994 (qv).

The Code is not concerned with any responsibility for design which may involve the main contractor. It assumes the use of standard forms of building contract after the second stage. If other forms of contract are to be used, some modification of detail may be necessary.

Where tenders are to be invited for public works contracts in excess of a particular value, the procedure must follow the EU Directive 71/305/EEC as amended by 89/440/EEC and the provisions of the code are so modified[131].

Two-stage tendering involves a first-stage competitive tendering procedure to select the contractor on the basis of pricing of documents related to the preliminary design. Thus a level of pricing is provided for use in subsequent negotiations. During the second stage, a tender is produced, using the

---

[131]Reference should be made to NJCC Procedure Note 19.

first-stage pricing on bills of quantities (qv) which properly document the finished design. The process is most suited to large or complex schemes where the involvement of the contractor at an early stage is desirable. It is important to remember that, although the system is often used when designs are fairly crude and time is short, the often long process of negotiation may not give any overall saving in time over single-stage tendering at a later point in the design process. During the first stage it is important to:

— Provide a competitive basis for selection.
— Establish the layout and design.
— Provide clear pricing documents which are flexible enough to be the basis for the pricing of the first-stage tender. Provision must be made for fluctuations between the first- and second-stage tenders.
— Clearly state the respective obligations and rights of the programme for the second stage and the conditions of contract.
— State the contract terms.

The exact nature of the first-stage documents will depend upon the circumstances. It is not intended that any contract for the execution of the work will be entered into at the end of the first stage.

The number of tenderers for the first stage should be restricted to six. In preparing a short list of tenderers the following must be borne in mind:

— The firm's financial standing.
— Recent experience of building over similar contract periods.
— General experience and reputation of similar building types.
— Adequacy of management.
— Health and safety competence.
— Quality assurance position.
— Adequacy of capacity.

Each firm on the short list should be sent a preliminary enquiry to determine if it is willing to tender.

The enquiry should contain:

— Job title.
— Description.
— Name of employer.
— Names of professional team.
— Name of planning supervisor.
— Location of site including plan.
— Approximate cost range.
— Number of tenderers.
— Principal nominated sub-contractors.
— Form of contract noting important additions or deletions.
— Correction of priced document.
— Second stage contract executed as a deed or under hand.
— Anticipated date for possession.
— Contract period.
— Anticipated date for despatch of tender documents.
— Length of tender period.

— Length of time tender must remain open.
— Guarantee requirements.
— Special conditions.

Once a contractor has confirmed his intention to tender, he should do so. If circumstances arise which make it necessary for him to withdraw, he should notify the architect before the tender documents are issued. If a contractor has expressed willingness to tender but is not chosen for the final short list, he must be informed immediately.

Note:
— Tender documents should be despatched on the stated date.
— Tenders must be submitted on the same basis together with the priced document.
— Standard forms of contract should not be amended.
— A time of day should be stated for receipt of tenders, and tenders received late should be returned unopened.
— The tender period should depend on the size and complexity of the job, but should be not less than five weeks.

If any tenderer required clarification of a point, he must notify the architect who should inform all tenderers of his decision. If a tenderer submits a qualified tender, he should be given the opportunity to withdraw the qualification without altering his tender figure, otherwise his tender should normally be rejected.

Under English law, a tender may be withdrawn at any time before acceptance (qv) which is why some tenders specify that the contractor has been paid a nominal sum (often £1) in consideration for keeping the tender open. Where there is consideration for keeping it open, the tender cannot be withdrawn. Under Scottish law, it cannot be withdrawn unless the words 'unless previously withdrawn' are inserted in the tender after the stated period of time that the tender is to remain open for acceptance.

After the tenders are examined, all the tenderers except the three adjudged most favourable should be informed immediately.

The quantity surveyor must keep the priced documentation (which should be submitted at the same time as the tender) strictly confidential. If there is an error in pricing, the tenderer should be given the opportunity to confirm the offer or correct genuine errors.

If he corrects and he is no longer the most favourable tenderer, the next tender should be examined. If he does not correct, an endorsement is required. Corrections must be initialled or confirmed in writing and if recommended for the basis of the second stage, reference must be made.

The tender considered to provide the best value should be recommended for acceptance. If one tender is not clearly the most favourable, two or more tenders may be given to the employer, together with recommendations, for his decision.

Acceptance of the first-stage tender is recommended to be confirmed in writing and the intentions of the parties clearly defined with regard to:
— Grounds for withdrawal from the second stage.
— Entitlement to costs and methods of ascertaining them if the parties fail to conclude second-stage negotiations to their mutual satisfaction.

— Reimbursement for any work done on site if second-stage procedures are abortive.

Acceptance of the first-stage tender is a particularly delicate operation. The employer, in particular, does not wish to find himself in the position of having accepted a contract sum at that stage. The terms of the letter of acceptance must be carefully worded to avoid such an eventuality. Depending upon the circumstances, it may be that a contract has been entered into. The question may be: What are the terms of the contract? There are two pitfalls:

— No contract exists. This is likely in many cases.
— A contract exists binding the employer to pay and the contractor to build. This would be far the worse of the two situations which could arise if insufficient care is given to the drafting of the invitation to tender, the tender and the acceptance.

The parties must carefully consider whether they wish to enter into a legally binding contract at all – even if restricted in scope.

The second stage is the completion of the design, production drawings and bills of quantities and the pricing of the bills from the first-stage tender prices.

The total of the priced bills will be recommended to the employer for acceptance as the contract sum. The Code states that no contract will have been entered into until the employer has accepted the sum. That may or may not be the case in practice. The parties must make clear, preferably in writing, their precise intentions in that respect.

If agreement cannot be reached, second-stage procedures may be restarted with the next most favourable tenderer or new first-stage tenders invited. After a contractor is appointed, all unsuccessful tenderers should be notified and, if feasible, a list of first-stage tender offers should be provided. If cost has not been the sole reason for acceptance, the fact should be stated. A contract must not be entered into until the successful firm has satisfied health and safety requirements.

A model preliminary enquiry for invitation to first-stage tender is appendixed to the Code, together with a model formal invitation to tender and form of tender. Notes are included for use in Scotland. The Code apparently works well in practice and this is probably because both parties, after the completion of the first stage, have an interest in bringing the procedure to a successful conclusion. The two stages can be summarised as reaching an agreement to try and a contract. The contractor and the employer's professional advisers should be aware of the possible legal and financial traps involved and should make due provision in the documentation.

See also: **Contract.**

Collateral contract/Collateral warranty An independent contract which is collateral to another contract can be created in several ways.

> 'Undertakings may be given that are collateral to another contract. They may be considered to be independent of that other contract either because they cannot fairly be regarded as having been incorporated therein, or because rules of evidence hinder

their incorporation, or because the main contract is defective in some way or is subject to certain requirements of form or is made between parties other than those by or to whom the undertaking is given. Such undertakings are often referred to as collateral contracts, or "collateral warranties".'[132]

Promises made by the employer to the contractor during pre-contractual negotiations may give rise to such a contract or warranty[133]. The classic case is *Shanklin Pier Ltd* v. *Detel Products Ltd* (1951)[134] where the employer contracted with a third party to paint the pier. The defendants induced the employer to specify their paint and gave assurances as to its quality. The paint was properly applied by the third party but did not live up to the defendants' promises. It was held that there was a collateral contract between the parties under which the employer could recover the amount it had to spend to put matters right. In *Greater London Council* v. *Ryarsh Brick Co Ltd* (1985)[135] it was held that where a supplier makes statements to a prospective purchaser about the quality of his goods, and because of those statements the purchaser causes a third party, such as a building contractor, to buy them, a collateral contract may arise between the supplier and the purchaser, so that if the goods prove defective, the purchaser can sue the supplier under the collateral contract.

In the construction industry, the use of formal collateral contracts between employer and a proposed sub-contractor is common. In the normal way, there is no privity of contract (qv) between employer and any sub-contractor, nominated or otherwise, and it is unlikely that third parties will be allowed to acquire rights under the Contracts (Rights of Third Parties) Act 1999 (qv), although circumstances may decree otherwise. But when a main contract is entered into in JCT 98 standard form, it is standard practice for a proposed nominated sub-contractor to be required to enter into the JCT Standard Form of Employer/Nominated Sub-contractor Agreement (NSC/W) which is collateral contract. This gives each party (employer and nominated sub-contractor) certain direct contractual rights against each other. Such collateral contracts are highly desirable in order to protect the employer as regards both nominated sub-contractors and nominated suppliers in three main areas:

— Where the nominated sub-contractor has carried out design work.
— Where the main contractor has a valid claim for extension of time under the main contract due to a failure by the nominated sub-contractor.
— Where the main contractor has a valid money claim under the main contract due to a failure by the nominated sub-contractor.

In these circumstances, delay by the nominated sub-contractor, or design failure, may be costly to the employer who, under JCT terms, has no claim against the main contractor. These and other defects are remedied by the collateral contract which gives the employer direct rights against the defaulting

---

[132]*Chitty on Contracts*, 28th edn, p. 957. Sweet & Maxwell.
[133]*Bacal Construction (Midlands) Ltd* v. *Northampton Development Corporation* (1976) 8 BLR 88, where statements about ground conditions were held to give rise to such a warranty.
[134][1951] 2 All ER 471.
[135](1985) 4 Con LR 85.

sub-contractor, and in return the nominated sub-contractor is given various rights against the employer, e.g. rights to direct payment.

A similar, but not identical, formal collateral contract is created by the RIBA/CASEC Form of Employer/Specialist Agreement ESA/1 for use under IFC 98 in respect of named persons as sub-contractors.

Warranties, sometimes referred to as 'duty of care agreements' are often required from architects and other members of the professional team. Although standard warranties have been produced which have the approval of a number of professional institutes and the British Property Federation, construction professionals are commonly asked to enter into warranties on forms which are specially drafted by their clients' legal advisors. Such forms usually have clauses dealing with the following matters:

— Reasonable skill and care of the warrantor.
— Design obligations.
— Prohibition on specifying certain materials.
— Licence to use copyright material.
— Obligation to continue professional indemnity cover.
— Assignment of the warranty.
— Take-over provisions (usually by a funder) in specified circumstances.

Warranties must be agreed by professional indemnity insurers before they are executed.

Commercial Court Part of the Queen's Bench Division of the High Court staffed by judges with special knowledge of commercial law and commercial matters. It deals largely with legal matters arising out of the financial and commercial activities of the City of London. The procedure is more flexible than the ordinary procedure and by consent the strict rules of evidence are often relaxed.

Many important questions arising from arbitration (qv) are determined in the Commercial Court, especially since the Arbitration Acts 1979 and 1996, but appeals in construction arbitrations are dealt with by the Technology and Construction Court (qv), formerly known as the Official Referees Court.

Commission (1) A body set up by the Crown or other authority, generally to enquire into and report upon something.

(2) An order, especially to an agent, to do something. Thus an architect is said to have received a commission when a client requests him to act on his behalf, for example, to prepare designs for a building.

(3) A form of remuneration which is related to the value or type of business generated. It is a common way of paying sales representatives. The theory is that if a man is paid in proportion to what he sells, he will sell more. An agent must not take any secret commission, i.e. one of which his principal is unaware. See also: **Agency.**

Common law The rules and principles expressed in judicial decisions over the centuries. It is unwritten and covers all law other than law made by statute (qv). Its essential feature is the doctrine of judicial precedent (qv) which is one of the most important sources of English law. Even where there is a comprehensive written contract, such as JCT 98, there may be implied terms (qv) which derive from common law[136].

At common law, unless the parties have agreed to the contrary, a building contractor impliedly undertakes that:
— He will do his work in a good and workmanlike manner.
— He will supply good and proper materials.
— The completed structure will be reasonably fit for its intended purpose[137].

Express terms of the contract will replace such implications and the third limb of this duty will not normally apply where the employer appoints an architect[138].

See also: **Equity.**

Commotion A term used in ACA 3, clause 11.5, alternative 2, as ground for awarding extension of time (qv). Various stages of violence are listed ranging from war (qv) to disorder (qv). In this context, it seems that the term refers to a violent disturbance between a riot (qv) and a disorder, although the dictionary allows 'violent disturbance', 'upheaval' and 'political insurrection' as definitions. In other contracts, commotion in this sense probably comes under the head of *force majeure* (qv).

See also: **Civil commotion; Civil war; Insurrection.**

Company See: **Corporation.**

Compensation event A term used in clause 6 of the NEC. The eighteen events are listed in clause 60.1 (the numbering system is rather curious). They include such things as variations (known as instructions to change the work information (qv)), tests, antiquities, etc. The effects of events on time and cost are dealt with together.

See also: **Engineering and Construction Contract.**

Competent Properly qualified. The word is used in a strictly legal context about a court, to denote the extent of its jurisdiction, or of a witness, to show that he is able to give evidence.

It is also used in contracts to stress that a particular person must be suitably qualified to do a particular job. So JCT 98, clause 10, refers to a 'competent person-in-charge'. The intention is clearly that such a person must be able to do his work with skill and care and also that he is the contractor's representative on the site. When considering 'competent' in the Quarries (General)

---

[136]*London Borough of Merton* v. *Stanley Hugh Leach Ltd* (1985) 32 BLR 51.
[137]*Hancock and Others* v. *B. W. Brazier (Anerley) Ltd* [1966] 2 All ER 901.
[138]*Test Valley Borough Council* v. *Greater London Council* (1979) 13 BLR 63.

Regulations 1956 the judge said:

> 'I am not prepared to hold either that "competent" means the most competent person available ... or that it means that he shall be so competent that he never makes a mistake. In my judgment, it means a man who, on a fair assessment of the requirements of the task, the factors involved, the problems to be studied and the degree of danger implicit, can fairly be regarded by the manager, and in fact is regarded at the time by the manager, as competent to perform ...'[139]

Clause 1.2 of ACA 3 requires the contractor to exercise 'all the skill, care and diligence to be expected of a properly qualified and competent contractor experienced in carrying out work of a similar scope, nature and size' to the project in hand. It is a question of fact whether or not a person is 'competent', i.e. has the necessary qualities and skills. GC/Works/1 (1998), clause 5 requires the contractor to employ a competent agent to supervise the execution of the works.

Completion In general, the point in time at which the contract works are finished. Different forms of contract qualify completion in various ways. Completion under JCT contracts has been held to be the same as practical completion (qv)[140]. For a fuller discussion see also: **Completion date; Taking-over.**

'Completion' is also used in connection with house purchase to mean the execution of the prescribed deed of transfer, when the purchase price is paid and the legal estate passes to the purchaser.

Completion date All standard forms of contract make provision for stating a specific date by which or a period within which the work is to be completed. Usually, failure by the contractor to so complete will result in his having to pay or allow the employer liquidated damages (qv) at a specified rate, subject to the contract provisions for extensions of time (qv). In the absence of such a contractual provision and where no completion date is expressly agreed, the contractor would be under an obligation to complete within a 'reasonable time'. In that case, the employer would be unable to recover liquidated damages if the works remained uncompleted after the elapse of a 'reasonable time' although he might, with difficulty, recover unliquidated or general damages on proof of loss, possibly subject to a ceiling on their amount equal to the failed liquidated damages clause[141].

JCT 98, clause 17 refers to practical completion (qv) as discharging the contractor's obligations with regard to the completion date. The completion date is referred to in clause 23.1 and the actual date is to be inserted in the Appendix (qv).

ACA 3 does not refer to a completion date but to a date on which the works are fit and ready for taking-over (clause 12), which clearly amounts to the same thing. The date for taking-over (qv) is to be inserted in the Time Schedule (qv).

GC/Works/1 (1998) refers to completion in clause 34 (1) and clause 1 (1) refers to the date for completion being calculated from the date of possession

---

[139] *Brazier* v. *Skipton Rock Co Ltd* [1962] 1 All ER 955 per Winn J at 957.

[140] *Emson Eastern Ltd (In Receivership)* v. *E. M. E. Developments Ltd* (1991) 28 Con LR 57.

[141] *Lorna P. Elsley, Executrix of the Estate of Donald Champion Elsley* v. *J. G. Collins Insurance Agencies Ltd* (1978) 4 Const LJ 318.

or the date of acceptance of tender as provided in the Abstract of Particulars (qv).
See also: **Essence of the contract.**

Composition with creditors A phrase used in some standard form contracts, for example in JCT 98, clause 27.3.1: 'If the Contractor makes a composition or arrangement with his creditors . . .' as a ground for determination (qv) of the contractor's employment under the contract. Essentially, it is an agreement between a debtor and his creditors on the basis that the creditors agree with the debtor and either expressly or by implication with each other, that they will accept less than the amounts due from the debtor in full satisfaction of their claims[142]. The agreement may be made orally or in writing or a combination of the two.

Compromise See: **Settlement.**

Conclusive evidence The final certificate (qv) under JCT 63 (until the 1976 revision) was final in two senses:
— The last occasion on which the architect could certify payment.
— The final certification that the works had been carried out in accordance with the contract.

The operative clause was 30 (7) which stipulated that the final certificate was to be conclusive evidence (unless proceedings had been commenced before issue or an arbitration request was made within 14 days of issue) in any proceedings that the works had been properly carried out and completed in accordance with the terms of the contract. This meant that the employer had no redress against the contractor if, for example, a month after the issue of the final certificate the employer discovered that all the ceilings were 150 mm lower than specified. Even if the employer sued through the courts, the final certificate was conclusive 'in any proceedings'. The employer could, of course, sue the architect, which was why architects waited as long as possible before issuing the final certificate. Sometimes the final certificate was never issued, a small sum being left outstanding in the hope that the contractor would not consider it worth his while to take legal action and the architect would never have to certify. Certain things were excluded from the conclusiveness of the final certificate – fraud, dishonesty or fraudulent concealment, any defect, including omissions, in the works which reasonable inspection (qv) at any reasonable time before the issue of the final certificate would not have revealed. The last are sometimes known as 'latent defects' (qv).

The position under JCT 98 (clause 30.9) and IFC 98 (clause 4.7) is different. The final certificate is conclusive evidence:
— That where the quality of materials or the standard of workmanship are expressly stated to be to the approval of the architect, they are to his reasonable satisfaction, but it is not conclusive that any other materials or workmanship comply with the contract.

---

[142]*Capes* v. *Ball* (1873) LR 8 Exch 186.

— That any necessary effect has been given to all the terms of the contract which require adjustment of the contract sum, except for accidental errors in arithmetic.

— That all and only such extensions of time as are due have been given.

— That reimbursement of any loss and/or expense is in final settlement of all and any claims arising from any matter referred to in clause 26 (JCT 98) or clause 4.12 (IFC 98) whether for breach of contract, duty of care, statutory duty or otherwise.

The proviso regarding fraud (qv) and proceedings remains. The difference is important because it means that if nothing is expressly stated in the contract to the approval of the architect, the conclusiveness affects only the financial aspects of the certificate. In any action brought by the employer after the issue of the final certificate, the contractor is not able to rely on the certificate as proof that the works are in accordance with the contract. However, even in relation to the matters listed in clause 30.9.1, the effect of the final certificate is merely to limit the evidence that might otherwise be called[143]. Architects will, no doubt, limit the situations in which they require work or materials to be to their approval. A discussion of the situation which arose after the judgment in *Crown Estates Commissioners* v. *John Mowlem & Co Ltd* (1994)[144] will be found under **Final certificate**.

ACA 3 (clause 19.5) expressly states that the final certificate does not relieve the contractor of any liability under the contract.

GC/Works/1 (1998) (clause 50 (7)) states that no certificate (qv) is to be conclusive and also that any certificate may be modified or corrected by any subsequent certificate. MW 98 does not make the final certificate conclusive.

See also: **Approval and satisfaction.**

Condition A term in a contract which is of fundamental importance to the contract as a whole. If such a term is broken by one party, the other party may accept the breach as repudiation[145]. He may elect to treat his obligations under the contract as at an end and sue for damages (qv). It is, therefore, crucial to appreciate which terms are conditions and which are simply warranties (qv) because breach of a warranty does not entitle the innocent party to rescind the contract. It is for the court to decide the question unless the parties have themselves specified that certain terms are to be treated as conditions.

The JCT and GC/Works/1 (1998) forms refer to the body of the printed contract form as 'conditions'. They are not all conditions in the legal and contractual sense; some of them are warranties or minor terms. The ACA 3 form refers to its terms as 'clauses', which is less liable to give rise to misunderstanding. Clause 2.1 of the JCT 98 is a good example of a true condition, since it sets out the contractor's fundamental obligations.

See also: **Condition precedent; Condition subsequent.**

---

[143] *P. & M. Kaye Ltd* v. *Hosier & Dickinson Ltd* [1972] 1 All ER 121.
[144] (1994) 70 BLR 1.
[145] *Photo Production Ltd* v. *Securicor Transport Ltd* [1980] AC 827 per Lord Diplock at 849.

Condition precedent A condition which makes the rights or duties of the parties depend upon the happening of an event. The right or duty does not arise until the condition is fulfilled. For example, under JCT 98, clause 26.1, the making of a written application by the contractor at the proper time is probably a condition precedent to payment under the contractual machinery for reimbursement of direct loss and/or expense. Similarly, before the employer can claim liquidated damages under many forms of building contract, the architect's certificate of delay is a condition precedent, e.g. JCT 98, clause 24; IFC 98, clause 2.7; ACA 3, clause 11.3.

There are many other examples in the standard forms. It should be noted, however, that it is sometimes open to question whether or not a term is a condition precedent unless it is expressly stated to be such and even then the courts will sometimes refuse to hold that a term is a condition precedent if to do so would be contrary to commercial sense in a special situation[146].

Take, for example, the notice provision found in JCT 98, clause 25. Although, at first sight, the requirement of written notice by the contractor appears to be a condition precedent to the awarding of an extension of time, that is not the case. The architect is under an independent duty to consider whether an extension of time is justified and to make any appropriate extension[147]. The provision under clause 25.3.3 requiring the architect to review and, if appropriate, make a further extension of time even if no notices have been given by the contractor, puts the matter beyond doubt. If the architect fails to carry out his duties under clause 25, the employer may lose his right to recover liquidated damages (qv).

It is probable that if a notice provision is to rank as a condition precedent, it must state a time for service and make clear that a failure to serve will mean loss of rights[148].
See also: **Condition; Condition subsequent.**

Condition subsequent A provision which terminates the rights of the parties upon the happening of an event, e.g. a contract clause providing for the termination of the contract on the outbreak of war (qv).
See also: **Condition; Condition precedent.**

Conditional contract Where an offer (qv) is made subject to a condition and is accepted by the other party, differing legal consequences may result:
— Where the parties have not settled all the terms, or the agreement is conditional on a further agreement, there is no contract. This interpretation is always adopted where the parties express their agreement as being 'subject to contract' (qv). Another possibility is that the agreement will be void for uncertainty, e.g. as in *Lee-Parker* v. *Izzet* (1972)[149] where agreement was reached 'subject to the purchaser obtaining a satisfactory mortgage'.

---

[146]*Koch Hightex GmbH* v. *New Millenium Experience Company Ltd* (1999) CILL 1595.
[147]*London Borough of Merton* v. *Stanley Hugh Leach Ltd* (1985) 32 BLR 51.
[148]*Bremer Handelsgesellscaft MBH* v. *Vanden Avenne-Izegem PVBA* [1978] 2 Lloyds Rep 109.
[149][1972] 2 All ER 800.

— Where there is complete agreement but it is suspended until the happening of a stated event (see: **Condition precedent)** such as the obtaining of an export licence. In some cases this may impose an obligation on one party to bring about the stipulated event or at least not to prevent it happening: *Mackay* v. *Dick* (1881)[150].

— Dependent on the wording used, the condition does not prevent the contract coming into existence, but merely suspends some aspect of contractual performance until the condition is satisfied.

Conditions of contract The clauses or terms in the main body of the contract, between the Recitals (qv) and the Appendix (qv). They are sometimes referred to as 'operative clauses'. The word 'condition' used in this sense must be differentiated from the same word used to denote a term of fundamental importance to the contract as a whole.

See also: **Condition; Condition precedent; Condition subsequent.**

Conditions of Engagement (RIBA) A document issued by the Royal Institute of British Architects for the benefit of clients and architects. It determined the minimum fees for which RIBA members could undertake work and the professional services which clients could expect to receive in return. The Conditions of Engagement were mandatory upon members of the RIBA. They were replaced in July 1982 by the Architect's Appointment (qv) and in 1992 by the Standard Form of Agreement for the Appointment of an Architect (SFA/92). This was supplemented in 1995 by the Conditions of Engagement (CE/95), probably intended for medium sized commissions, and in 1996 by the Small Works Agreement (SW/96). SFA/92 had a special design and build version for 'employer client' and for 'contractor client'. There were also supplements, e.g. for historic building works.

In 1999, new versions were produced as SFA/99, CE/99 and SW/99 together with design and build supplements and sub-consultancy terms. SFA/99 is divided into the following sections:

— Articles of agreement (the basic agreement between architect and client).
— Appendix (some variable matters to be inserted).
— Schedule 1 Project description.
— Schedule 2 Services (with a selection of 'other activities').
— Schedule 3 Fees and expenses.
— Schedule 4 Other appointments (other consultants and elements to be designed by others).
— Services supplement: Design and management.
— Conditions of engagement.
— Attestation.

Confidence, breach of See: **Confidentiality.**

Confidential communications See: **Privilege.**

---

[150](1881) 6 App Cas 251.

Confidentiality The law recognises that certain relationships give rise to a duty to maintain confidentiality and will award damages (qv) or an injunction (qv) as appropriate for breach or threatened breach of that duty.

> 'The obligation to respect confidence is not limited to cases where the parties are in contractual relationship ... If the defendant is proved to have used confidential information, directly or indirectly obtained from the plaintiff without consent ... he will be guilty of an infringement of the plaintiff's rights.'[151]

This is a developing area of the law and protection is not confined to business relationships. Some standard form contracts deal expressly with the matter, e.g. ACA 3, clause 3.3 which is in very plain terms, but even where the contract is silent it is clear that the relationships in contracting give rise to a duty to maintain confidentiality. Clause 29 (2) of GC/Works/1 (1998) also deals explicitly with confidentiality of information.

The principle is that someone who has received information in confidence should not take unfair advantage of it, but it is now established that the courts can take the public interest into account[152].

It is now common practice for fax cover sheets to contain a warning to any person who may receive it in error that the fax may contain confidential and/or privileged information and that it is not to be copied or otherwise distributed. Similarly warnings are sometimes placed on the ubiquitous e-mail. In appropriate cases an injunction may be obtained to restrain use[153]. The Court of Appeal has recently allowed publication of material covered by the Official Secrets Act 1989 once the information was in the public domain[154]. It remains to be seen what impact this will have on the concept of confidentiality and the ability to restrain its breach by injunction.

Consequential loss Many supply contracts contain terms purporting to exclude the supplier's liability for 'consequential loss or damage' caused by such matters as late delivery, defects in materials supplied and so on. The use of the word 'consequential' causes much debate but, in the context of building and related contracts, its meaning is quite clear.

In *Croudace Construction Ltd* v. *Cawoods Concrete Products Ltd* (1978)[155] the Court of Appeal decided that 'consequential loss or damage' means the loss or damage which does not result directly and naturally from the complained breach of contract. Damages are not consequential if they result directly and naturally from the breach or event on which reliance is put. Loss which directly and naturally results in the ordinary course of events from a breach of contract is recoverable as 'direct loss and/or expense' under JCT 98, clause 26, and similar provisions in other contracts.

---

[151] *Saltman Engineering Co Ltd* v. *Campbell Engineering Co Ltd* [1963] 3 All ER 413 per Lord Greene MR at 414.
[152] *Lion Laboratories Ltd* v. *Evans* [1984] 2 All ER 417.
[153] *Peter Pan Manufacturing Corporation* v. *Corsets Silhouette Ltd* [1963] 3 All ER 402.
[154] *Attorney General* v. *Times Newspapers Ltd* (2001) The Times 25 January.
[155] (1978) 8 BLR 20.

'Consequential loss' clauses merely protect suppliers etc. 'from claims for special damages which would be recoverable only on proof of special circumstances and for damages contributed to by some supervening cause'[156].

In *Millar's Machinery Co Ltd* v. *David Way & Son* (1934)[157] a contract provided that suppliers did 'not accept responsibility for consequential damages'. It was held that this clause did not exclude liability for the buyer's expenses in obtaining other machinery to replace the defective machine.

See also: **Causation; Damages; Direct loss and/or expense; Foreseeability.**

Consideration Something which is given, done or foreborne by one party in return for some action or inaction on the part of the other party. It must have some legal value. It is a vital part of a simple contract (but not of a contract executed as a deed (qv), i.e. a specialty contract (qv)).

There are some general rules which apply to consideration. It must:

— Be genuine; it must not be a vague promise or one in which there is no legal benefit to the other party or legal detriment suffered by the promising party.

— Be legal; it must not be unlawful.

— Be possible; it must be capable of fulfilment at the time the contract is made or at the time stipulated for performance. This must be distinguished from the consideration becoming impossible during the course of the contract (see: **Frustration**).

— Be present or future; it cannot be something already done or given at the time the contract is made.

— Move from the promisee; the parties entering into the contract must provide the consideration.

Consideration need not be adequate. If two parties have entered into a genuine contract where what is given by one of the parties does not appear to be equivalent to what is given by the other, the courts will rarely intervene. There are exceptions to some of these general rules and in some instances the existence of consideration may be difficult to prove. In the case of building contracts the consideration will be the carrying out of the works by the contractor and the payment by the employer.

It used to be said that a promise on the part of one party to do something which he was already obliged to do was not good consideration[158]. In modern times, the courts are more ready to find the existence of consideration so as to reflect the true intention of the contracting parties than was the case in the nineteenth century. So, in *Williams* v. *Roffey Brothers & Nicholls (Contractors) Ltd* (1990)[159], the Court of Appeal held that a main contractor's promise to pay extra to a sub-contractor to complete on time was an enforceable contract and did not fail for lack of consideration. The court treated the practical benefits gained by the contractor as sufficient to render it enforceable even though

---

[156] *Saint Line Ltd* v. *Richardsons, Westgarth & Co Ltd* [1940] 2 KB 99 per Atkinson J at 103.
[157] (1934) 40 Com Cas 204.
[158] *Stilk* v. *Myrick* (1809) 2 Camp 317.
[159] [1990] 2 WLR 1153.

there was no detriment to the promisee. Lord Justice Russell said: 'A gratuitous promise, pure and simple, remains unenforceable unless given under seal[160]. But where, as in this case a party undertakes to make a payment because by doing so it will gain an advantage arising out of the continuing relationship with the promisee the new bargain will not fail for want of consideration'.

Executed consideration exists where the consideration on one side consists of the doing of an act, the doing of which brings the contract into existence. A good example is a typical estate agent's contract to sell a house. The client says 'I will pay you $2\frac{1}{2}$% if you sell my house'. There is no contract until the house is sold, and so the estate agent is not liable if he does not try to sell the house and the client can withdraw the agency before the house is sold.

Executory consideration exists where the consideration consists of an exchange of promises to be performed in the future.
See also: **Contract.**

Construction The term has very different meanings in legal and building contexts.

Legal The terms of a contract (qv) are construed so as to arrive at their precise meaning and effect. Where a term is ambiguous and reference is made to other terms within the contract this is called interpretation rather than construction.

Erection of a structure The common-sense meaning within the industry is the process of erection of a structure which may be a building or it may be a dock or a road. This simple approach has been adopted in relation to the concept of 'construction phase' (qv) which is used within the Construction (Design and Management) Regulations 1994 (qv). Additionally, there is now a statutory definition of another related term, 'construction operation' (qv)[161].

Construction Confederation The new name for the Building Employers' Confederation. This is the body which looks after the interests of its contractor members in matters such as wages, working rules and contract conditions.

Construction contract A term found in the Housing Grants, Construction and Regeneration Act 1996 (qv), part II. It is defined in s. 104 as an agreement for carrying out, arranging for the carrying out, or providing labour for carrying out construction operations (qv). It is expressly stated to include agreements for architectural design or surveying work or for providing advice on building, engineering, interior or exterior decoration or the laying out of landscape.

Construction Contracts (Northern Ireland) Order 1997 The Order which came into force in Northern Ireland on 1 June 1999. It is substantially the same as part II of the Housing Grants, Construction and Regeneration Act 1996 (qv).

---

[160]At that time a seal was necessary to form a contract. Now, not only is it unnecessary, the mere existence of a seal does not, of itself, create a deed.
[161]Section 105, Housing Grants, Construction and Regeneration Act 1996.

A Scheme for Construction Contracts (Northern Ireland) Regulations Northern Ireland 1998 has also been produced which is virtually identical to the Scheme for Construction Contracts (England and Wales) Regulations 1998. See: **Housing Grants, Construction and Regeneration Act 1996.**

Construction (Design and Management) Regulations 1994 Regulations

which place particular duties on clients, their agents, designers and contractors to take account of health and safety, and to co-ordinate and manage it effectively during all the stages of a project from inception to eventual repair and maintenance procedures. They came into force on 31 March 1995 throughout Great Britain and in certain circumstances elsewhere (see Regulation 20). In Northern Ireland, the Regulations are dated 1995. The Regulations are detailed and they are administered by the Health and Safety Executive under the Health and Safety at Work Act 1974. The key parts of the Regulations are:

— Clients and their agents must be reasonably satisfied that they are using competent persons to fill the crucial roles and that sufficient resources are devoted to the project.
— A planning supervisor must be appointed to take responsibility for co-ordinating the health and safety elements of the project at design stage. The planning supervisor is responsible for the health and safety plan and health and safety file preparation.
— A designer must perform his duties to avoid, reduce or control risks during construction and maintenance.
— A principal contractor (usually the contractor on site) must develop the health and safety plan and ensure that all on site comply with the plan and all relevant health and safety legislation including providing necessary information.
— Other contractors involved in the project must co-operate with the principal contractor.
— A health and safety file is produced on completion of the project, which not only satisfies the usual requirements for a maintenance manual, but also warns of particular risks and dangers.
— Domestic householders do not have duties under the Regulations.

Civil liability is excluded by Regulation 21 with the exception of duties under Regulations 10 and 16 (1) (c). Consequently, an employer would in general have no recourse against a contractor who was in breach of the Regulations. Most standard forms, therefore, make express provision for the employer and the contractor to comply with the Regulations so that failure to do so will amount to a breach of contract and it will be actionable in arbitration or the civil courts as appropriate (see, for example, JCT 98, clause 6A).

The Health and Safety Executive have produced an excellent 'Approved Code of Practice' referring to the Regulations. The Code has a special legal status. If a person is prosecuted for breach of health and safety law and it is proved that the person has failed to comply with the Code, he will be found to be at fault by a court unless he is able to show that he has complied with the law in some other way.

Construction Industry Council Model Adjudication Procedure (CIC Procedure) These are rules for the conduct of adjudication (qv) in compliance with the Housing Grants, Construction and Regeneration Act 1996 (qv). They were first issued in February 1998 and were the result of a task force consisting of representatives of all the major professional and contracting institutes, the Official Referees Solicitors Association and the CIC.

Construction Industry Model Arbitration Rules (CIMAR) Rules for the conduct of construction arbitration. They were first issued in February 1998 and originated by the Society of Construction Arbitrators in response to the Arbitration Act 1996. At the time of publication, endorsement of the Rules was indicated by the following:
— The Association of Consulting Engineers.
— The British Institute of Architectural Technologists.
— The British Property Federation.
— The Chartered Institute of Arbitrators.
— The Chartered Institute of Building.
— The Chartered Institution of Building Services Engineers.
— The Civil Engineering Contractors' Association.
— Construction Confederation.
— The Construction Liaison Group.
— The Institution of Mechanical Engineers.
— The Institution of Electrical Engineers.
— The Royal Institute of British Architects.
— The Royal Institution of Chartered Surveyors.
— The Specialist Engineering Contractors' Group.

The Joint Contracts Tribunal (qv) provides for the use of the CIMAR in the arbitration provisions in all JCT contracts in lieu of the formerly prescribed JCT Arbitration Rules (qv).

These rules, together with the Arbitration Act 1996 amount to a significant overhaul of the arbitration process. The Rules comprise the following:
(1)   Objective and application.
(2)   Beginning and appointment.
(3)   Joinder.
(4)   Particular powers.
(5)   Procedure and evidence.
(6)   Form of procedure and directions.
(7)   Short hearing.
(8)   Documents only.
(9)   Full procedure.
(10)   Provisional relief.
(11)   Default powers and sanctions.
(12)   Awards and remedies.
(13)   Costs.
(14)   Miscellaneous.

Appendix I Definitions.

Appendix II Sections referred to within the body of the rules but not reproduced therein.

Construction Industry Scheme (CIS) The scheme commenced on 1 August 1999. It replaced the statutory tax deduction scheme provided for in JCT forms of contract. The governing legislation is the Income and Corporation Taxes Act 1988 and The Income Tax (Sub-Contractors in the Construction Industry) (Amendment) Regulations 1998 SI 2622. Essentially the scheme provides that no party designated a 'contractor' (qv) under the Act may make a payment under a contract for construction operations (qv) unless the sub-contractor has either:

— a tax certificate; or
— a certifying document; or
— a valid registration card.

In the case of a tax certificate or certifying document, the contractor may pay without deduction. In the case of a card, the contractor may pay, but he must make the statutory deduction[162].

Construction management A system of procurement, the essential features of which are that a construction manager (who may well be a contractor) is appointed to act solely in a management capacity for which the employer pays a fee. Each professional and each trade contractor is contracted directly to the employer. In this respect it is significantly different to a management contract (qv). The individual trade packages are tendered separately in sequence to obtain the overall best price for the project.

The construction manager is one of the professional team and he acts as leader and co-ordinator of all the consultants and trade contractors. The particular contractual structure is more conducive to this relationship than the traditional management contract where the contractor is rarely a permanent part of the design team. The precise details of his role may vary, but commonly he will carry out all administrative functions in respect of the trade contractors. For example, he may issue certificates of all kinds and give instructions to the trade contractors. In order to be able to do this effectively, it is important that each contract contains a set of powers and duties which interlock with all other contracts and that each contract, in addition, makes express provision for the co-ordinating role of the construction manager so that each member of the team knows exactly the extent of their roles.

Advantages are:

— Input from the contractor (construction manager) at an early stage.
— Early start on site.
— Should be lowest overall cost.
— Employer has total control.
— The system is ideally suited to fast track projects.

---

[162]Reference can be made to JCT Practice Note 1 (Series 2) published by RIBA Publications (1998).

Disadvantages are:
— No certainty of final cost.
— Employer takes most of the risk.
— No guaranteed completion date.

There is no standard form of contract although the Joint Contracts Tribunal (qv) is reputed to be finalising one. Many construction management projects proceed under specially drafted contract documents and, currently, this is the best option. Suggestions that existing forms such as ACA 3 and IFC 98 can be easily adapted are wishful thinking.

Construction operations A term used in the Housing Grants, Construction and Regeneration Act 1996, Part II. A detailed explanation of its meaning is to be found in s. 105. The first part of the section defines the term, the second part clarifies processes which are not included in the term. Broadly, construction operations include construction, alteration, repair, maintenance and demolition of building and civil engineering work; supply and installation work; cleaning if carried out with other construction operations, ancilliary work such as foundations, excavations, clearance, etc. and painting and decorating.

Processes not included are drilling or extraction of oil or gas; extraction of minerals; assembly, installation or demolition of plant or machinery, etc. where the main activity is nuclear processing, power generation, water or effluent treatment, production, transmission or bulk strorage of chemicals, oil, food etc.; manufacture or supply only of components or materials; and the making or installation of purely artistic works. The Act makes provision for the Secretary of State to add to or change any of the operations which are to be treated as construction operations (s. 105 (3)).

The term is also used in the Construction Industry Scheme (CIS) (qv) and it is defined in s. 567 of the Income and Corporation Taxes Act 1988 in similar terms to those noted above.

Construction phase A term used in the Construction (Design and Management) Regulations 1994 (qv) and defined in Regulation 2 as 'the period of time starting when construction work in any project starts and ending when construction work in that project is completed'. The definition is important, because, among other things, the client must ensure as far as reasonably practicable that the health and safety plan has been prepared before the construction phase begins.

Construction Sites Directive The common abbreviation for the European Council Directive 92/57/EEC on the implementation of minimum safety and health requirements at temporary or mobile construction sites. Effect is given to this directive, except for certain details, by the Construction (Design and Management) Regulations 1994 (CDM Regulations) (qv).

Constructive acceleration See: **Acceleration.**

Consultant Literally, a specialist who gives expert advice or assistance. None of the standard forms of building contract mentions consultants specifically. It is important to remember, therefore, that a consultant will have no express authority to issue instructions under any of the standard forms unless a suitable clause is written into the contract. It is not advisable to do this because it is essential that the control of the work rests in the hands of one person: the architect or contract administrator under JCT contracts or whoever is designated to fulfil the role under other standard forms. It must be recognised, however, that the dubious practice of consultants visiting site and giving instructions directly to a sub-contractor does exist.

Under the BPF system (qv), consultants are defined as persons or firms contracted by the client (qv) to produce design and cost services additional to those provided by the design leader (qv). They may be experts in any relevant field and are paid a fixed fee to cover all costs and expenses. Consultants under that system work under the terms of a model consultancy agreement prepared by the BPF and are responsible only for their own part of the work. The design leader (qv) is responsible for the co-ordination of the work of all consultants.

The RIBA have prepared standard sub-consultancy agreements for use by all disciplines sub-consulting to an architect.

See also: **Conditions of Engagement.**

Consultant switch A procedure whereby a consultant engaged by the employer ceases acting for the employer and commences acting for a contractor on a project where the procurement system is design and build or one of its derivatives. The 'switch' usually takes place on the appointment of the design and build contractor. It is essential that two separate agreements are executed, each containing different, but appropriate, terms to suit the very different situations. The first contract must come to an end before the second is executed. The process is commonly, but inaccurately, referred to as 'novation' (qv) which is a completely different process. True novation is sometimes operated instead of consultant switch, but neither novation nor consultant switch is to be recommended.

One of the supposed advantages is that the design team on being switched to the contractor to complete the design and production drawings will be able to develop them while remaining true to the original concept in a way which would not otherwise happen. This exceedingly optimistic view ignores the fact that each consultant, on being switched, will thereupon owe a duty to the contractor in respect of the completion of the production information (qv). This may simply mean that the consultants are obliged to comply with the contractor's instructions to produce a scheme which is inexpensive to construct, but perhaps not what the design team originally had in mind. It is virtually impossible for the procedure to operate without creating a conflict situation for the consultant. Some bespoke terms of appointment aggravate the situation by requiring the consultants to report back to the employer in specific instances after the switch. Consultants should resist being 'switched'.

Consultation A word used in a number of standard forms, notably in JCT 98
clause 13.4.1.2, alternative A, and IFC 98 clause 3.7.1.2, option A, where the
quantity surveyor is to act 'after consultation with the architect'. Consultation
is also required between architect and contractor in JCT 98 clause 8.4.2 and
elsewhere. It is the act of seeking information, opinion or advice. The consultor
must provide sufficient information and give sufficient opportunity to allow the
advice to be given[163]. There is no implication that the advice or information
need be acted upon although it is probably implied that where the contract
requires one party to consult another, the information and advice should
inform the action.

Contingency An unexpected event. The architect normally arranges for a contingency sum to be inserted in the bills of quantities (qv). The amount is usually
about 3% of the expected contract sum. The purpose of the sum is to cover the
cost of unforeseen items.

    If, unusually, there are no such items, the whole of the sum is deducted from
the contract sum and represents a saving to the employer. A contingency sum is
not intended to cover additional work to that originally envisaged or the
correction of specification errors. In certain types of building, e.g. old or
complex existing structures, the contingency sum may be increased to reflect
the fact that there is more chance that unforeseen situations (hidden rainwater
pipes, eccentric structure, rot) may be discovered.

Continuous improvement A term used in connection with prime contracting
(qv). An essential facet of the system is that work processes and methods must
be mapped out and programmes must be in place to improve them. Because
this is an ongoing process, it cannot take place unless the other essential
element is in place: long-term relationships between prime contractor and
suppliers. Improvement must be achieved in terms of what the client receives
and the profitablility of the whole supply chain.

See also: **Right first time; Supply chain partners; Supply clusters.**

Contra proferentem A principle or rule of contract construction.

> 'If there is an ambiguity in a document which all the other methods of (interpretation)
> have failed to resolve so that there are two alternative meanings to certain words, the
> court may construe the words against the party who put forward the document and
> give effect to the meaning more favourable to the other party.'[164]

    The rule does not seem to apply to 'negotiated' standard form contracts,
such as the current editions of JCT forms, where the document is prepared by
representatives of actual and potential users[165]. Probably, however, the rule
would apply where the employer makes *substantial* amendments to the printed

---

[163]*Fletcher* v. *Minister of Town and Country Planning* [1947] 2 All ER 496.
[164]May, A, *Keating on Building Contracts* (1995) 6th edn, p. 47. Sweet & Maxwell.
[165]*Tersons Ltd* v. *Stevenage Development Corporation* (1963) 5 BLR 54, a decision on the 4th edition of
the ICE Conditions of Contract.

text so that it ceases to be a 'negotiated document' and is put forward by him as his own. Probably, too, it applies to manuscript or typewritten insertions, e.g. in the Appendix to the JCT forms, where these are inconsistent with the printed conditions[166].

The best known example of the application of the *contra proferentem* principle in the construction industry is the decision of the Court of Appeal in *Peak Construction (Liverpool) Ltd* v. *McKinney Foundations Ltd* (1970)[167] which involved Liverpool Corporation's own form of contract.

See also: **Unfair Contract Terms Act 1977.**

Contract A binding agreement between two or more persons which creates mutual rights and duties and which are enforceable at law. There must be an intention to create a legal relationship. Thus, a simple promise to do something for a person is not legally binding. For example, if A agrees to give £5 to B and in return B agrees to clean A's car, a legally binding contract is in existence. If B simply promises to clean A's car, there is no contract and A can do nothing if B fails to keep his promise.

There are two basic types of contract:
— Specialty contracts (qv) or contracts executed as deeds. This type of contract is often used by local authorities and corporations (qv).
— Simple contracts (qv) or contracts made in writing or orally. If written, they may be recorded in correspondence or may be a document(s) signed by the parties. This type of contract is the most common.

Figure 5 illustrates the major differences between specialty and simple contracts.

A number of features are essential in order to enter into a valid contract:
— There must be an *offer* (qv) by one party.
— There must be an unqualified *acceptance* (qv) by the other party.
— There must be *consideration* (qv) except in the case of deeds.
— The parties must have *capacity to contract* (qv).
— There must be an *intention to create a legal relationship*[168].
— There must be *genuine consent*. For example: there must be no duress involved.
— The object of the contract must be *possible*.
— The object of the contract must be *legal*. For example an agreement to defraud the Inland Revenue would not be a binding contract[169].

It is common in the construction industry for a contract to be formed without it being possible to identify a formal offer and acceptance due to the volume or type of correspondence between the parties. In such circumstances, the important thing is whether it can truly be said that the parties eventually came

---

[166] *Bramall & Ogden Ltd* v. *Sheffield City Council* (1983) 1 Con LR 30.
[167] (1970) 1 BLR 114.
[168] *Harvey* v. *Facey* [1893] AC 552.
[169] For example in *Taylor* v. *Bhail* (1996) 50 Con LR 70, the Court of Appeal dismissed a contractor's claim for monies owed, where the contractor and employer had inflated the insurance quotation, as this amounted to obtaining monies by deception.

Specialty	Simple
(i) Form	
The contract is created by the deed itself	Normally writing is merely evidence of the contract, which exists apart from and in the absence of writing
(ii) Consideration	
Need not be present	Must always be present
(iii) Limitation	
12 years from the date on which the cause of action arises, i.e. breach	6 years (Limitation Act 1980)
(iv) Estoppel*	
Statements in a deed are conclusive against the parties to it, except where there is a latent ambiguity, or fraud, duress or mistake is proved	Statements in a simple contract are only *prima facie* evidence of their truth

* Estoppel is a rule of evidence which precludes a person from denying the truth of some statement made by him, or the existence of facts which by words or conduct he has led others to believe in.

Figure 5   Simple and specialty clauses compared.

to an agreement on the essential terms[170]. Determining whether a contract has come into existence and the precise nature of its terms can be a difficult task. Contracts may be:

— *Valid:* they satisfy all the requirements for a legally binding contract.
— *Void:* they are not contracts at all because they are lacking in some important respect, e.g. lack of proper acceptance.
— *Voidable:* a contract which is not void but which can be made void at the instance of one of the parties.
— *Unenforceable contracts:* contracts which are valid but whose terms cannot be enforced because of some special reason, e.g. the operation of the Limitation Act 1980 (see also: **Limitation of actions**).

Contracts for the erection of buildings are normally entered into by using one of the standard forms available. They have the following advantages:

— Designed specially for construction work.
— Comprehensive and continually updated in the light of experience and developments in the law.
— The contents are generally understood by the industry.
— Certain contracts are negotiated documents and, therefore, not to be construed *contra proferentem* (qv) against either party.

See also: **ACA Form of Building Agreement; Agreement for Minor Building Works; Anticipatory breach of contract; Breach of contract; Burden of a**

---

[170]*G. Percy Trentham Ltd* v. *Archital Luxfer Ltd* [1993] 1 Lloyds Rep 25 per Steyn LJ at 29–30.

contract; Change of parties; Contracts (Rights of Third Parties) Act 1999; Discharge of contract; Divisible contract; Entire contract; Essence of the contract; Formalities of contract; Fraudulent misrepresentation; GC/Works/1 contract; Illegal contract; Implied contract; Innocent misrepresentation; JCT contracts; Misrepresentation; Mistake; Performance; Privity of contract; Quasi-contract; Rectification; Repudiation; Rescission; Standard forms of contract.

Contract bills An expression used to refer to the bills of quantities (qv) if they are, as is usual, to become part of the contract documents (qv). Invariably such bills will have been priced by the contractor. The contract bills are defined in JCT 98 clause 1.3 as the priced bills of quantities referred to in the first Recital and signed by the parties to the contract. The contract makes frequent reference to them, for example in clause 5.1: 'The Contract Drawings and the Contract Bills shall remain in the custody of the Employer. . .'.

Contract data A term used in the NEC. It is similar in use to the Appendix (qv) to JCT contracts and to the abstract of particulars in GC/Works/1 (1998). It is in two parts: data provided by the employer and data provided by the contractor. The variable parts of the contract, such as the names of the parties, the starting date, etc.
See also: **Engineering and Construction Contract.**

Contract documents A document is anything on which marks have been made with the intention of communicating information. Such things as writing, printing, typescript, computer printout, drawings and photographs are documents. The documents which are brought together to form the evidence of a contract, agreed by the parties and signed as such, are termed the 'contract documents'. Most of the standard forms of contract define what are to be the contract documents: JCT 98, clause 2.1; ACA 3, Recital C; GC/Works/1 (1998), clause 1 (1); MW 98, 1st Recital; IFC 98, 2nd Recital. WCD 98, strangely, does not define them although it is not difficult to deduce what they comprise.

The printed form, drawings, specification (qv), bills of quantities (qv), schedules and schedules of rates (qv) are commonly included, depending on the type of contract desired. It is important, although rarely completely achieved in practice, that the documents are consistent with one another. In the case of inconsistencies, most standard forms provide that the printed conditions must override any other provisions if there is conflict. This reverses the general rule, that specially written terms take precedence over printed terms, and sometimes leads to unwelcome results.

Thus, under JCT terms, if the employer by a clause in the bills of quantities was given 21 days to honour the architect's certificates, it would have no effect unless the corresponding clause in the printed conditions had been properly amended and initialled by the parties.

All the contract documents should be signed by the parties and identified as being contract documents. Some such endorsement as 'This is one of the contract documents referred to in the Agreement dated ...' and signed by the parties should suffice.

See also: **Bills of quantities; Contract bills; Contract drawings; Priority of documents; Specification.**

Contract drawings

The drawings specifically referred to in the contract. They are usually identified by drawing number together with any revision number. The drawings must be signed by the parties and bound in with the rest of the contract documents (qv).

Although all the standard forms make provision for the architect to issue 'such further drawings or details which are reasonably necessary to explain and amplify the Contract Drawings ... to enable the contractor to carry out and complete the Works in accordance with the Conditions' (JCT 98, clause 5.4.2), such additional drawings cannot modify the contractor's obligations as contained in the contract documents. What that means is that the architect cannot, without a variation in cost to the contract, change anything contained in the contract drawings or contract bills. The contract drawings are usually small-scale drawings: plans, elevations, sections, site plan. It is important that they are as accurate as possible and they must be the same drawings on which the contractor submitted his tender. It is not unknown for drawings to be revised between the date of invitation to tender and the signing of the contract, but it must be the original tender drawings which are signed and bound into the contract[171].

See also: **Contract documents.**

Contract sum

The amount or consideration (qv) which the employer agrees to pay to the contractor for carrying out the works. It is written into the contract documents.

All the standard forms contain provision for adjusting the contract sum and, therefore, the amount of the final account (qv) may well be greater or less than the contract sum. The 'contract sum' itself, however, is clearly defined as a precise amount of money. The contract sum, therefore, can never change, although it can be adjusted. When adjusted, it is known as the 'adjusted contract sum'. The issue may be very important under certain circumstances, e.g. if an architect bases his fees on a percentage of the contract sum. No matter how much additional work is instructed or contractor's claims for loss and/or expense certified, the architect's fee would remain the same. To overcome that problem, a percentage fee should be based on the 'adjusted contract sum', the 'final account' or some such phrase. Architects who are engaged on the standard RIBA Conditions of Engagement (qv) (SFA/99, CE/99 SW/99 etc.) should not have this problem.

The contract sum is generally stated to be exclusive of VAT.

---

[171]A fuller discussion of this point may be found in Chappell's *Contractual Correspondence for Architects and Project Managers* (1996) 3rd edn, p. 82, Blackwell Science.

Contract sum analysis An analysis of the contract sum provided by the contractor in accordance with the stated requirements of the employer. It is referred to in JCT 98 Standard Form Without Quantities, IFC 98 and WCD 98. It should be noted that the contract sum analysis is not a contract document except under WCD 98 although in all circumstances where it is used, it is the priced document for valuation purposes.

JCT Practice Note 23 (1987) explains the purpose, use and content of the contract sum analysis in detail. Uses of the contract sum analysis include:
— Valuation of variations and provisional sum work.
— To enable the calculation of fluctuations in accordance with the formula rules.
— To help the calculation of interim certificates and payments.

If the formula rules are to be operated, it is important for the invitation to tender to state clearly what form the contract sum analysis must take to enable the calculation to take place.

Contractor One who enters into a contract with another. Generally, the person or firm who contracts with the employer and undertakes to construct the project. The word is used to make the distinction between a person who enters into a contract to carry out work and services, often called an independent contractor, and a person who is a servant or employee of the person for whom he does the work. The contractor, unlike the employee, is not subject to detailed control. In the construction industry, a contractor is invariably the person, partnership or company which carries out construction work. All the standard forms of contract refer to the contractor in this sense.

'Contractor' is defined in the Construction (Design and Management) Regulations 1994 (qv) as 'any person who carries on a trade, business or other undertaking (whether for profit or not) in connection with which he (a) undertakes to or does carry out or manage construction work, (b) arranges for any person at work under his control (including, where he is an employer, any employee of his) to carry out or manage construction work'.

An entirely different definition of 'contractor' is referred to in the Construction Industry Scheme (CIS) (qv). Reference should be made to IR14/15 (CIS). A contractor for these purposes embraces not only construction companies, but also many employers. That is why the employer under a JCT contract must state whether it is a 'contractor' for the purposes of the Scheme. Local authorities and government departments are included and also businesses who do not include construction operations among their trading activities, but carry out or commission construction work on a regular basis on their own property. It is a contractor if:
— its average annual expenditure on construction operations in the period of three years ending with its last accounting date exceeds £1 million; or
— if it has not been trading for the whole of the last three years its total expenditure on construction operations for the part of the three year period exceeds £3 million.

A business which becomes a contractor will continue to be treated as a contractor until the Inland Revenue is satisfied that, during each of three successive years, its construction expenditure has been below £1 million. Domestic householders having work carried out on their own premises are not contractors for these purposes, neither is a business which does not have any trade as a contractor and its average expenditure on construction averages less than £1 million in recent years.

Contractor's skill and care In the absence of any express terms in the contract, the law will always imply that the contractor:
— Will carry out his work in a workmanlike manner.
— Will supply good and proper materials.
— Will ensure that the completed structure is reasonably fit for its intended purpose. In the case of a dwelling it must be reasonably fit for human habitation. This limb is normally inapplicable where there is an architect when the contractor's obligation is simply to comply with the specification[172].

These implied terms may be excluded (subject to the provisions of the Unfair Contract Terms Act 1977 (qv)) by an express term to that effect in the contract.

Other factors, also, may operate to reduce the liability of the contractor. For example, if the employer has the services of an architect on whose advice he relies[173]. ACA 3 makes the position quite clear, so far as that contract is concerned, by including a special clause (1.2) which expressly refers to and preserves all implied warranties or conditions and puts on the contractor the duty to perform his obligations under the contract with 'all the skill, care and diligence to be expected of a properly qualified and competent contractor experienced in carrying out work of a similar scope nature and size to the Works'.

Contracts (Rights of Third Parties) Act 1999 The Act came into force on 11 May 2000 and it applies throughout the UK. It interferes with the principle of privity of contract (qv) by giving the entitlement to third parties, who are not parties to the contract in question, to enforce certain rights under the contract. Specific criteria must be satisfied:
— The contract must give the third party a right.
— The terms must confer a benefit (unless it is clear that the parties did not intend a benefit to be conferred).
— The third party must be identified in the contract. That can be by name, by class or by description. (It should be noted that the third party may not have existed at the time the contract was entered into, e.g. a newly formed limited company).

Such a right may only be enforced in accordance with the terms of the contract, and the party against whom the third party seeks to enforce the terms

---

[172]*Lynch* v. *Thorne* [1956] 1 All ER 744.
[173]*Rotherham Metropolitan Borough Council* v. *Frank Haslam Milan & Co Ltd* (1996) 59 Con LR 33.

may use any defences and remedies available under the contract and may raise any set-off or counterclaim. In some instances, the third party may be treated as a party to an arbitration agreement in the contract.

The parties can rescind or vary the contract in order to remove the right, but not if:

— the third party has communicated his agreement to the term; *and* the parties know that the third party has relied on the term; *or*

— It was reasonably foreseeable that the third party would rely on the term and he has relied on it.

To overcome that, the Act allows parties to include a term in the contract by which they agree to rescind or vary without the consent of the third party or setting out circumstances for the third party's consent. Most usefully, parties to a contract may expressly exclude third party rights under that contract. That seems to be the simplest approach and it is the approach favoured by the Joint Contracts Tribunal (qv) which, by amendment, has inserted such an excluding clause in all the standard forms.

Certain contracts are excluded from the operation of the Act, including promissory notes and other negotiable instruments, contracts of employment and agency contracts.

It was forecast that the advent of the Act would put an end to the use of collateral warranties (qv). In view of the opportunities to negate the operation of the Act in respect of particular contracts, it is unlikely that the demise of collateral warranties is imminent.

Contributory negligence Governed by the Law Reform (Contributory Negligence) Act 1945. An action for negligence (qv) against one party cannot be defeated merely by proving that the other party contributed to the damage by reason of his own negligence. In such circumstances, if the negligence of both parties is proved, the court will reduce the damage payable by the defendant (qv) by a proportion which has regard to the 'contributory negligence' of the plaintiff (qv). Contributory negligence may have limited application in a contractual dispute[174]. The contractual term in question must be one to take 'reasonable care' and there must be a concurrent duty of care (qv) in tort (qv)[175]. It is generally advisable for the defendant (qv) to join in the proceedings any other party from which a contribution is demanded.
See also: **Civil Liability (Contribution) Act 1978.**

Copyright Rights relating to creative work of an artistic, dramatic, literary or musical nature. They usually belong to the originator or creator. Generally, copyright remains with the creator of the work for his lifetime and for 50 years thereafter. No one may produce, reproduce or copy his work without his express permission. Ownership of copyright may be transferred from the creator or a licence (qv) may be given to someone to reproduce the work while the

---

[174]For example, *Barclays Bank plc* v. *Fairclough Building Ltd (No. 2)* (1995) 76 BLR 1.
[175]*Forsikringsaktieselskapet Vesta* v. *Butcher* [1989] 1 All ER 402.

creator retains the ownership of the right. In published works it is usual, though not essential, to show that copyright is claimed thus: © Alice Davis (2000).

Copyright is governed by the Copyright Designs and Patents Act 1988. An important innovation is the introduction of 'moral rights'. The author of the copyright work in general has the right to be identified as the author. Sections 77 (4) to (5) refer to works of architecture. The creator has the right to be identified whenever any kind of copy of the work is issued to the public. He also has the right to be identified on the building. The creator must assert his right before it can be infringed. If there is a delay in asserting the right, the court can take it into account in awarding any remedy. The right not to have the work subjected to derogatory treatment is also recognised. Treatment is said to be derogatory if it is a distortion or mutilation of the work or is otherwise prejudicial to the author's honour or reputation. When applied to a building, the author has the right to require his identification to be removed.

An author may take action for breach of statutory duty if a moral right is infringed. The court may grant an injunction (qv) on terms that the infringing act must cease unless an appropriate disclaimer is made dissociating the author.

Architects have copyright in their designs. An architect commissioned to design a building retains the copyright in his design but, normally, the client has a licence to reproduce that design as a building, provided the client has agreed the matter with the architect or paid a sufficient fee such that the architect's agreement to the use of his design is implied[176]. In the RIBA Standard Agreement for the Appointment of an Architect (SFA/99) (qv), the position is clearly set out:

— Copyright in all documents and drawings prepared by the architect and in work executed from them remains the property of the architect.

— The client has a licence to reproduce and use the designs for any purpose provided that it relates to the project on the particular site and may allow consultants and contractors providing services for the project to do likewise. The purposes are broadly drafted, but they expressly exclude any right to reproduce the design to extend the project or, of course, any other project.

— The architect is not liable if the copyright material is used for a purpose other than that for which it was intended. That would be the ordinary common law position in any event.

— If the copyright material is used after the architect has finished performance of his services (which stops short of work stage D) and before practical completion (qv), the client must obtain the architect's consent and/or pay any agreed licence fee before he may proceed to execute the work, provided that the architect shall not withhold his consent unreasonably. This is clearly to overcome the problem faced by many architects when a client will only agree to engage them for limited work, but wish to take the benefit of the architect's designs.

If the architect suspects that a client, or anyone else, is about to use his designs without consent, express or implied, he can apply to the court for an

---

[176]*Stovin-Bradford* v. *Volpoint Properties Ltd and Another* [1971] 3 All ER 570.

injunction to restrain them. Note, however, that the courts will not grant an injunction if the work has been commenced because they consider that damages in the form of a suitable fee for reproduction will amply recompense the architect and stopping expensive building work is not justified in such circumstances. What constitutes commencement of building work may be a difficult matter to decide[177].

It may also be difficult to prove infringement of copyright. It is easy to show that a design has been copied if every detail is exactly the same as the original, but the position is not so straightforward if portions only of the design have been copied. Small alterations to a design will not overcome the rights of the original designer. Similarly, if a substantial and recognisable feature of the original design is copied, the original architect will have a good case. The issue is a matter of degree and very uncertain in many instances. The architect should try to negotiate a suitable fee rather than resort to the courts in such instances.

In an action for breach of copyright, the court may have regard to the flagrancy of the infringement and any benefit accruing to the defendant by the infringement when considering whether to award additional damages 'as the justice of the case may require'.

Section 107 of the 1988 Act makes certain infringements of copyright a criminal offence. The penalty may be a fine or imprisonment or both.

Corporation An artificial legal person having a distinct legal existence, a name, a perpetual succession and a common seal. Corporations are classified as:
  — *Corporations sole*, which consist of only one member at a given time and are the successive holders of certain offices, e.g. the Bishop of Exeter. It is an office or function as opposed to its holder in his private capacity.
  — *Corporations aggregate*, which consist of many members. They come into existence either by grant of a royal charter or by or under authority of an Act of Parliament, e.g. a limited liability company. The corporation is a separate legal entity distinct from the individuals who are its members for the time being.

Contracts made beyond the powers of the corporation, as laid down in its charter or limited by statute, are *ultra vires* (qv) and void. This is not of great importance as regards building contracts. Corporations can make contracts in the same form as is available to private individuals. The age-old requirement of the common law that corporations had to contract under seal (qv) was abolished by the Corporate Bodies Act 1960.

Corporeal property Tangible property such as land or goods which has a physical existence in contrast with incorporeal property (qv) which consists of intangible legal rights. A *corporeal hereditament* is a tangible interest in land — the land itself and things which are annexed to or form a part of it ('fixtures' (qv)) while an *incorporeal hereditament* is a right over land, such as a right of way or

---

[177] *Hunter* v. *Fitzroy Robinson & Partners* (1978) 10 BLR 84.

other easement (qv). The word 'hereditament' denotes that the property is inheritable.

See also: **Personal property; Real property.**

Corroboration Evidence (qv) which tends to strengthen other evidence. It is not strictly necessary in English law but it is always desirable. The court may act on the testimony of one witness alone, but in certain specified cases, e.g. perjury, corroboration is required.

See also: **Hearsay; Parol evidence.**

Corrupt practices Many standard form contracts contain clauses forbidding the contractor from indulging in corrupt practices, such as the giving of bribes or the taking of secret commissions. GC/Works/1 (1998), clause 24 is such a provision and entitles the employer to determine the contract and/or to recover from the contractor the amount or value of the bribe, etc. JCT 98, clause 27.4 (IFC 98, clause 7.4) confers a similar right to determine the contractor's employment 'under this or any other contract' for such practices, which are, in any case, a criminal offence under the Prevention of Corruption Acts 1889 to 1916.

Very strict legal rules at common law enable the employer to rescind a contract tainted by corrupt practices and to recover any secret bribes or commission in any case[178].

See also: **Bribery and corruption; Fraud.**

Cost reimbursement contract A type of contract by which the contractor receives all his costs together with a fee. There are four common variations:
— *Cost plus percentage:* The contractor is paid the actual cost of the work reasonably incurred plus a fee, which is a percentage of the actual cost, to cover his overheads and profit. This form of contract is often used for maintenance work or for work where it is difficult to estimate the work to be done or for emergency work. It is possible to invite tenders on the basis of the percentage but there is no incentive for the contractor to make good progress or to save money because his fee rises with the total cost of the job. The Joint Contracts Tribunal (qv) has produced a suitable form of contract – the Standard Form of Prime Cost Contract (PCC 98).
— *Cost plus fixed fee:* Similar to the cost plus percentage contract and used for similar situations. The important difference is that, because the fee is a fixed lump sum, the contractor has more incentive to finish quickly and maximise his profit as a percentage of turnover. It is usual for some indication of the total cost to be given to tenderers. The Standard Form of Prime Cost Contract (PCC 98) is applicable.
— *Cost plus fluctuating fee:* Similar to the fixed fee contract and used for similar situations. An estimate is made of the total cost. The amount of the fee received by the contractor varies inversely to the costs actually

---

[178] *Salford Corporation* v. *Lever* (1891) 63 LT 658.

achieved. Thus, if the costs are less than the estimated costs, the contractor receives a greater fee calculated in accordance with an agreed sliding scale and vice versa. It is to the contractor's advantage to reduce costs and finish the work quickly. PCC 98 can also be used for this type of procurement.

— *Target cost:* Used in similar situations to the contracts previously discussed, it can also be used for a wide variety of conditions. Priced bills of quantities (qv) or a priced schedule are agreed and a target cost is obtained for the project. The contractor's fee is usually quoted as a percentage of the target cost. Provision is made for the target cost to be adjusted to take account of variations and fluctuations. The contractor is paid the actual costs reasonably incurred. The total of these costs is compared with the adjusted target cost. If they show a saving, the fee is increased in accordance with a pre-agreed formula, and vice versa. The disadvantage of this type of contract lies in the complex measurement procedures involved and the difficulty of agreeing targets and percentages. The NEC has this as one of its options.

See also: **Engineering and Construction Contract; Management contract; Prime contracting; Target cost (contract) (BPF); Value cost contract.**

Costs After litigation (qv) or arbitration (qv) the general rule is that the unsuccessful party has to pay the reasonable costs of the successful party. This general position may change where one of the parties has offered to settle, made a payment into court or made a *Calderbank* (qv) offer. Under the Civil Procedure Rules (CPR) (qv), it is now possible for a court to make partial costs orders so that where a party wins on four out of five issues, the court may order that both parties pay costs on an issue by issue basis.

Costs are generally ordered on the standard basis which usually equates to between 60% and 85% of the actual costs incurred. If the costs are not agreed, they will be determined in front of a costs judge by means of a costs assessment (qv). In this procedure the costs must be justified as being both reasonable and proportionate to the issue(s) in dispute. There are no clear guidelines on the question of proportionality.

Costs may be awarded occasionally on an indemnity basis. This means the costs are assumed to be reasonable unless shown to the contrary. It is rare for all costs to be recovered, even on the indemnity basis.

The irrecoverable costs (the difference between the actual costs and those recovered from the paying party) are an important factor in considering the question of settlement (qv) as this element grows as the proceedings progress such that going to trial may no longer be cost-effective.

The costs position for statutory adjudication is wholly different as each party bears its own costs. The adjudicator has no power to award costs, unless the parties have expressly given him power to make such an award[179].

See also: **Calderbank offer; Commercial Court; Official Referees; Sealed offer.**

---

[179] *Northern Developments (Cumbria) Ltd* v. *J. & J. Nichol* [2000] BLR 158.

Costs assessment Procedures under the Civil Procedure Rules (qv) which used to be known as taxation of costs (qv). Costs are principally dealt with under Parts 43 to 48 and the accompanying practice directions. Essentially, it is the process by which the court determines the amount of costs to be paid by a party following the giving of an order for costs if the parties cannot agree on the amount.

Costs judge See: **Taxation of costs.**

Counsel A barrister or group of barristers.

Since 3 April 1989, arbitrators and members of certain professional bodies such as the RIBA, RICS and ASI have been able to obtain counsel's opinion direct without the necessity to engage a solicitor to instruct counsel on their behalf. This new facility is called 'direct professional access' and is limited to obtaining opinions or seeking advice in conference; construction professionals cannot instruct counsel direct to appear in court proceedings, and for conducting litigation they must employ a solicitor. There is no such bar in respect of arbitration proceedings.

Counterclaim In legal proceedings, a defendant may respond to a claim for damages by serving a defence and a claim for damages against the claimant (qv). This latter claim is termed a 'counterclaim' or 'crossclaim'. The counterclaim is not part of the defence; it may, indeed, have no relevance to the original claim. It may, however, be properly described as a set-off (qv) and so be a defence as well as a counterclaim. It may simply be a claim which the defendant intended to pursue in any event. It is for the court to decide whether it is convenient to deal with both claim and counterclaim at the same time. If the court decides that it is not convenient, the counterclaim may be struck out and it is for the defendant then to bring a separate action, as claimant, on the substance of the counterclaim. It follows from s. 14 of the Arbitration Act 1996 that the respondent in an arbitration may not raise an unrelated counterclaim in that same arbitration against the claimant unless the claimant agrees to that counterclaim being heard. Otherwise, such an unrelated counterclaim can only be brought by serving a further notice of arbitration. A similar situation is to be found in statutory adjudication, because the adjudicator may only consider the questions referred to him by the party seeking the adjudication and the respondent's defence to those questions.

Architects will be familiar with the device of counterclaiming if they have been involved in suing for outstanding fees. In many cases, a client will counterclaim, alleging negligence, in order to prevent the architect obtaining summary judgment (qv). It is difficult to show that a counterclaim is entirely frivolous and the architect may drop his original claim or face long delays before, possibly, obtaining judgment.

See also: **Abatement; Pleadings; Set-off.**

Counter-offer For a contract to come into existence there must be an offer (qv) by one party and an unqualified acceptance (qv) by the other. If the second party signifies 'acceptance' with qualifications, this is not true acceptance, but merely a counter-offer, which the first person is free to accept or reject. A counter-offer destroys the original offer and the second party may not subsequently purport to accept the original offer[180].

A counter-offer must be distinguished from a mere request for further information[181]. For example, if contractor A requests a quotation from supplier B, B's quotation is the offer. It may contain special terms of business. If A writes purporting to accept the offer subject to his own contract terms, this is a counter-offer. The process may continue and is known to lawyers as the 'battle of forms'.

In these circumstances, if there is a contract it is often the set of terms last in time which were acted upon which is decisive. The correct approach is to see whether one party has accepted the other's terms by express words or conduct, e.g. by acting upon them[182]. However, in some cases there will be no contract at all because neither party has accepted the other's offer or counter-offer.

Countryside and Rights of Way Act 2000 A new Act which makes new provision for access to countryside. Many of the provisions of the Act came into force on 31 January 2001. The Act introduces the so-called right to roam, giving effective rights of way (qv) over land in the absence of any easement (qv) or other established right.

Course of dealing Where parties have dealt with each other consistently using certain terms in a substantial number of previous transactions of a particular type and they enter into a further agreement of the same type, the previously used terms may be incorporated into their contract[183]. The situation is relatively rare and the criteria are strictly examined. It certainly does not entitle a contractor to rely on the terms of a single previous contract as governing the parties in a subsequent contract.

See also: **Contract.**

Courts *The Concise Oxford Dictionary* defines a court as an 'assembly of judges or other persons acting as tribunal' as well as a 'place or hall in which justice is administered'.

Courts can be classified in several ways. **Figure 6** represents diagrammatically the major courts in England and Wales. They are divided into *superior* and *inferior* courts. Inferior courts are those which are subject to control by the High Court. Only the decisions of superior courts play any part in the development of judicial precedent and it is only decisions of superior courts which have any binding authority in later cases (see: **Judicial precedent**). Some

---

[180]*Hyde* v. *Wrench* (1840) 3 Beav 334.
[181]*Stevenson* v. *McLean* (1880) 5 QBD 346.
[182]*Butler Machine Tool Co Ltd* v. *Ex-Cell-O Corporation (England) Ltd* [1979] 1 All ER 965.
[183]*J. Spurling Ltd* v. *Bradshaw* [1956] 2 All ER 121; *McCutcheon* v. *David McBrayne Ltd* [1964] 1 WLR 125.

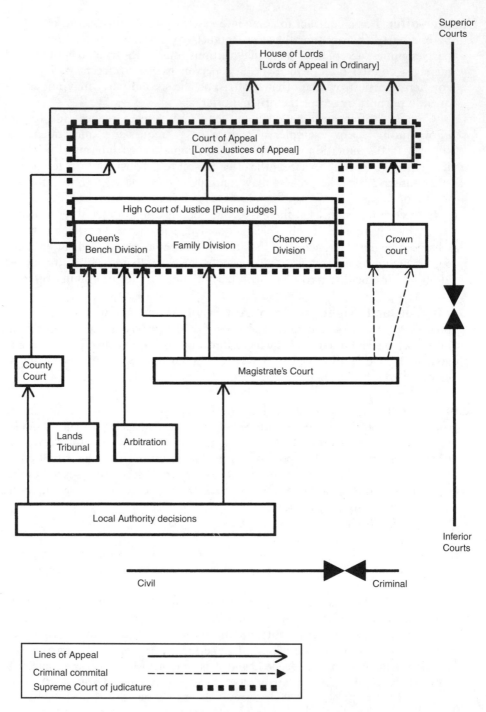

Figure 6 Diagram showing the organisation of the English court system.

courts have only criminal jurisdiction, while others hear civil matters only. Some are hybrid and can hear both types of case.

Magistrates' courts deal mostly with minor criminal matters and are normally staffed by Justices of the Peace who have no formal legal qualifications, but must undergo some training. Paid District Judges (formerly Stipendary Magistrates) − who have formal legal qualifications − are appointed in London and major centres. Magistrates' courts have some limited jurisdiction, e.g. hearing certain appeals against local authority decisions.

Crown courts deal with serious criminal matters and also hear appeals from decisions of magistrates' courts. They are part of the Supreme Court of Judicature and are served by High Court judges and circuit judges.

County courts deal with the bulk of civil litigation; there is no monetary limit on the amount of the claim in both contract and tort claims, but higher value or complex matters are litigated in the High Court. They are staffed by circuit judges appointed by the Queen on the advice of the Lord Chancellor from among practising barristers and solicitors of experience. There is no jury.

The High Court of Justice (which is part of the Supreme Court of Judicature) is divided into three:
— Queen's Bench Division.
— Family Division.
— Chancery Division.

High Court judges are appointed by the Crown on the advice of the Lord Chancellor from the ranks of eminent practising barristers and possibly solicitors of long standing. Building contract disputes are normally dealt with in the Queen's Bench Division, generally by the judges of the Technology and Construction Court (qv).

The Court of Appeal consists of a criminal and a civil division, is presided over by the Lord Chief Justice and the Master of the Rolls respectively and consists of Lords Justices of Appeal, who are usually promoted from the High Court bench. They sit in courts of two or three and hear appeals (qv) from both county courts and the High Court. The criminal division hears appeals against conviction and sentence.

The House of Lords as a judicial body consists of Lords of Appeal in Ordinary, together with the Lord Chancellor. Appeals are heard by the Appellate Committee of the House of Lords, usually sitting as a committee of five, but in very important cases there may be seven members. Before an appeal can be heard the appellant must obtain the permission of the Court of Appeal or the Appeal Committee of the House of Lords itself. In practice, only matters of the greatest importance proceed this far.

The structure of the courts in Northern Ireland is very similar to England and Wales although it is a separate jurisdiction. The decisions of the respective High Courts and Courts of Appeal are not strictly binding on each other, but usually they will be followed if in other respects the law on a particular matter is the same. The structure in Scotland is different. The House of Lords is the final appellate court for Northern Ireland and Scotland as well as for England and Wales. The Judicial Committee of the Privy Council hears appeals from such Commonwealth countries as retain the right of appeal, colonial territories, ecclesiastical courts, the Isle of Man and Channel Islands and certain professional tribunals.

See also: **Commercial Court.**

Covenant A promise or an agreement executed as a deed. A covenant can also be implied by law, in certain cases, e.g. leases.

Restrictive covenants most directly affect the construction industry. They restrict the use of freehold land according to the original agreement. They are attached to the land not the person, so that a person buying land also takes on the benefit or burden of any restrictive convenant which applies to the land: s. 56, Law of Property Act 1925. Thus, a covenant may restrict the building of anything on land A for the benefit of the owner of land B. New owners may purchase the land but the restrictive covenant remains unless, of course, the two owners (who must be the only ones affected) agree that the covenant may be removed. In the case of a restrictive covenant imposed on all the owners of land in a particular area for the benefit of that area (i.e. a 'building scheme'), such as a housing estate, the covenant can be enforced by any of the owners.

To enforce a restrictive covenant, the following conditions must be satisfied:
— The covenant must confer a benefit on the land or other land.
— The covenant must be preventive, i.e. to stop something occurring, and must not require the expenditure of money. A covenant to build and keep a boundary wall in good condition is not restrictive.
— The person seeking to enforce the covenant must show that he has been assigned the benefit of the covenant or that it attaches to his land.

Outmoded restrictive covenants may be modified or discharged by the Lands Tribunal (qv).

An *express covenant* may be made, usually between landlord and tenant, in written form. It often covers such things as the tenant's duty to repair, insure against fire and pay the rent.

An *implied covenant* is one that is not written down but is implied by law. Common implied covenants relating to landlord and tenant, if not expressly stated, are that the tenant will have 'quiet enjoyment' of the land (no other party can question his right to the land) and that the tenant will pay the rent.

See also: **Restrictive covenant.**

CPR See: **Civil Procedure Rules.**

Criminal liability Liability which arises under the criminal law, as opposed to civil law. Conviction of a criminal offence may result in a fine, imprisonment or some other punishment. A crime is an offence against the state. The law has declared various kinds of conduct criminal. For the most part, criminal liability in the building industry will result from breach of some specific statutory provision or requirement of regulations, e.g. Regulation 7 (notification of project) of the Construction (Design and Management) Regulations 1994.

Cross-examination The second stage in the examination of witnesses in judicial or arbitral proceedings when the witness is cross-examined by or on behalf of the opposing party. Leading questions may be put, and a very wide range of questions is allowed. The object of cross-examination is to shake the witness's testimony and establish matters which are favourable to the cross-examining party. The witness can be asked questions the answers to which tend to discredit him by showing that he is a person not to be believed.
See also: **Examination-in-chief, Re-examination; Witness.**

Crown The term 'the Crown' may mean the Queen acting as head of state on the advice of her ministers and is largely synonymous with the term 'the State'.

In the context of building contracts the term means the various Government departments. In general, the Crown has the same power to make contracts as local authorities, companies in the private sector or individuals, but the following should be noted:
— There are limits on the contractual capacity of the Crown, although their extent is not entirely clear. In practical terms, building and other contracts with Crown departments can be enforced by and against the Crown under the Crown Proceedings Act 1947.
— There are special Crown contracting procedures which have been developed over the years.
See also: **GC/Works/1 contract.**

Crown privilege The Crown has a right to object to producing a document in court on the ground that it is contrary to the public interest to do so. This is also known as Public Interest Immunity (PII). The privilege (qv) is claimed by an affidavit (qv) sworn or a certificate signed by the appropriate minister which states that he has examined the document personally and objects to its being produced. The courts may, however, question a claim of Crown privilege.

Custom Long-standing practice or usage is binding on those within its scope. It is a subsidiary source of law, though largely unimportant today. Evidence of trade custom or usage may be given and proved to show that words in a contract are to be interpreted in a particular way, e.g. in *Myers* v. *Sarl* (1860)[184] where evidence was allowed to show that 'weekly account' was a term of art well known in the building trade at that time. Implied terms (qv) may also be established by proving trade custom or usage, e.g. 'reduced brickwork' as meaning brickwork 9 inches thick[185].

---

[184](1860) 3 E & E 306.
[185]*Symonds* v. *Lloyd* (1859) ER 622.

It must be established, however, that the custom relied on is:
— Open and notorious, i.e. generally accepted and acted on.
— Not contrary to law[186].
— Reasonable and certain in its operation.

In general, customs will only be implied if they are not expressly excluded and where they do not contradict any other terms implied by the general law. They are difficult to prove in practice.

---

[186]*Crowshaw* v. *Pritchard & Renwick* (1899) 16 TLR 45.

Damage; Damages Damage is any harm suffered by a person. For an action (qv) to lie, it must be wrongful damage. Damages are also the compensation awarded by the court or claimed by the claimant (qv).

See also: **Consequential loss; Direct loss and/or expense; Remoteness of damage; Resitutio in integrum; Special damages.**

Dangerous premises A local authority (qv) has the power to deal with premises which are in a dangerous or defective condition under ss. 76 to 81 of the Building Act 1984. The procedure is by way of complaint to the magistrate's court (qv) which may make an order requiring works to be carried out where any building or structure or part thereof is in such a condition as to be dangerous to a person in the street, in the premises themselves, or in adjoining premises. Section 78 in fact contains an emergency procedure which can be invoked where immediate action is necessary. The local authority may take any necessary action to abate the nuisance (qv) and recover expenses from the person in default.

See also: **Abatement; Occupiers' liability.**

Day A 24 hour period extending from midnight to midnight is called a natural day. The period between sunrise and sunset is called a civil day. Contracts commonly refer to day in the first sense; they may also refer to working days. In the absence of any special definition in the contract, a working day is any day other than Sundays, Good Friday, Christmas Day, a bank holiday or a day declared to be a non-working or non-business day. The term working day must be expressly stated in the contract if that is what is meant; it will not be implied unless to do otherwise would make nonsense of the particular provision.

ACA 3 refers throughout to 'working days' and defines this expression (article G) as meaning Monday to Friday inclusive, excluding any day which is a public holiday in the country in which the works are to be executed and any day which is a holiday under the Building and Civil Engineering Annual and Public Holiday Agreements.

The JCT contracts, in conformity with part II of the Housing Grants, Construction and Regeneration Act 1996 (qv), set out which days will be excluded where something is to be done within a specified period of days. For example, IFC 98 clause 1.14 states that public holidays are to be excluded and clause 8.3 defines a public holiday as meaning Christmas Day, Good Friday and any day which is a bank holiday under the Banking and Financial Dealings Act 1971. The effect is that most weekends will be counted as days for the purpose of reckoning a notice period. This will be particularly difficult for construction firms who tend to close down for two weeks over the Christmas and New Year periods. Unless they make arrangements to check their mail during the holiday period, they may return to find that the period within which they had to do something has expired without them ever knowing that it had started.

Under rule 2.8 of the Civil Procedure Rules referring to court procedure, a period of time expressed in days refers to clear days. Part 6 states that Saturday, Sunday, Christmas Day, Good Friday or a bank holiday will be excluded from the calculation of any period of five days or less. It also sets out deemed days of service for documents served by first class post (the second day after posting), document exchange (the second day after being left at the document exchange), actual delivery (the day after delivery), fax (on the same day as a business day if before 4 PM, otherwise on the next business day) or other electronic method (the second day after the day of transmittal).

If the contract requires 14 days' notice, the notice expires on the fifteenth day. However, if 14 clear days' notice is required, the notice does not expire until the sixteenth day.

See also: **Month; Notices; Year.**

Dayworks If works are carried out by the contractor and the works cannot properly be valued by measurement, they may be valued on a prime cost (qv) basis. The amount of work done and materials used are recorded and a percentage is added.

JCT 98 makes provision (clause 13.5.4) that vouchers (commonly called 'daywork sheets') must be delivered to the architect or his representative not later than the end of the week following that in which the work was carried out. The valuation must comprise either the prime cost of the work (as defined in the 'Definition of Prime Cost of Daywork carried out under a Building Contract', current at date of tender and issued by the RICS and CC) plus percentage additions as set out by the contractor in the contract bills (qv) or if the work is of a specialist nature and the body representing the employers in that trade has issued a definition, the prime cost calculated in accordance with that definition plus the percentage additions as before.

GC/Works/1 (1998) provides for daywork (clauses 42 (5) (d)) to be valued by the value of materials used and plant and labour employed in accordance with the basis of charge for daywork described in the contract. Clause 42 (12) requires the contractor to give the quantity surveyor reasonable notice of the commencement of daywork and to deliver vouchers to him by the end of the week following that in which the work was done.

Neither ACA 3 nor MW 98 expressly provides for daywork. Clauses 17 and 3.6 respectively provide instead for valuations to be agreed. In practice, daywork calculations in respect of work and materials will take place.

Where it has been decided by the quantity surveyor that valuation of a variation is to be undertaken using daywork sheets and they have been signed by a properly authorised person, the quantity surveyor has no power to substitute his own estimate of the hours it should have taken to do the work[187].

---

[187]*Clusky (trading as Damian Construction)* v. *Chamberlain* (1994) April BLM 6.

It is, of course, quite possible to carry out the whole of a contract using dayworks as a basis for valuation and payment.

De minimis Very minor. The term is generally used in relation to the issue of the certificate of practical completion (qv) under JCT contracts. The certificate may be issued when there are *de minimis* items still to be executed[188]. In this sense, it seems sensible to interpret the phrase in the context of the particular work included in the contract. It is unlikely that the amount of work to be considered *de minimis* under a multimillion pound complex use contract will be the same as under the contract for a private dwelling house. It must have some relation to the total work to be done and particularly to whether the completion of such work is likely to cause the employer any inconvenience after occupation. That appears to be the inference to be drawn from the judgment in *Westminster Corporation* v. *J Jarvis & Co Ltd* (1969)[189].

Death The death of a person may end some claims and liabilities. For example, a contract for personal services ends on the death of the person contracted to give those services. This situation may occur in respect of individual architects or contractors and *would* apply, for example, if the employer had engaged a sculptor to embellish some part of the building and the sculptor died. In general, claims for negligence against a party do not lapse on the death of that party but may be pursued against his heirs. Death is important in respect of many situations, most notably of course, wills. A partnership ends with the death of any one partner although the terms of the partnership usually provide for the remaining partners immediately to form a new partnership to continue the business – often retaining the deceased partner's name as part of the title. See also: **Frustration.**

Debenture A document, issued by a company, which acknowledges a loan and provides for repayment with interest. It also contains a charge which is fixed on property which is definite or ascertainable, or, a floating charge over property which is subject to change. A debenture holder has the right to make an immediate appointment, without notice, of a receiver (in the case of a floating charge – a receiver (qv) or receiver and manager) if:
— There has been a default in repayment of interest.
— The security is in jeopardy.
There is usually a provision in the debenture to the effect that the receiver or receiver and manager shall be deemed to be an agent of the company.
The company's assets do not vest in the receiver, but he has power to realise the assets by sale. The receiver does not become a party to contracts in existence with the company, and it follows that he cannot vary them[190]. Furthermore, the receiver 'must fulfil company trading contracts entered into

---

[188] *H. W. Neville (Sunblest) Ltd* v. *Wm Press & Sons Ltd* (1981) 20 BLR 78.
[189] [1969] 1 WLR 1448 (CA) per Salmon LJ at 1458 whose view was not expressly disapproved by the House of Lords.
[190] *Parsons* v. *Sovereign Bank of Canada* [1911] AC 160.

before his appointment or he renders it liable to damages if he unwarrantably declines to do so'[191].

See also: **Insolvency; Liquidation.**

Debt A sum of money owed by one party to another, and recoverable by means of legal action. Liquidated damages due to the employer are often stated to be 'recoverable as a debt'. A speedy way to do this, if the debtor has no defence or counterclaim, may be to apply for summary judgment (qv).

If a party cannot pay his debts as they fall due, he is insolvent (qv) which may result in bankruptcy (qv) in the case of an individual, or liquidation (qv) in the case of a company registered under the Companies Acts.

Deceit A tort (qv) consisting essentially of a fraudulent misrepresentation (qv) made with the intention that the other person should rely on it and which causes damage to him.

See also: **Fraud.**

Deed See: **Specialty contract.**

Deed of Arrangement See: **Arrangement, deed of.**

Deemed To be treated as. The word is used not only in statutes (qv) but also in building contracts. The 'deemed' thing must be treated for the purposes of the statute or contract as if it were the thing in question. For example, clause 10 of JCT 98 states that instructions given by the architect to the person-in-charge 'shall be deemed to have been issued to the Contractor', i.e. such instructions shall be treated as though they have been issued to the contractor. What is 'deemed' is conceded not to be true[192]. The preliminaries to bills of quantities may state that the contractor will be deemed to have inspected the site and made all necessary enquiries about the nature of the ground. That means that he is to be treated as having done so even if it is perfectly clear to all parties that the contractor has never set foot on site.

Deemed variation Generally, an architect's instruction which is treated as being an instruction requiring a variation even though the instruction may not specifically state as much. Deemed variations are provided for in the standard forms, e.g. JCT 98, clause 2.2.2 where there are errors, etc. in the contract bills and clause 6.1.3 in relation to divergences (qv) between statutory requirements and contract documents.

See also: **Deemed.**

---

[191]*George Barker (Transport) Ltd* v. *Eynon* [1974] 1 All ER 900.
[192]*Re Cosslett (Contractors) Ltd, Clark, Administrator of Cosslett (Contractors) Ltd in Administration* v. *Mid Glamorgan County Council* [1997] 4 All ER 115.

Defamation A tort (qv) which consists of the publication to a third party of false and derogatory statements about another person without lawful justification. A statement is defamatory if it exposes the person defamed to 'hatred, ridicule or contempt'. Defamation in a permanent form, e.g. in writing, is called *libel* (qv) while in an impermanent or transitory form, e.g. the spoken word, it is called *slander*. Each repetition of a defamatory statement, whether oral or written, amounts to a separate publication and each person repeating it is liable as well as the person who originated the statement.

Defamation is of little importance in the context of building contracts, save as regards 'reasonable objections' made to a proposed nominated sub-contractor (qv) under, e.g. clause 35 of JCT 98. Provided such objections are made reasonably, they will be given privilege (qv) unless the maker was actuated by malice (spite or ill will) or published his objection beyond those who have an interest to receive it, i.e. the architect and (possibly) the employer. The same principle applies to references about the character and abilities of a former employee. There are particular duties owed to employees to exercise due care and skill in the preparation of references. It is not sufficient that an employer believes what he says is true, he must exercise reasonable care in checking the truth[193].

Default Failure to act, especially a failure to meet an obligation. The word is used frequently in building contracts, especially in indemnity clauses (qv). JCT 98, clause 20.2 thus refers to '... negligence, omission or default of the contractor ...'. An earlier version of that clause was considered by the High Court in *City of Manchester* v. *Fram Gerrard Ltd* (1974)[194] where it was held that for there to be a 'default' does not necessarily require that the injured party should be able to sue the defaulter. The judge cited the decision of Parker J in *Re Bayley Worthington & Cohen's Contract* (1909)[195] where it was said: 'Default must ... involve either not doing what you ought to do or doing what you ought not, having regard to your relations with the other parties concerned in the transaction; in other words, it involves breach of some duty you owe to another or others. It refers to personal conduct and is not the same thing as breach of contract'.

On the facts before him, Kerr J held that 'default' is established 'if one of the persons covered by the clause either did not do what he ought to have done, or did what he ought not to have done in the circumstances, provided ... that the conduct in question involves something in the nature of a breach of duty...'. On the facts he held that the conduct of sub-contractors in applying and using a waterproof coating which contained a phenolic substance and misinforming the plaintiffs about the curing period amounted to 'default' in the context of the indemnity clause.

More recent authority, in the Court of Appeal, has considered whether there is a difference between breach of contract and default and concluded that

---

[193] *Spring* v. *Guardian Assurance PLC and Others* (1994) 3 All ER 129.
[194] (1974) 6 BLR 70.
[195] [1909] 1 Ch 648.

119

'default' in a contractual document means a breach of contract especially if damages are said to be incurred[196].

Defect correction period A term used in the NEC (qv). The period is to be inserted by the employer into the contract data (qv). It is referred to in the core clauses in clause 4. It is not otherwise defined. Clause 43.1 (a sub-section of clause 4) states that the period begins 'at Completion for Defects notified before Completion and when the Defect is notified for other Defects'.

Defective Premises Act 1972 The construction of dwellings is subject to the provisions of this Act which came into force on 1 January 1974. The Act does not apply to Scotland or Northern Ireland[197] and is limited to dwellings (including blocks of flats) as well as dwellings which are created by conversion or enlargement. Where an 'approved scheme' is in operation in relation to the dwelling, the Act does not bite.

Section 1 (1) provides: 'Any person taking on work for or in connection with the provision of a dwelling ... owes a duty to see that the work which he takes on is done in a workmanlike or, as the case may be, professional manner, with proper materials and so ... that the dwelling will be fit for human habitation when completed'.

The provision is 'in addition to any duty a person may owe apart' from the Act and extends the common law duties owed to the buyer of a house in the course of erection in a number of ways. It applies to 'conversions' and not just the erection of a dwelling. It extends its benefits to every person acquiring an interest in the dwelling, i.e. subsequent purchasers, subject to the limitation period (see: **Limitation of actions**) which arises 'at the time when the dwelling was completed' or, in the case of rectification work, 'at the time when the further work was finished'.

A builder who carries out work in compliance with instructions given by or on behalf of the person for whom the dwelling is being built, e.g. under JCT 98 or JCT Minor Works contracts, is given a defence. He has no liability under the Act 'to the extent that he does (the work) properly in accordance with those instructions ... except where he owes a duty to (the client) to warn him of any defects in the instructions and fails to discharge that duty': s. 1 (2).

Section 6 (3) outlaws clauses excluding or restricting liability under the Act, and probably extends to such provisions as JCT 98, clause 30.9, in so far as it makes an architect's final certificate (qv) conclusive under certain circumstances.

It is important to appreciate that the Act is very widely drawn. The duty imposed by s. 1 (1) extends not only to builders and developers but also to architects and other designers. It also extends to local authorities, housing associations, etc. when exercising their powers under the Housing Acts: s. 1 (4) (b).
See also: **National House Building Council.**

---

[196]*Perar BV* v. *General Surety & Guarantee Co Ltd* (1994) 66 BLR 72.
[197]In Northern Ireland the equivalent is the Defective Premises (Northern Ireland) Act 1972.

Defective work In the context of all standard forms of building contract defective work is work which is not in accordance with the contract. The architect may have a degree of discretion in accepting or rejecting work, but he has no power to insist upon higher standards than those laid down in the contract documents. There is, of course, an implied term (qv) in every building contract to the effect that the contractor will do the work in a good and workmanlike manner[198]. A contractor who complies precisely with a detailed specification may not be liable if the specification is inadequate[199]. However, the better view probably is that a contractor who discovers that a particular detail, if constructed, would lead to seriously defective work has a duty to point out the defect to the employer[200].

JCT 98 deals with defective work, by implication, in many clauses requiring the contractor to carry out the work properly (notably clause 2.1) and, expressly, clauses 8.4, 17.2, 17.3 and 27.2.1.3. These clauses give the architect power to have defective work removed from the site and to have defects which appear during the defects liability period (qv) made good, and they give the employer power to determine the contractor's employment if the contractor neglects to remove defective work and thereby the works as a whole are materially affected.

ACA 3 similarly carries the implication that the contractor will not produce defective work and provides for dealing with it in clauses 8.1 (a), 12 and 20.1 (d). These clauses give the architect similar powers to those in JCT 98 and give the employer similar powers of termination if the contractor persistently neglects to remedy the defective work at the request of the architect.

GC/Works/1 (1998) empowers the project manager, in clause 40 (2) (d), to require the removal or re-execution of any work, while clause 21 (l) obliges the contractor to make good at his own cost any defects in the works notified to him by the employer as having appeared during the maintenance period (qv). This is backed up by default powers including the power of determination for failure to comply with an instruction requiring the rectification or making good of defective work: clause 56 (6) (a).

The employer also has his common law rights in respect of defective work whether before or after completion.

The architect has no duty to the contractor to discover defective work[201]. However, clause 8.2.2 of JCT 98 comes perilously close to requiring just such a duty when it provides that, where materials, goods or workmanship are to be to his reasonable satisfaction, the architect must express any dissatisfaction within a reasonable time of its execution.

See also: **Contractor's skill and care; Latent defect; Patent defect.**

Defects clause A clause in a contract to permit the contractor, for a specified period, to return to the site in order to remedy defective work at his own cost.

---

[198] *Test Valley Borough Council* v. *Greater London Council* (1979) 13 BLR 63.
[199] *Lynch* v. *Thorne* [1956] 1 All ER 744.
[200] *Equitable Debenture Assets Corporation Ltd* v. *William Moss Group and Others* (1984) 2 Con LR 1; *Victoria University of Manchester* v. *Hugh Wilson and Lewis Womersley* (1984) 2 Con LR 43.
[201] *Oldschool* v. *Gleeson (Construction) Ltd* (1976) 4 BLR 103.

Its purpose is to remove the necessity for the employer to bring an action for damages at common law in respect of defective work. If work is defective he will be able to bring an action (within the limitation period) even though the defects liability period (qv) has expired. This is subject to the effect, if any, of the final certificate (qv).

See also: **Defects liability period; Maintenance clause.**

Defects liability period A period of time after the works are completed and during which the contractor must make good any patent or other defects. The start of the period is signalled by the date on which the architect certifies the works:

— Have achieved practical completion (JCT 98, clause 17.1; IFC 98, clause 2.9; MW 98, clause 2.5).
— Are completed in accordance with the contract (GC/Works/1 (1998), clause 39).
— Are fit and ready for taking-over (ACA 3, clause 12).

Or when, dependent on the form of contract, partial possession has taken place.

GC/Works/1 (1998) and ACA 3 refer to it as the 'maintenance period'. Many contractors and architects use the same terminology which is misleading, maintenance (qv) having a rather different meaning to defects liability. The length of the period is a matter for the contracting parties. Usually a period of six months is inserted by the architect for general work and three months for minor works. There is nothing to prevent much longer periods being specified provided the contractor is aware at the time of tender and can price accordingly. It is common for mechanical and engineering works to have a twelve months' period in order to allow defects to appear during the full range of seasonal variations of temperature and humidity. This is sensible, but it should be noted that most standard forms make no provision for defects liability periods of different lengths for different elements in the same contract. Most contracts refer to 'period' in the singular and provide for only one certificate of completion of making good defects, one release of retention at that stage and one schedule of defects. The specification of two periods, although common, is incorrect. The answer is to decide on the longest period required and apply it to the whole of the work.

All the main forms of contract incorporate the phrase 'which (may, shall) appear' during the period to indicate the extent of the contractor's liability. It makes practical sense that defects which are present at the time of completion of the works are included[202]. Although it seems reasonable and the contractor has, in any case, liability to carry out the work in accordance with the contract, it is wise to note any outstanding matters at the time of completion to avoid disputes. ACA 3 makes express provision to do this in clause 12.1.

All the forms make reference to 'defects, shrinkages and other faults' (MW 98 refers to 'excessive shrinkages') except GC/Works/1, clause 21, which refers to defects only. The phrase must be interpreted *ejusdem generis* (qv) so that

---

[202] *William Tomkinson* v. *Parochial Church Council of St Michael* (1990) 6 Const LJ 319.

'other faults' must be similar to defects and shrinkages (qv). Under JCT 98, the contractor's obligation is to make good defects arising from:

— Workmanship or materials not being in accordance with the contract documents (qv).

— Frost occurring before the date certified for completion.

Under GC/Works/1, clause 21, the contractor's obligation is to make good at his own cost any defects in the work resulting from what the employer considers to be default by the contractor or his agents or sub-contractors and which appear during the maintenance period.

The architect has the whole of the defects liability period and, in the case of the JCT 98 and ACA 3 forms, 14 and 10 days respectively after the end of the period in which to notify the contractor of defects. The contractor has a reasonable time (qv) in which to make good the defects at his own cost. When all the defects have been made good, JCT 98, MW 98 and IFC 98 require the architect to issue a certificate to that effect.

It is sometimes thought that the end of the defects liability period signals the end of the contractor's liability for defects in the works. That view is quite wrong. The defects liability period is primarily for the contractor's benefit so that he can rectify defects and put the works in accordance with the contract. It enables him to deal with these matters at minimum cost, but it does not remove the employer's common law rights to sue for breach of contract within the limitation period (see: **Limitation of actions**). Sometimes, the employer will be justified in refusing to allow the contractor back on site to make good defects, but the reason for refusal must be more than trivial[203]. If the employer authorises the architect to instruct the contractor not to make good defects and employs others to do so, or if the architect does not notify the contractor of defects until after the end of the defects liability period, the amount deductible from money payable to the contractor is confined to what it would have cost the contractor to make good if he had been allowed to make good or been notified respectively[204]. Under JCT 98 and IFC 98, this is referred to as an 'appropriate deduction'. Of course, if the contractor is notified of defects at the appropriate time and simply refuses to make good, the employer is entitled to employ others to carry out the remedial work and to charge all costs involved against the contractor.

Defence In statement of case (qv), formerly pleadings (qv), it is a set of reasons put forward by the defendant (qv) to show why a claim made by the claimant (qv) should not succeed. They are carefully drafted and couched in formal language. They may range from a complete denial of the claimant's allegations, possibly coupled with a counterclaim (qv), to an admission of the claim while raising matters in justification. There are many variations in the form of defence, depending upon the ingenuity of the defendant's legal advisers. Defence is dealt with by Part 15 of the Civil Procedure Rules.

---

[203] *City Axis* v. *Daniel P. Jackson* (1998) CILL 1382.
[204] *Pearce & High* v. *John P. Baxter and Mrs A. Baxter* [1999] BLR 101.

Defendant The person against whom legal proceedings are brought and called, in Scotland, the defender. In arbitration, he is referred to as the respondent.

Defined terms A phrase used in the Engineering and Construction Contract (NEC) (qv) in clause 11.1. Defined terms have capital initials in the text of the contract, which is the usual drafting convention. The definitions are contained in clause 11.2.
See also: **Contract data; Identified terms.**

Delay In the context of building contracts the term 'delay' is used to indicate that the works are not progressing as quickly as intended and, specifically, that as a result completion may not be achieved by the completion date (qv) specified in the contract documents (qv).

Most standard forms provide that the employer is entitled to deduct liquidated damages (qv) if the contractor does not achieve completion by the due date. In order to preserve the employer's right to deduct such damages, provision is also made for the contractor to be given extensions of time (qv) in certain circumstances. In the absence of an extension of time clause, there is no power to extend time[205]. JCT 98, in clause 25.2.1, and IFC 98, in clause 2.3 lay an obligation upon the contractor to notify the architect of all delays which may affect the progress of the work. ACA 3, in clause 11.5 (alternatives 1 and 2), is not absolutely clear on the point and it may well be that the contractor is obliged to notify only those delays for which he is seeking extension of time, although this was not the intention of the compilers. Under GC/Works/1 (1998), clause 36 (1) (and MW 98), it seems that the contractor is only obliged to notify delays for which he is seeking an extension of time, but under clause 35 his agent is to attend monthly progress meetings and three days before each meeting is required to submit a report to the project manager which must, among other things 'explain any new circumstances arising since the previous meeting which in his opinion have delayed or may delay completion'.
See also: **Acceleration of work; Extensions of time.**

Delay damages A term used in the NEC (qv). It is roughly equivalent to liquidated damages (qv) under other forms of building contract.

Delegated legislation Byelaws, rules and regulations made by local authorities, Secretaries of State, etc. under powers delegated to them by Parliament. It is sometimes referred to as secondary legislation.

Today, Parliament tends to pass Acts (see: **Act of Parliament**) of a general character and entrusts to particular ministers the power of giving effect to these general provisions by means of specific regulations. The characteristic of all delegated or subordinate legislation is that power to make it must be derived from Parliament. Once validly made, however, these byelaws and regulations

---

[205] *Peak Construction (Liverpool) Ltd* v. *McKinney Foundations Ltd* (1970) 1 BLR 114.

have statutory force and effect, e.g. The Scheme for Construction Contracts (England and Wales) Regulations 1998 (see: **Housing Grants, Construction and Regeneration Act 1996**).

Delegated legislation can be challenged in the courts on the ground that it is *ultra vires* (qv), i.e. that the person making it has acted beyond his powers. Regulations and byelaws so made are void.

Delegatus non potest delegare Literally, a delegate cannot delegate. A general principle that someone to whom powers have been delegated cannot delegate them to someone else. The same rule applies to duties. In general, an architect has no power whatever to delegate his duties to anyone unless his contract with the client expressly empowers him to do this[206].

Delict Broadly speaking, delict is the Scottish equivalent of the English law of tort (qv). Most actions in delict are based on negligence (qv).

Deposition A statement on oath (qv) of a witness in judicial proceedings, duly signed by the maker. Depositions are common in criminal courts and statute allows them to be used in civil proceedings in certain circumstances.

Depositions are dealt with in Part 34 of the Civil Procedure Rules. The party from whom evidence is to be obtained after an order under rule 34.8 is termed a 'deponent'. The order may also require the production of any document. See also: **Affidavit; Discovery.**

Derogation Taking away something which is already granted. Thus it also means prejudicing or evading what is already granted; for example, where a landlord has granted a lease and he later purports to create a right of way over the leased land in favour of a third party. The basic principle is that nobody can derogate from his own grant.

Design A rather vague term denoting a scheme or plan of action. In the construction industry, it may be applied to the work of the architect in formulating the function, structure and appearance of a building or to a structural engineer in determining the sizes of structural members.

In general terms in relation to building contracts, the architect will be responsible for the design of the building and the contractor is responsible for the materials and workmanship in putting the design together on the site. This generality is often qualified in practice, however, depending upon the circumstances. The contractor may take all or some responsibility for design, for example in a design and build contract (qv), in the Designed Portion Supplement (qv) to JCT 98 or under clause 42 (performance specified work (qv)), or design responsibility may be thrust upon him, for example, where the architect does not undertake any inspection[207]. The contractor may also become liable

---

[206] *Moresk Cleaners Ltd* v. *Thomas Henwood Hicks* (1966) 4 BLR 50, where an architect, without his client's permission, employed a contractor to design a structure.
[207] *Brunswick Construction Ltd* v. *Nowlan* (1974) 49 DLR (3rd) 93.

for a part of the design if the employer can show that he relied on the contractor rather than the architect or where the contract makes clear that the contractor must fill in the gaps in a specification[208].

The professional designer such as an architect is under a duty to exercise reasonable care in his design. 'The test is the standard of the ordinary skilled man exercising and professing to have a special skill. A man need not possess the highest expert skill at the risk of being found negligent .... it is sufficient if he exercises the ordinary skill of an ordinary competent man exercising that particular art'[209]. However, by the terms of a particular contract the designer may in effect be guaranteeing the result and undertaking that the designed structure is reasonably fit for its intended purpose[210].

For a discussion of the complex problems involved in design liability see *Design Liability in the Construction Industry,* by D. L. Cornes, 4th edn, 1994, Blackwell Science.

Design and build contract Sometimes known as a 'package deal contract' (qv). In the classic form of this type of building contract the contractor takes full responsibility for the whole of the design and construction process from initial briefing to completion. In practice, the design and build contractor is often engaged only after the employer's design team have done a substantial amount of design work. The greater the amount of design carried out by the employer's designers before tendering, the smaller the contractor's design responsibility tends to be. The JCT have produced a standard form of contract to cover this kind of work where the employer is not employing an architect in the traditional way (Standard Form of Building Contract With Contractor's Design, 1998 Edition (WCD 98)). It appears to follow JCT 98 quite closely but not only is the philosophy completely different, there are considerable differences in the wording of the clauses. It is contractor driven. Naturally, it omits all references to 'architect' and inserts 'employer' instead where necessary. The main headings are as follows:

Recitals.
Articles.
(1)   Contractor's obligations.
(2)   Contract sum.
(3)   Employer's agent.
(4)   Employer's requirements and contractor's proposals.
(5)   Dispute or difference − adjudication.
(6A)   Dispute or difference − arbitration.
(6B)   Dispute or difference − legal proceedings.
(7.1)   Planning supervisor.
(7.2)   Principal contractor.
(8)   Modifications to the contract − sectional completion.

---

[208] *Basildon District Council* v. *J. E. Lesser Properties* (1987) 8 Con LR 89; *Bowmer & Kirkland Ltd* v. *Wilson Bowden Properties* (1996) 80 BLR 131.
[209] *Bolam* v. *Friern Hospital Management Committee* [1957] 2 All ER 118.
[210] *Greaves & Co (Contractors) Ltd* v. *Baynham, Meikle & Partners* (1975) 4 BLR 56.

Conditions.
(1) Interpretation, definitions, etc.
(2) Contractor's obligations.
(3) Contract sum – additions or deductions – adjustment – interim payments.
(4) Employer's instructions.
(5) Custody and supply of documents.
(6) Statutory obligations, notices, fees and charges.
(6A) Provisions for use where Appendix 1 states that all the CDM Regulations apply.
(7) Site boundaries.
(8) Work, materials and goods.
(9) Copyright, royalties and patent rights.
(10) Person-in-charge.
(11) Access for employer's agent, etc. to the works.
(12) Changes in the employer's requirements and provisional sums.
(13) Contract sum.
(14) Value Added Tax – supplemental provisions.
(15) Materials and goods unfixed or off-site.
(16) Practical completion and defects liability period.
(17) Partial possession by employer.
(18) Assignment and sub-contracts.
(19) (Number not used).
(20) Injury to persons and property and indemnity to employer.
(21) Insurance against injury to persons and property.
(22) Insurance of the works.
(23) Date of possession, completion and postponement.
(24) Damages for non-completion.
(25) Extension of time.
(26) Loss and expense caused by matters affecting the regular progress of the works.
(27) Determination by employer.
(28) Determination by contractor
(29) Execution of work not forming part of the contract.
(30) Payments.
(31) Statutory tax deduction scheme – CIS.
(32) (Number not used).
(33) (Number not used).
(34) Antiquities.
(35) Fluctuations.
(36) Contributions, levy and tax fluctuations.
(37) Labour and materials cost and tax fluctuations.
(38) Use of price adjustment formulae.
(39) Settlement of disputes – adjudication – arbitration – legal proceedings.
Supplementary provisions.
(S1) (Number not used).

(S2) Submission of drawings, etc. to employer.

(S3) Site manager.

(S4) Persons named as sub-contractors in Employer's Requirements.

(S5) Bills of quantities.

(S6) Valuation of change instructions – direct loss and/or expense – submission of estimates by the contractor.

(S7) Direct loss and/or expense – submission of estimates by the contractor.

Appendices.

Supplemental provisions (VAT).

Supplemental provisions for EDI.

It is anticipated that the employer will normally nominate an architect or clerk of works to be his agent (termed 'employer's agent' (qv)) for contract purposes. ACA 3 may also be used as a design and build contract.

See also: **Design.**

Design leader The term used under the BPF System (qv) to describe the person with overall responsibility for the pre-tender design and for sanctioning the contractor's design. He may be an employee of the client or an independent consultant and is usually, though not necessarily, an architect or an engineer. The design leader co-ordinates the work of all consultants and obtains statutory approvals, etc. He provides design advice on variations (qv) as the project proceeds, and the limits of his authority are clearly defined in the BPF manual. He cannot issue orders to consultants which would vary the work from the brief or lead to increased cost or delay and he cannot give instructions to the contractor except in an emergency.

The design leader's duties may vary from project to project, but in essence he assumes total contractual responsibility for pre-tender architectural and engineering design for a fixed fee.

Designed Portion Supplement A supplement produced by the JCT for use with JCT 98 where the employer wishes the contractor to take some design responsibility for a specific element of the building such as foundations, mechanical engineering services, etc. It inserts additional clauses into the standard form and amends many of the existing clauses. The effect is to give the contractor similar responsibility in respect of the particular element as if it was carried out under WCD 98. In carrying out the design element, the contractor is responsible for integrating his design into the architect's design as evidenced in the tender documents. However, if the architect subsequently issues instructions to vary any of the 'designed portion', he must also include in his instruction the integration of any change into the overall design of the building.

See also: **Performance specified work.**

Details Small subordinate items. In building contracts, the term is used to denote the large-scale drawings of the architect or consultants (qv). It may also be used to refer to schedules giving minute particulars, e.g. a bar bending schedule could come under the general heading: steelwork details.

Determination A decision or, more commonly, the bringing to an end of something, for example, the determination of a dispute. The word is used in the context of building contracts to refer to the ending of the contractor's employment. Both parties have a common law right to bring their obligations under the contract and sometimes the contract itself to an end in certain circumstances (see: **Contract**) but most standard forms give the parties additional and express rights to determine upon the happening of specified events. Some of these rights are similar under different contracts.

It should be noted that in some instances the giving of notice is required, while in others determination may be automatic (e.g. bankruptcy). Some contracts distinguish between determination which is the fault of one party or the other or which is the fault of neither party. GC/Works/1 (1998) gives no contractual right to the contractor to determine. This contract also refers to determination of the *contract*, while others refer to the determination of the *contractor's employment* under the contract. In practice, it makes little difference, since all contracts make express provision for what is to happen after determination. Although it may be argued that putting an end to a contract also removes any obligations under clauses purporting to deal with subsequent events, the courts have taken the view that a party referring to bringing the contract to an end actually means bringing the parties' primary obligations to an end[211].

In ACA 3, 'termination' is used instead of 'determination' but the effect is the same. In all cases the procedure prescribed by the relevant clause should be followed exactly. Determination which is not carried out strictly in accordance with the contract provisions may amount to repudiation. However, it is probably not repudiation if a party honestly, albeit mistakenly, relies on a determination provision[212]. It is worth noting that the rights of the parties to seek adjudication (qv) or arbitration (qv) continue after determination[213].

Deviations Departures from prescribed contractual standards.
See also: **Extra work.**

Difference See: **Dispute.**

Diligently See: **Regularly and diligently.**

Direct loss and/or expense The phrase used in JCT 98, clause 26, and IFC 98, clause 4.11 to describe the reimbursement to which the contractor is entitled under the claims (qv) provisions of the contract in respect of both disruption (qv) and prolongation (qv).

---

[211] *Photo Production Ltd* v. *Securicor Transport Ltd* [1980] AC 827.
[212] *Woodar Investment Development Ltd* v. *Wimpey Construction UK Ltd* [1980] 1 All ER 571.
[213] *Heyman* v. *Darwins Ltd* [1942] All ER 337.

After a good deal of controversy, it is now clearly settled law that this phrase – or similar phrases such as 'direct loss and/or damage' – extends to those heads of claim which would be recoverable at common law as damages for breach of contract[214].

In practice, this requires precise and exact calculation. Figures cannot be plucked out of the air and it is up to the contractor to prove that he has in fact suffered or incurred the loss or expense which he is claiming[215]. Such calculation should specify each causative event and the loss or expense attributed to it. The claim should not be a single global figure[216].

See also: **Claims; Compensation event; Consequential loss.**

Direct payment clause Where nominated sub-contractors (qv) are involved in the work, JCT 98, clause 35.13.5 provides that the employer may pay a nominated sub-contractor directly if the contractor has failed to discharge sums due on the previous certificate. The procedure is as follows:

(1) Before the issue of each certificate, the contractor must furnish reasonable proof to the architect that any sums directed to be paid to the sub-contractor have been paid.

(2) If the contractor fails to provide reasonable proof, the architect must issue a certificate to that effect stating the amount in question. (If the architect is satisfied that absence of proof stems solely from failure on the part of the sub-contractor, these provisions do not apply.)

(3) If the certificate is issued, the employer must pay the amount direct to the sub-contractor and deduct an equal sum from future payments due to the contractor (including VAT), provided that the employer is not obliged to pay more than is available to him by means of deduction from the contractor.

(4) If two or more sub-contractors are to be paid and the sum available is insufficient, the employer is to divide the amounts pro rata owing or in some other fair way.

There are no provisions in standard forms for direct payment other than to nominated sub-contractors.

The nomination clauses of GC/Works/1 (1998) are 63 and 63A, but they do not contain any provision which entitles the employer to pay the nominated sub-contractor directly. However, clause 63 (2) does allow the employer to order and pay directly for any prime cost (qv) items provided the contractor's profit is retained by the contractor in the contract sum.

Directions A term used in construction contracts, particularly JCT 98, clause 12, usually to mark a distinction from 'instruction' (qv). The clerk of works' directions are said to be of no effect unless confirmed in writing by the architect.

---

[214]*Wraight Ltd* v. *P. H. & T. (Holdings) Ltd* (1968) 13 BLR 26; *F. G. Minter Ltd* v. *Welsh Health Technical Services Organisation* (1980) 13 BLR 1.
[215]*Alfred McAlpine Homes North Ltd* v. *Property and Land Contractors Ltd* (1995) 76 BLR 65.
[216]*Wharf Properties Ltd* v. *Eric Cumine Associates (No 2)* (1991) 52 BLR 1.

A direction might thus be defined as a provisional instruction pending confirmation. In ordinary language, however, the distinction between direction and instruction is not clear except that 'instruction' has more force. In the nominated sub-contract conditions (NSC/C), clause 4.2, the contractor is empowered to issue any instructions of the architect and 'may issue any reasonable direction in writing to the Sub-Contractor in regard to the Sub-Contract Works'. A clear distinction is drawn between an 'instruction' and a 'direction'. An instruction comprises those matters about which the architect is empowered to issue an instruction under the main contract, while a direction is concerned with the regulation 'for the time being [of] the due carrying out of the Works' (NSC/C clause 4.1.1). A similar, although not identical distinction is to be found between instruction issued under MC 98 and WC/2.

In law, a judge may issue a direction to a jury. In this case, he is clarifying a point of law. A summons for direction asks the court to decide various procedural matters, for example: the dates for exchange of particular documents. Directions are also given by arbitrators and adjudicators (qvv) for such things as the service of documents, times and dates of hearings and the like.

Directly If an action is to be carried out 'directly', it must be done quickly or as soon as possible[217] (qv). It is stricter than 'forthwith' (qv).
See also: **Immediately.**

Discharge of contract Release from contractual obligations. This may occur in a number of ways:
— Performance: when both parties have fulfilled their obligations under the contract, e.g. the builder has completed the building and the employer has paid for it.
— Agreement: where both parties agree to treat the contract as at an end (see: **Accord and satisfaction**).
— Frustration (qv).
— Breach (qv): the breach must be of some fundamental term of the contract in order to allow the injured party to treat it as repudiation (qv).
— Operation of law: examples are: the contract falling under the Limitation Act, bankruptcy of one party or the object of the contract becoming illegal during its currency.
— Replacement of one contract by another (novation (qv)) usually accompanied by a change in the identity of one of the parties. In the case of a simple contract (qv) for a lump sum, if one party issues instructions to vary the contract works, the other party is entitled to consider the original contract at an end and a new contract, incorporating the variation, in being. Severe financial repercussions may result. The effect is avoided in the standard forms of building contract by the insertion of a variation clause to allow variations of the original contract works.

---

[217] *Duncan* v. *Topham* (1849) 8 CB 225.

Part release from obligations under a contract may be obtained by:
— Waiver: where one party agrees to waive his rights to have the other party fulfil some obligation.
— Estoppel (qv).

Disclaimer A technical phrase referring to the power of a trustee in bankruptcy (qv) or the liquidator (qv) to renounce any kind of onerous property, including contracts. Thus, in the case of a liquidator, s. 178 of the Insolvency Act 1986 confers this right on him in the case of unprofitable contracts (among other things), and a similar power is conferred on the trustee in bankruptcy by s. 315.

'Disclaimer' is also used colloquially to refer to notices or to contract terms which purport to limit liability for breaches of contract, etc.

See also: **Exemption clause; Unfair Contract Terms Act 1977.**

Disclosure/inspection of documents The term 'disclosure' has been introduced by the Civil Procedure Rules (CPR) (qv) replacing the former procedure of discovery. The CPR has altered the principles governing the production of documents. It is dealt with under Part 31. In both arbitration and litigation, disclosure of documents is the procedure under which one party provides to the other not only the documents which he will produce at the hearing but all other documents bearing on the issue. Each party serves on the other a list of all documents which are or have been in his possession or control relating to the matters in dispute. All the documents listed must be made available for inspection by the other party who may take copies of them. This is so, no matter how prejudicial to the disclosing party's case the documents are, e.g. internal memoranda commenting on the validity of a claim, etc. The list is not confined to a selection. The only exception is that certain documents are privileged, e.g. Counsel's opinions, correspondence with one's own solicitor about the dispute, etc.

In arbitration, an order for discovery is made at the preliminary meeting. For the purpose of CPR Part 31, a document is in the control of a party when he has possession or the right of possession or a right to inspect and take copies and a document has been 'disclosed' when a party states that it exists or has existed.

The list of documents is usually prepared in a standard form (N265) and includes three parts:

Part 1 Relevant documents which are listed numerically in date order and which the party has in his control and to the inspection and copying of which he has no objection. For example, the contract documents, correspondence between the parties, invoices, etc.

Part 2 Relevant documents which the party has in his control, but which he objects to produce, and which must contain a statement of the grounds on which privilege is claimed.

Part 3 Relevant documents listed as above which have been, but are no longer in the control of the party, e.g. originals of correspondence. He must say when they were last in his control, what has become of them and who has possession of them.

Inspection of documents is usually followed by the preparation of an agreed 'bundle' of documents which both parties are prepared to admit as evidence without the need for strict proof.

See also: **Arbitration; Pleadings; Privilege.**

Discovery See: **Disclosure.**

Discrepancies

Differences or inconsistencies. Thus, if a contract drawing (qv) showed bricks for a particular situation to be rustic facings and the contract bills (qv) gave the bricks for the same situation to be smooth-faced engineering bricks, there would be a discrepancy between the drawings and the bills. It is quite possible, in fact quite common, for there to be discrepancies of various kinds among the many constituent parts of the contract documents (qv). One drawing may not agree with the rest of the drawings or it may be in conflict with the information in the bills or specification.

All the standard forms make provision for the treatment of discrepancies. JCT 98, clause 2.3 states that if the contractor finds any discrepancy in or divergence between two or more of the:

— Contract drawings;
— Contract bills;
— Architect's instructions, except instructions requiring a variation;
— Any further drawings, etc. issued by the architect;
— The numbered documents (qv) attached to NSC/A;

he must notify the architect in writing and the architect must issue an instruction resolving the difficulty. A similar provision, in clause 6.1, refers to the finding of a 'divergence' between statutory requirements and the contract documents.

There is some dispute as to the precise meaning of the word 'if' in these clauses, i.e. 'If the contractor shall find ...'. It is generally assumed among architects that the average contractor using normal skill and care should find discrepancies in good time so as to avoid costly mistakes, the word 'if' indicating that there might not be any discrepancies, not that the contractor may not find them. Contractors usually take the clause to indicate that their obligation is only to report discrepancies *if* they find them; the contractor's view is correct[218].

ACA 3 removes such disputes (clause 1.5) by expressly making the contractor responsible for using his 'skill, care and diligence' to ensure that there are no discrepancies at the date of the contract. If he subsequently finds a discrepancy, he must notify the architect who shall issue an instruction. Only if the contractor could not reasonably have found the discrepancy at the date of the contract will he be entitled to payment.

---

[218]*London Borough of Merton* v. *Stanley Hugh Leach Ltd* (1985) 32 BLR 51.

MW 98 provides in clause 4.1, that inconsistencies shall be corrected and such corrections be treated as variations under clause 3.6. The contractor is not made specifically responsible for finding inconsistences but, reading this clause in conjunction with clause 1.1, it seems probable that he would be. GC/Works/1 (1998) states (clause 2 (1)) that in the case of any discrepancy, the supplementary conditions and annexes prescribed by the abstract of particulars (qv) prevail over the conditions which, in turn, prevail over any other document forming part of the contract. The written specification (even if part of the bills of quantities (qv)) prevails over the specification content of the drawings. Clause 40 (2) (b) empowers the PM to issue an instruction. It does not, however, resolve the question of responsibility for finding discrepancies.
See also: **Inconsistency.**

Discretion The ability to decide something in the light of what is fair and reasonable in all the circumstances. Discretionary power is vested in judges in certain cases and some contracts may appear to give the architect discretionary powers, but it has been remarked that:

> 'The occasions when an architect's discretion comes into play are few, even if they number more than the one which gives him a discretion to include in an interim certificate the value of any materials or goods before delivery on site ... The exercise of that discretion is so circumscribed by the terms of that provision of the contract as to emasculate the element of discretion virtually to the point of extinction.'[219]

Modern JCT contracts give the architect no discretion about the inclusion of off-site materials (qv) and goods. It is for the employer to decide at tender stage whether he is prepared to make such payment, and if the appropriate list of uniquely identified or non-uniquely identified goods and materials is not attached to the contract, no payment can be certified for them.

Disorder A rather loose term included in ACA 3, clause 11.5, alternative 2, as a ground for awarding an extension of time. It may be considered as a serious disturbance of public order, probably involving an element of violence, rather than the lesser sorts of disorder which can nonetheless amount to a breach of the peace.
See also: **Civil commotion; Civil war; Commotion; Insurrection; Riot.**

Dispute Strictly, a calling into question. In the context of construction contracts it is usually associated with adjudication (qv) or arbitration (qv) when it refers to a disagreement between two parties. Usually, that will take the form of a proposition by one party which the other has rejected, although sometimes long silence or delay in responding on the part of the other party may also be considered a rejection. A dispute may exist for the purpose of adjudication or arbitration even though it may be obvious which party is correct[220].

---

[219]*Partington & Son (Builders) Ltd* v. *Tameside Metropolitan Borough Council* (1985) 5 Con LR 99 per Judge Davies at 108.
[220]*Hayter* v. *Nelson* [1990] 8 Lloyds Rep 265.

Adjudication and arbitration clauses usually refer to 'dispute or difference', but they probably have essentially the same meaning in this context. The Arbitration Act 1996 s. 82 (1) says that 'dispute' includes 'difference'. No adjudication or arbitration can take place unless a dispute or difference exists[221].

Disruption A term commonly used by contractors making a claim for additional money. The ordinary meaning of disruption is 'violent destruction or dissolution'. Therefore, with reference to a construction contract it cannot cover minor interferences with progress. A claim for disruption may be distinguished from a prolongation claim (qv) in that it does not depend upon the completion date being exceeded to be successful. An architect's instruction may cause the contractor severe disruption of his programme. But by efficient re-organisation, he may be able to complete the contract on time. Alternatively, the time taken to carry out certain activities may be extended, but there is no effect on the completion date, because the activity is not critical and the extended time does not use up the whole of the available float (qv). Despite having completed on time, he will have incurred considerable administrative costs for which he is entitled to be reimbursed over and above any value of the instruction. Labour, materials, plant and the contractor's planned sequence of operations may also be affected. In all cases it is for the contractor to prove the loss and/or expense incurred as a consequence of disruption.

The term is used in the ACA 3 and GC/Works/1 (1998) forms of contract, clauses 7.1 and 46 respectively, to describe severe breaking down of the orderly progress of the works.

See also: **Acceleration of work; Claims; Extension of time; Loss and expense.**

Distress A summary remedy under which someone may take possession of the personal goods of another person and hold them to compel performance of a duty or the satisfaction of a debt or demand. Distress is often used by the Inland Revenue to enforce payment of income tax. The most common example of distress is the right of a landlord to distrain on his tenant's goods for non-payment of rent.

Disturbance A word often used in connection with the regular progress (qv) of the works. It means an interruption or disruption (qv) and usually forms grounds for a contractual claim.

See also: **Claims; Loss and expense.**

Divergence A separating or differing. The word is found in clauses 2.3 and 6.1 of JCT 98. It is used in conjunction with the word 'discrepancy' (qv) in clause 2.3 and appears to add little to the meaning. In clause 6.1, it is used alone because it better expresses the sense that the requirements of the contract documents and statutory requirements may differ.

See also: **Discrepancies; Inconsistency.**

---

[221]For example, see *Cruden Construction Ltd* v. *Commission for The New Towns* (1994) 75 BLR 134.

Divisible contract One in which payment is due for partial performance, in contrast to an entire contract (qv). Most construction contracts of any size are divisible in this sense in that they provide for payment in instalments. Part II of the Housing Grants, Construction and Regeneration Act 1996 (qv) requires such provision in every construction contract (qv) which is expected to last more than 45 days. But the intended result is often avoided by the drafting of bespoke contract terms. Otherwise, whether a contract is entire or divisible is a matter which depends upon the intentions of the parties as interpreted by the courts.

Documentary evidence Evidence in recorded form, normally written, printed or drawn, but electronic forms are also considered to be documents[222]. Examples are: letters, drawings, contract documents (qv), deeds, wills, books and reports. Before documentary evidence is admissible in court, it must be proved authentic. That is not to say that the contents of the document must be proved to be correct, but that the document must be what it is purported to be. For example, a document put forward as being a report on a specific topic written by one person for the benefit of another must be shown to be about the topic, written by that person for the benefit of another. The contents of the report may later prove to be in error. The burden of proving documentary evidence is removed if both parties to the dispute agree. In most building disputes, much of the documentary evidence can be agreed in advance, leaving only key documents or points of law to be decided by the court or arbitrator.
See also: **Admissibility of evidence.**

DOM/1 The form of domestic sub-contract for use with JCT 98. It is approved by the Construction Confederation (qv) and published by them. Although still in the 1980 edition, it incorporates amendments 1 to 3 and 5 to 10 (amendment 4 is only suitable for use where the main contract is without quantities). Amendment 10 makes it suitable for use with JCT 98. It is in two parts:
— Articles of agreement.
— Sub-contract conditions.
The articles can be used without the sub-contract conditions, which they incorporate. The articles have an appendix which is divided into parts which are to be completed to show:
Part 1 – details of the main contract and the main contract appendix.
Part 2 – details of the sub-contract.
Part 3 – insurance cover.
Part 4 – time periods.
Part 5 – daywork rates.
Part 6 – VAT clauses applicable.
Part 7 – retention.
Part 8 – adjudication details.

---

[222]See CPR Rule 31.4, *Derby & Co Ltd* v. *Weldon (No 9)* [1991] 1 WLR 652 (computer database records) and *Alliance & Leicester Building Society* v. *Ghahremani* [1992] 32 RVR 198 (word processing file).

Part 9 – particular attendance.
Part 10 – fluctuations.
Part 11 – materials list, etc.
Part 12 – basic transport, etc. prices.
Part 13 – formula details.
Part 14 – arbitration details.
Part 15 – performance specified work items.
Part 16 – other matters agreed.
The sub-contract conditions are very much in standard form.

DOM/2 The form of domestic sub-contract for use with WCD 98. It is substantially based on DOM/1 (qv) with the necessary amendments to step down the provisions of WCD 98 and in particular the design responsibility. It is approved by the Construction Confederation (qv) and published by them. Although still in the 1981 edition, it incorporates amendments 1 to 7. It is in two parts:
— Articles of agreement.
— Sub-contract conditions.
The articles can be used without the sub-contract conditions, which they incorporate. The articles have an appendix which is divided into parts which are to be completed to show:
Part 1 – details of the main contract, the main contract appendices and any special obligations imposed by the employer's requirements.
Part 2 – details of the sub-contract.
Part 3 – insurance cover.
Part 4 – time periods.
Part 5 – daywork rates.
Part 6 – VAT clauses applicable.
Part 7 – retention.
Part 8 – adjudication details.
Part 9 – particular attendance.
Part 10 – fluctuations.
Part 11   materials list, etc.
Part 12 – basic transport, etc. prices.
Part 13 – formula details.
Part 14 – arbitration details.
Part 15 – other matters agreed.
The sub-contract conditions are very much in standard form.

Domestic sub-contractor A term found in the JCT 98 contract, principally in clause 19.2. It refers to any person or firm, other than a nominated sub-contractor (qv), to whom the contractor (qv) sub-lets any portion of the works. If the contractor wishes to sub-let the plastering work, he must first obtain the architect's written permission. The contactor does not have to obtain the architect's consent to the actual sub-contractor to be used (although it is good practice to do so), only to the fact of sub-letting.

Of course, it might be reasonable for the architect to refuse consent until he is informed of the name of the sub-contractor. JCT 98 also enables the employer to specify domestic contractors by means of a list in the contract bills: clause 19.3. The employer details in the contract documents work which the contractor is to price, but which in fact is to be executed by a domestic sub-contractor chosen by the contractor from a list provided by the employer.

Provided that the contractor has the choice of at least three persons named in the contract bills (qv) by the employer, the chosen sub-contractor will be a domestic sub-contractor (see: **Named sub-contractors**). If the list falls below three for any reason and is not increased, the work is to be carried out by the contractor who may sub-let it to a domestic sub-contractor. There is no contractual relationship between the employer and the domestic sub-contractor. Claims between them must pass through the contractor. Thus, if the domestic sub-contractor's work is defective, the employer will seek redress from the contractor. It is then for the contractor to seek redress, in turn, from the domestic sub-contractor. In those circumstances it is important that the terms of the main contract and sub-contract interlock so that the respective rights and obligations are stepped up and down the contractual chain.

See also: **DOM/1; DOM/2; Privity of contract; Sub-contractor; Vicarious performance.**

Drawings and details The usual means of communicating information from the architect to the builder. Reference is made to both in the JCT 98 and ACA 3 forms of contract. No reference is made to 'details' in the GC/Works/1 (1998) form. In practice, it probably makes little difference because the provision of drawings and details would be covered in such references as 'drawings ... or other design information' (clause 46 (2) (a)). Strictly, a drawing is not always a detail, neither is a detail always a drawing. A drawing might best be described, in this context, as a visual representation of a building or some part of a building usually drawn to a designated scale. A detail would normally be a drawing of some small part of a building so as to show, to a large scale, the important features of construction. 'Details', plural, may also mean a written description going into some depth. For example, the architect may furnish *details* of concrete lintels by providing the contactor with schedules giving bar lengths and diameters, lintel sizes and number, and describing the position of the bars in the lintels; but he would provide a *detail* of a concrete lintel by producing a drawing to full-size or half full-size. In general, when the architect or contractor refers to drawings they mean all the drawings irrespective of size and scale; when they refer to details they mean large-scale drawings. If reference to schedules is intended, the word 'schedule' is usually used.

See also: **Contract drawings.**

Due date The correct date by which some action should be commenced or completed. Thus the due date for completion is the date stated in the contract document (qv) by which the works must be complete. The date when payment is due is referred to in s. 110 (1) of Part II of the Housing Grants, Construction and Regeneration Act 1996 (qv).

Due time The correct period of time. In building contracts, the due time for completion is the length of time between the date for commencement and the date for completion, i.e. the contract period.

Duty of care See: **Care, duty of.**

Duty of care agreement See: **Collateral warranty.**

Early warning meeting A phrase encountered in the NEC. Part of the philosophy of the contract is that each party must give early warning of problems to the other. Where, under clause 16, an early warning meeting is convened, any proposals discussed and considered and any decisions made must be minuted by the project manager (qv) and those minutes must then be copied to the contractor for his records and action as appropriate.

See also: **Engineering and Construction Contract; Minutes of meeting.**

Easements and profits An easement is a right, held by one person to use the land belonging to another or to restrict the use by another. Examples are right of way (qv), right of drainage and right to discharge water on to neighbouring property. These are known as *positive easements* as compared to right of light (qv) and right of support, which are known as *negative easements*.

An easement is attached to land, not to a person. The land which enjoys the benefit is known as the *dominant tenement*; the land on which the easement is exercised is known as the *servient tenement*. For an easement to exist, the two pieces of land must have different owners. A *profit à prendre* is the right to remove something from another's land, for example, turf or gravel and where several people enjoy the right communally it is known as 'a right of common' and must be registered under the Commons Registration Act 1965. Both easements and profits may be created by:
— Act of Parliament.
— Express grant, normally by deed.
— Express reservation, when land is sold.
— Prescription (qv).

See also: **Wayleave; Countryside and Rights of Way Act 2000.**

Economic duress Economic pressure put on a party to enter into a contract or vary the terms of a contract which may form grounds for relief to a party in that situation[223].

Egan Report The short name for *Rethinking Construction*, the report of the Construction Task Force chaired by Sir John Egan, published by the Department of the Environment, Transport and the Regions (1998). The report contends that the UK construction industry at its best is excellent, but there is concern that the industry is under-achieving. The task force is convinced that radical change and improvement in quality and efficiency can be spread throughout

---

[223]*D & C Builders Ltd* v. *Rees* [1965] 3 All ER 837; *Universal Tankships of Monrovia* v. *International Transport Workers Federation* [1982] 2 All ER 67; *Atlas Express Ltd* v. *Kafco (Importers and Distributors) Ltd* [1989] 1 All ER 641.

the construction industry. Five key drivers are identified:
— Committed leadership.
— Focus on the customer.
— Integrated processes and teams.
— A quality driven agenda.
— Commitment to people.

Ambitious targets and effective measurement are said to be essential for improvement. The targets include an annual 10% reduction in construction cost and time together with a yearly reduction in defects of 20%. The industry should create an integrated project process around product development, project implementation, partnering the supply chain and the production of components. Waste must be eliminated and customer value increased. The industry must provide decent and safe working conditions, improve management and supervisory skills, and design projects for ease of construction making maximum use of standard components and processes. Competitive tendering must be replaced with long-term relationships.

The report has been criticised by some as being couched in jargon which amounts to little more than pious hopes in practice.

See also: **Prime contracting.**

Eichleay formula A USA formula for calculating the 'head office overhead' percentage of a contractor's money claim for delay. It is widely used in Federal Government contracts but has also been adopted in non-government contract cases although it is not universally accepted even in the USA.

This formula has sometimes been used in this country as an alternative to the Hudson or Emden formulae (qvv). The Eichleay formula is a three-step calculation:

$$(1) \quad \left[\frac{\text{Contract billings}}{\text{Total contractor billings for contract period}} \right] \times \left[\begin{array}{c} \text{Total HO overhead} \\ \text{for contract period} \end{array} \right] = \left[\begin{array}{c} \text{Allocable} \\ \text{overhead} \end{array} \right]$$

$$(2) \quad \left[\frac{\text{Allocable overhead}}{\text{Days of performance}} \right] = \left[\begin{array}{c} \text{Daily contract HO} \\ \text{overhead} \end{array} \right]$$

$$(3) \quad \left[\begin{array}{c} \text{Daily contract HO} \\ \text{overhead} \end{array} \right] \times \left[\begin{array}{c} \text{Days of} \\ \text{compensable delay} \end{array} \right] = \left[\begin{array}{c} \text{Amount of} \\ \text{recovery} \end{array} \right]$$

The formula can be subjected to a number of criticisms and, at best, gives a rough approximation. In particular, the formula does not require the contractor to prove his actual increased overhead costs from the delay, which is an essential requirement in English law, e.g. under JCT 98, clause 26.

Moreover, as set out above there is the possibility of double recovery, to allow for which it is at least necessary to deduct any head-office overhead recovery allowed under normal valuation rules in respect of variation orders. It is unusual, too, in applying daily rates.

See also: **Emden formula; Hudson formula.**

Ejusdem generis rule A rule used in the construction and interpretation of contracts (qv) to the effect that where there are words of a particular class followed by general words, the general words must be treated as referring to matters of the same class as those listed. It is important to note that the rule applies only where the specific is followed by the general and not where the general is followed by the specific. For example, in *Wells* v. *Army & Navy Co-operative Society Ltd* (1902)[224] an extension of time clause in a building contract allowed the architect to grant an extension of time to the contractor if the works were 'delayed by reason of any alteration or addition ... or in case of combination or workmen, or strikes, or by default of sub-contractors ... *or other causes beyond the contractor's control*'.

The 'other causes' were held to be limited to those *ejusdem generis* with the specific causes listed and therefore did not include the employer's own default in failing to give the contractor possession of the site.

IFC 98, clause 2.10 refers to 'defects, shrinkages or other faults'. The phrase 'other faults' is to be interpreted *ejusdem generis* to mean faults of the same class as defects and shrinkages.

The modern tendency of the courts is to restrict the operation of the rule[225]. The rule will not apply if the parties establish that the words used are to be given an unrestricted meaning. In any event, as was remarked in the *Henry Boot* case, while the rule is ordinarily applied in the case of deeds (qv), wills and statutes (qv), 'it is of less force when one is dealing with a commercial contract'[226].

Electronic data interchange (EDI) Some standard forms of contract now have provision for electronic data interchange or the interchange of information between the parties by e-mail. For example, JCT 98, clause 1.11 provides that supplemental provisions annexed to the contract will apply if the appendix (qv) so states. The supplemental provisions set out the requirements and the following should be noted:

— The parties must enter into the EDI agreement no later than the coming into existence of a binding contract between employer and contractor.

— Except where expressly stated in the supplemental provisions, nothing in the EDI agreement may override or modify the contract. The contract in this context presumably refers to the printed form although it may refer to the printed form together with the other contract documents (qv).

— It is essential that all parties connected with the project, including the whole of the professional team, know what data may be transferred by EDI and which participants will be using the facility. It is unlikely that

---

[224](1902) 86 LT 764.
[225]*Henry Boot Construction Ltd* v. *Central Lancashire New Town Development Corporation Ltd* (1980) 15 BLR 1.
[226]See also: *Chandris* v. *Isbrandtsen-Moller Co Inc* [1951] 2 All ER 613.

any professional firm has no e-mail facility, but it is quite common for firms to dislike doing all their business in this way, because of the abuses to which the system is prone. It is probably pointless to put the agreement in place if very few of the participants wish to use it.

— Adopted protocols must be stated in the contract documents.
— Communications which the contract requires to be in writing (qv) may be exchanged in accordance with the EDI agreement with certain very specific exceptions:

  • Determination of the contractor's employment
  • Suspension of performance of the contractor's obligations
  • The final certificate
  • Notices in connection with any of the dispute resolution procedures. For example, notices of intention to seek adjudication or notices referring a dispute to arbitration must be given in writing
  • Agreements amending the EDI agreement.

However, it is clear that there is nothing to prevent the parties agreeing that all certificates, other than the final certificate, may be delivered by EDI. So the architect may issue financial certificates and the certificate of practical completion by e-mail in appropriate cases. The issue of an architect's instruction could become instantaneous.

A useful publication from the Joint Contracts Tribunal is *Electronic Data Interchange in the Construction Industry*.

Elemental bills of quantities A system of classification of the contents of the bills of quantities (qv) into elements instead of the more usual trade or constructional section divisions. In practice, it means that the lists of work and materials are grouped under headings which reflect the parts of a building, for example: floors, roofs, windows, staircases, rather than carpentry, joinery, finishes, etc.

The principal benefit of elemental bills is in cost analysis where the various parts of a building may be accurately costed and comparisons of costs made with the use of differing materials or with other similar buildings. A quantity surveying office which uses this method will, in time, build up a very useful set of comparative costs to aid cost estimating for new buildings. Some architects and contractors, used to traditional bills, find it difficult to locate items quickly in elemental bills and put up resistance to their use. On the other hand, many architects and contractors find them more logical than traditional bills. Work items are easy to locate once the principle has been grasped.

Emden formula Another formula approach to the controversial topic of overhead and profit recovery in a claims situation under standard form contracts. Unlike the Hudson formula (qv) this one takes a percentage from the contractor's overall organisation, i.e. on costs and profit expressed as a percentage of annual turnover.

It is so called because it appeared in Emden's *Building Contracts & Practice* 8th edn, 1990, vol. 2, p.N/46, Butterworths:

'When it is desired to claim extra head-office overheads for a period of delay a calculation is adopted as follows:

$$\frac{h}{100} \times \frac{c}{cp} \times pd$$

where h = the head-office percentage, c = the contract sum, cp = the contract period and pd = the period of delay (cp and pd should be calculated in the same units, e.g. weeks).

The head-office percentage is normally arrived at by dividing the total overhead cost and profit of the organisation as a whole by the turnover ... The formula ... notionally ascribes to the contract in question an amount in respect of overheads and profit proportional to the relation which the value of the contract in question bears to the total turnover of the organisation.'

Although this approach is more realistic than that of some other formulae it is of limited value in practice and is simply a rough and ready approximation of the situation. In principle, it is necessary for the contractor to prove that there was an increase in overhead costs attributable to the delay or disruption and this is something which any formula method of calculation ignores. The Emden formula was however applied (although wrongly referred to as the Hudson formula) somewhat uncritically and apparently without argument in *J. F. Finnegan & Son Ltd* v. *Sheffield City Council* (1989)[227], but this should not be taken as judicial approval of this or any other formula.

See also: **Eichleay formula; Hudson formula.**

Emergency powers Those powers which may be invoked by the Government in cases of emergency, national danger or other wholly exceptional circumstances, and now derived almost entirely from Act of Parliament, e.g. the Emergency Powers Acts 1920 and 1964, which give the Government a permanent reserve of power for use in peacetime emergencies, such as during a major strike. They are seldom invoked in practice. Many building contracts provide for what is to happen in the case of exercise of emergency powers. For example, JCT 98, clause 25.4.9 allows 'the exercise after the Base Date by the United Kingdom Government of any statutory power which directly affects the execution of the works' by, e.g., restricting the availability or use of labour, as a ground for extension of time, and the exercise of such powers might well fall within the meaning of *force majeure* (qv).

See also: **Base date; Government action.**

Employer In building contracts, the word does not have the legal 'master and servant' connotation of employment law. It is used to refer to the building owner, the person or body which commissions building work and enters into a contract with the building contractor. The JCT 98, ACA 3, IFC 98, MW 98 and GC/Works/1 (1998) contracts use the word 'employer' throughout in this sense.

---

[227](1989) 43 BLR 124.

See also: **Master.**

Employer's agent The person appointed to act for the employer in WCD 98. Although only mentioned twice in the Conditions (in clauses 5.4 and 11), article 3 provides that he is to act for the employer in receiving or issuing applications, consents, instructions, notices, requests or statements or for otherwise acting for the employer under any condition. If the employer wishes some other arrangement to apply and to reserve some powers to himself, he must give appropriate written notice to the contractor. He does not act for the employer in other than this limited capacity, but he is bound by the normal rules of agency (qv).

Employer's representative A term first introduced with the publication of JCT 98. It occurs in clause 1.9. The employer may issue a written notice to the contractor stipulating a date from which all the functions of the employer under the contract will be exercised by someone appointed as the employer's representative. In the notice, the employer may specify exceptions to the functions exercisable by the employer's representative. A footnote makes clear that, to avoid possible confusion, neither the architect nor the quantity surveyor should be appointed to this role. Although not expressly stated, it seems that a project manager (qv) may fill the role.

Employer's Requirements One of the contract documents in WCD 98, although not expressly noted as such. The intention is that the employer, probably with professional advice, produces this document and sends it to the contractor or contractors who are to tender for the design and construction. JCT Practice Note CD/1A summarises the main points to be borne in mind in preparing the document. It may range from little more than a description of the accommodation required to a full scheme design.

It is essential from the employer's point of view that his requirements are as comprehensive as necessary. If the employer wishes to keep a very tight control on the contractor, he will produce a very full and detailed performance specification. Despite what may sometimes be thought, the preparation of a proper performance specification is a skilful and time-consuming task.

Before the contract is executed, the Employer's Requirements must be made to match the contractor's proposals exactly. This is because there is no express provision in the contract for dealing with any discrepancies between the two documents although discrepancies in the Employer's Requirements are dealt with in clause 2.4.1. JCT recognise the problem in Practice Note CD/1, but opt simply to advise that discrepancies should be eradicated. Reading the contract as a whole, it is our view that the Employer's Requirements would take precedence in the event of a discrepancy.

Encroachment Intruding gradually or by stealth on to another person's land. Minor encroachments in neighbouring property are quite frequent when fences are erected or rebuilt, and boundaries are frequently varied in this way. The process is commonly called 'squatter's rights' or, more accurately, acquiring title by adverse possession (qv).

Engineering and Construction Contract (NEC) This contract, which was the subject of warm recommendation in the Latham Report (qv), introduces a number of new concepts into the construction contract arena. The contract consists of a number of core clauses, to which other clauses can be added to produce variations to suit different procurement requirements. The idea is not new, but this contract is the first to put the idea into operation. JCT and other contract families produce a different contract to suit each procurement route.

Other useful provisions include:
— Pre-estimation of the effects of instructions, to include time as well as financial effects.
— Acceleration subject to acceptance of a contractor's quotation.
— Option to include a bonus clause.
— Something called low performance damages may be used to enable the employer to recover if equipment does not function in accordance with a specified performance level.

Almost the whole of the contract is written in the present tense and very odd sentence construction is used on occasion. The use of the present tense also makes it difficult to interpret whether actions are powers or duties. It also seems as if a deliberate attempt has been made to avoid the use of any words or phrases which have been considered and interpreted by the courts. Therefore, precedents of interpretation from decided cases may have little value in relation to this contract. The intention is to make the contract simple. One can expect that contractors' claims will be plentiful. It should not be forgotten that the contract claims provisions (called 'compensation events' (qv) in this contract) do not prevent the contractor from mounting a claim for breach of contract at common law if he feels so inclined.

Among other points it should be noted:
— The contract refers to giving the contractor possession of the site, but elsewhere that he must share working areas with others. Working areas appear to be the site or at least parts of it and other places stated by the contractor. Therefore, it is not clear that the contractor has exclusive possession of the site so as to be the occupier for the purposes of the Occupiers Liability Acts (qv).
— The contract has something which it refers to as 'delay damages' (qv). It may be assumed that they are the same as liquidated damages and subject to the same rules, but it may take a test case to clarify the position.
— There is an early warning system concept which is a good idea, but it could just as easily and a great deal more clearly have been expressed as a duty on both parties to warn.

See also: **Contract data.**

Ensure Effectively, a guarantee that something will be done. It involves more than simply using best endeavours (qv). The obligation to 'ensure' or 'secure' the doing of something imparts an absolute liability to perform the duty set out[228]. Professional terms of appointment, if drafted by a client's solicitors, will often require the professional to ensure proper performance of the works. Professionals can gain some comfort from a recent decision which makes the requirement to ensure dependent on other clauses and surrounding circumstances before it can be given its full effect[229].

Entire completion See: **Performance.**

Entire contract A contract in which 'complete performance' by one party is a condition precedent (qv) to the liability of the other party[230]. For example, where the carrying out and completion of work by one party is necessary before payment by the other party is due[231]. Whether or not a contract is an entire one is a matter of interpretation of the contract; it depends on what the parties agreed. A lump sum contract (qv) is not necessarily an entire contract. The test for complete performance is in fact 'substantial performance' (qv). What is substantial is not determined by a comparison of cost of work done and work omitted or done badly[232].

Equities The right to invoke equitable remedies for fraud, mistake, etc. Equities are the lowest kind of interest in property, etc.

Equity Literally, fairness or natural justice. A body of rules which grew up alongside the common law as a supplement to it and formerly administered in separate courts. In time, the principles became systematised and equity supplemented and sometimes prevailed over common law. Equity must not be confused with ethical or moral concepts. Originally there was a moral aspect to the system, but the modern attitude is summed up by the following statement, made in dismissing a claim against a company director guilty of sharp practice: 'if we were sitting in a court of honour, our decision might be different'[233].

Towards the end of the nineteenth century, legislation fused the administration of law and equity and so both legal and equitable rules and remedies are now applied throughout the legal system. It is expressly laid down that whenever there is any conflict between common law (qv) and equity, the latter is to prevail. Such cases of conflict are now rare.

Errors Mistakes (qv). In the context of building contracts, errors are usually made in regard to fact or to law. Errors of fact may be sufficient to allow one party to apply to the court to have the contract put aside. Errors of law can also have the effect of bringing the contract to an end[234].

---

[228] *John Mowlem & Co* v. *Eagle Star & Others* (1995) 62 BLR 126.
[229] *Department of National Heritage* v. *Steenson Varming Mucahy* (1998) 60 Con LR 33.
[230] *Cutter* v. *Powell* (1795) 6 Term Rep 320.
[231] *Sumpter* v. *Hedges* [1898] 1 QB 673.
[232] *Hoenig* v. *Isaacs* [1952] 2 All ER 176.
[233] *Re Cawley & Co Ltd* (1889) 42 Ch D 209 per Fry LJ at 212.
[234] *Kleinwort Benson Ltd* v. *Lincoln City Council* [1999] 1 AC 153.

Most forms of contract make some provision for the correction of errors. JCT 98 refers, in clause 2.2.2.2, to the correction of errors in the contract bills; in clause 7 to errors arising from the contractor's inaccurate setting out; in clause 14.2, to the acceptance of errors in the computation of the contract sum (qv); in clause 30.9.1.2, to accidental inclusion or exclusion of any work, materials, goods or figure in any computation or any arithmetical error, all of which are excluded from the conclusiveness of the final certificate.

ACA 3 refers (clause 1.4) to the correction of errors in the contract bills and, in clause 3.1, to mistakes, inaccuracies, discrepancies and omissions in drawings for which the contractor is responsible.

GC/Works/1 (1998) refers, in clause 3 (3), to the correction of errors in the bills of quantities.

All such references to errors are intended to prevent the contract being vitiated by providing an agreed remedy for them. Errors in bills of quantities submitted in connection with tendering procedures are often dealt with by the Codes of Procedure for Single- and Two-stage Selective Tendering (qvv).

Obvious *clerical* errors in a contract will be read by the courts as corrected when interpreting a contract, but this does not apply to mistakes made by the contractor in his tender price. Such mistakes are binding on the contractor unless, before the tender is accepted, the employer or the architect discovers the difference and realises that it was not intentional. If the error is discovered the position is different[235].

Mistakes in bills of quantities are not infrequent and give rise to problems. In lump sum contracts (qv) errors not discovered by the employer or architect before acceptance clearly bind the contractor in relation to the original work[236].

Where valuation of additional quantities is necessary and the rates have errors in them, because they are either too high or too low, the rates are not to be corrected, but are used as the basis of the valuation of the additional work[237].

ESA/1 The collateral warranty produced for use with IFC 98. It was prepared by the Royal Institute of British Architects and the Committee of Associations of Specialist Engineering Contractors. Its object is to put a contractual relationship in place between each named person as sub-contractor (qv) and the employer principally to give the employer a route of redress against the sub-contractor if he carried out design work. There are also other clauses dealing with the sub-contractor's liability to the employer if he fails to provide information on time and provision for advance ordering of work and materials.

Escrow A written legal undertaking to do something which is delivered to a third party and released only after a stipulated condition has been fulfilled.

---

[235] *Webster* v. *Cecil* (1861) 30 Beav 60.
[236] *Riverlate Properties Ltd* v. *Paul* [1974] 2 All ER 656.
[237] *Dudley Corporation* v. *Parsons & Morrin Ltd*, CA, unreported, 8 April 1959; *Henry Boot Construction Ltd* v. *Alstom Combined Cycles Ltd* [1999] BLR 123.

Essence of the contract A term, the breach of which by one party gives the other party a right to treat it as repudiatory, is sometimes said to be of the essence of the contract. It must be a term so fundamental that its breach would render the contract valueless, or nearly so, to the other party. A term may be of the essence because it is stated to be so by the contract itself or it may be judged to be of the essence by the court. Where a term is not originally of the essence it may be made of the essence by one party giving the other a written notice to that effect. In that case, failure to comply with the notice would be evidence of a repudiatory breach rather than a repudiatory breach itself.

The phrase is often used in connection with time. If a party unreasonably delays his performance, time may be made 'of the essence' if the other party serves a notice on the party in breach setting a new and reasonable date for completion[238].

In building contracts, time will not normally be of the essence unless expressly stated to be so. This is because the contract makes express provision for the situation if the contract period is exceeded.

See also: **Delay; Fundamental term; Repudiation.**

Essential Indispensable. The word occurs in standard form contracts. JCT 98, clauses 25.4.9, 25.4.10.1 and 25.4.10.2 refer to such labour, goods or materials or fuel or energy as are 'essential' to the proper carrying out of the works. Before the architect may consider any of the circumstances in the clauses as relevant events (qv), he must be certain that the labour, goods or materials, etc. really are indispensable. It is not enough that they would have been useful or helpful or quite important. 'Essential' is not a word which lends itself to half measures and it seems that a strict interpretation is intended.

Establishment charges Otherwise known as 'establishment costs', these are the cost to the contractor of his site administration. They include such things as purely supervisory or administrative staff, site accommodation, water, light, heat, electricity charges, canteen, welfare, etc. The costs are important to a contractor who is framing a claim for loss and expense.

See also: **Claim.**

Estate A technical term used in connection with the ownership of land. It describes the extent of the proprietor's interest in the land, e.g. a freehold or leasehold estate. In common parlance it also refers to the land itself, e.g. 'the Whiteacre Estate'. The same term is also used as the equivalent of property, e.g. the estate of a deceased person.

Estimate A term widely used in the building industry. It has two possible meanings:
— Colloquially and in the industry generally it means 'probable cost' and is then a judged amount, approximate rather than precise.

---

[238] *Behzadi* v. *Shaftsbury Hotels Ltd* [1992] Ch 1.

— A contractor's estimate, in contrast, may, dependent on its terms, amount to a firm offer, and if this is so, its acceptance by the employer will result in a binding contract[239].

As regards its first connotation, architects and quantity surveyors are frequently required to provide an estimate of cost to a client at an early stage of a project in order that he can decide whether to proceed. It is generally accepted that the estimate will be higher or lower than the final figure. An initial estimate may be as much as 15% astray and it is, therefore, essential that the architect or quantity surveyor inform the client of the possible margin of error. Other factors should be stated, such as whether inflation has been taken into account, exclusion of VAT and the currency of prices. The final cost estimates, produced before the tender date, may have a very small margin of error, say 5%. Realistically, this is too small because variations in tender price may easily be in the order of 15%, excluding those prices which are clearly not intended to be competitive. Clients tend to expect accuracy and, therefore, architects will often err on the high side in order to avoid unpleasant surprises when tenders are opened. It is certainly a mistake to pitch any estimate too low simply to 'sell' a scheme because subsequent failure to achieve the figure is unfair to the client and he may sue for the return of his fees − at the very least[240]. Where estimates, in the colloquial sense, are put forward, they should always be qualified as 'rough estimate' or 'approximate estimate' to avoid any possible suggestion that it is a firm offer or figure.

It is not unusual, on small works, for contractors to produce an estimate of the cost of carrying out work. In the absence of a professional adviser, the employer may not realise that the final figure may exceed the estimate. If the word 'estimate' is used, it must have some proper foundation in calculations and facts.

See also: **Budget price; Offer; Quotation; Target cost; Tender.**

Estoppel A principle which precludes a person from denying the truth of a statement made by him or from alleging that a fact is otherwise than it appeared to be from the circumstances.

Several kinds of estoppel have been identified:

— *Estoppel by deed* A statement of fact in a deed (qv) cannot be disputed by either party to it. Thus a party to a deed cannot deny the truth of the recitals (qv) it contains.

— *Proprietary estoppel* Where a party has done acts in reliance that he will acquire rights over another's land.

— *Promissory estoppel* A promise that strict legal rights will not be enforced.

— *Estoppel by record* Where a party is barred from pursuing a cause of action (sometimes referred to as 'action estoppel') or raising an issue (sometimes referred to as 'issue estoppel') which has already been the subject of judgment in the courts.

---

[239]*Crowshaw* v. *Pritchard & Renwick* (1899) 16 TLR 45.
[240]*Nye Saunders & Partners* v. *Alan E. Bristow* (1987) 17 Con LR 73.

— *Estoppel by representation* Where someone expressly or impliedly by conduct has made a factual statement or conducted himself so as to mislead another person he cannot afterwards go back on the representation. For example, allowing another person to appear to be one's agent or to have an authority wider than he in fact has.

— *Estoppel by convention* Where parties to a contract have acted on the assumption, mutually agreed, that certain facts can be regarded as being true for the purposes of that contract. The parties are thus precluded from denying the truth of those assumed facts.

Estoppel can be used only as a defence, it cannot be used as the basis of an action.

See also: **Agency.**

Evasion See: **Avoidance.**

Evidence Information tending to establish facts, the facts themselves or opinions based on the facts. In court, there are rules of evidence, which must be observed, as to what evidence may be produced. In civil cases the burden of proof lies with the person asserting the facts. The standard of proof is the balance of probability, i.e. it is more likely to be as the person asserting states than otherwise.

See also: **Admissibility of evidence; Expert witness; Hearsay; Parol evidence.**

Ex contractu Arising out of contract. The term is used to refer to claims which arise out of the *express* provisions of the particular contract in contrast to other types of claim (qv). For example, JCT 98, clause 26 confers on the contractor a right to claim for loss and expense caused by matters materially affecting regular progress of the works and similar *ex contractu* claims arise under ACA 3, clause 7 (employer's liability) and GC/Works/1 (1998), clause 46 (prolongation and disruption expenses). The architect has power under the terms of his appointment to quantify or agree *ex contractu* claims, but not other types of claim.

All the current forms of contract allow additional or alternative claims for breach of contract based on the same facts, see e.g. JCT 98, clause 26.6, provided that the facts do amount to a breach of contract. Not all the 'matters' under clause 26.2 amount to breaches of contract. The contractor can recover his loss or expense only once, but a claim for breach of contract may avoid some of the restrictions under the particular contract clause. For example, if the contractor has neglected to make application within a reasonable time in accordance with JCT 98, clause 26.1.1, it is probable that a late claim will be rejected by the architect. There is nothing to prevent the contractor making the same claim at common law for breach of contract, but he can only recover his loss once.

Ex gratia claim or payment A claim or payment met or made 'as a matter of grace'. It is sometimes called a 'sympathetic' claim and the essential point is that the employer is under no legal obligation to meet it. *Ex gratia* payments are sometimes made to settle or compromise a claim rather than go to the expense of contesting it in litigation or arbitration.

Under most standard form contracts the architect or his equivalent has no authority to settle such claims or to authorise *ex gratia* payments. He must be given express authority by the employer if he is to settle such claim, and none of the standard contracts endows him with that authority.

GC/Works/1, Edition 2, clause 63 (4) gave the employer (the 'Authority') power to make 'such allowance, if any, as in [its] opinion is reasonable' where the employer has exercised the special power of non-default determination contained in the clause. This provision in effect merely enabled the employer to make an *ex gratia* allowance which it could do in any event and the allowance is to cover only the contractor's 'unavoidable losses or expense (excluding loss of profits) directly due to the determination' for which he has not been fully reimbursed.

See also: **Claim.**

Ex officio By virtue of one's office.

Examination-in-chief The first stage in the examination of a witness (qv) in judicial or arbitral proceedings. It is carried out by the party calling the witness, generally through counsel or a solicitor. There are many strict rules which must be observed in examination-in-chief, e.g. leading questions may not be asked. A leading question is one which suggests its own answer, e.g. 'Did the site agent tell you that he could not care less?' However, a witness may be led in the introductory part of his testimony, e.g. 'Are you the project architect?' This process is now discouraged by the Civil Procedure Rules (CPR) (qv)[241] where examination-in-chief is generally replaced by the contents of the relevant witness statement.

See also: **Cross-examination; Re-examination.**

Examination of site Under the general law the employer does not warrant the suitability of the site for the works[242]. The precise conditions of contact may emphasise the position or they may amend it so that the contractor is entitled to additional payment if the ground conditions are not as represented to him. It is quite common for a clause to be inserted in the bills of quantities (qv) requiring the contractor to satisfy himself regarding all matters in connection with the site. GC/Works/1 (1998), clause 7 (1) is in similar vein, but clause 7 (2) states that the provision does not apply to any information given or referred to in bills of quantities which is required to be given in accordance with the method of measurement expressed in the bills. Thus, if the bills have been prepared in accordance with the Standard Method of Measurement (qv) the contractor may well have a claim if the ground is not as described[243].

---

[241]CPR Rule 32.5(2).
[242]*Appleby* v. *Myers* (1867) LR 2 CP 651.
[243]*C. Bryant & Son Ltd* v. *Birmingham Hospital Saturday Fund* [1938] 1 All ER 503.

GC/Works/1 clause 7 (5) also gives the contractor a right to claim for unforeseeable ground conditions (qv) in specified circumstances.

JCT 98 also refers to the SMM in clause 2.2.2.1 and the contractor is entitled to claim additional payment if he is put to more expense in excavation than he was led to expect. ACA 3 provides, where the contractor is responsible for the provision of drawings (clause 2.6), that ground conditions are his responsibility but he is entitled to receive payment for any measures he needs to take after encountering 'adverse ground conditions or artificial obstructions' unless he should have foreseen them or an adjustment is made to the contract sum (qv) under clause 1.4 (bills of quantities). Whether the contractor is able to claim for site conditions is a matter of interpretation of the contract (qv).

Subject to anything agreed to the contrary, an architect will normally have a duty to inspect the site before carrying out design work[244]. Where an independent structural engineer is appointed to investigate the sub-soil conditions, the architect is probably entitled to rely on the structural engineer's advice, but in any event the architect will be protected if the terms of engagement include an appropriate clause[245].

See also: **Misrepresentation.**

Excepted risks The term is used in JCT 98 to describe those risks which are carried by the employer and which may affect the execution of the works although they are outside the contractor's control. The definition reflects the exceptions commonly to be found in 'All Risks' policies of insurance.

The definition covers ionizing radiations or contamination by radioactivity from any nuclear fuel or from nuclear waste from the combustion of nuclear fuel, radioactive, toxic, explosive or other hazardous properties of any explosive nuclear assembly or nuclear component thereof, and pressure waves caused by aircraft or other aerial devices travelling at sonic or supersonic speeds.

See also: **Accepted risks.**

Exceptionally adverse weather See: **Adverse weather conditions.**

Execute work To carry out the work specified in a contract.

Execution A word with several meanings in a legal context. 'To execute a contract' means to render it effective by signing it, or by completing it as a deed. It may also mean to carry out its terms.

Execution is also the process by which judgments of the court may be enforced, hence 'Writ of Execution' or 'Warrant of Execution' which directs

---

[244] *Eames London Estates Ltd* v. *North Hertfordshire District Council* (1980) 259 EG 491.
[245] *Investors in Industry Commercial Properties* v. *South Bedfordshire District Council* (1985) 5 Con LR 1, referring to clauses 1.20, 1.22 and 1.23 of the RIBA Conditions of Engagement (qv). The current SFA/99 does not have these clauses, but it does contain clauses which probably have the same effect.

the sheriff (or county court bailiff) to seize the judgment debtor's personal property to satisfy the judgment, costs and interest.

Exemption (exception or exclusion) clause A clause in a contract with attempts to exclude liability or limit it in some way.
See: **Unfair Contract Terms Act 1977.**

Expense The term used in GC/Works/1 (1998) (qv), clause 46, in reference to contractor's claims for disruption and prolongation. It is narrowly defined in clause 46 (6) as meaning 'money expended by the contractor, [and does] not include any sum expended, or loss incurred, by him by way of interest or finance charges however described'.

Expert Someone with special skill, knowledge or professional qualifications. If a person is deemed to be acting as expert and not as arbitrator, the effect is:
— The provisions of the Arbitration Act 1996 are not applicable to their decisions.
— There is no requirement for the 'expert' to hold a hearing.
— An expert is liable for negligent decisions (*Sutcliffe* v. *Thackrah* (1974)) even though an arbitrator is immune from an action for negligence[246].
Clause 31 (5) of GC/Works/1 (1998) refers to testing by an 'independent expert'.
See also: **Adjudication; Arbitration.**

Expert witness A witness who appears for one party at an arbitration hearing or in court proceedings and who gives evidence based upon his expert knowledge of some facet of the case. His duty is to assist the court or tribunal. An expert witness may and usually does give his opinion. Expert evidence is given by a person with the requisite skill and experience about the opinion that he holds on the basis of facts related to and/or perceived by him. Other witnesses may only give evidence as to facts, i.e. what they saw or heard. Thus, in a building case, a labourer may be called upon to give evidence as to what he saw before a brick wall collapsed — cracking, leaning, etc. He would not be asked what, in his opinion, caused the collapse. An experienced engineer, architect or surveyor, however, may be asked to give his expert opinion on the cause of the collapse. Anyone may be an expert witness provided only that he has the necessary expertise in the field in dispute. Thus a bricklayer would be entirely suitable, if properly experienced, to give an opinion on, say, standards of brickwork.

Part 35 of the Civil Procedure Rules (qv) has considerably tightened the rules concerning expert witnesses and evidence. Such evidence must be restricted to what is reasonably required to resolve the proceedings. Rule 35.3 makes clear that the expert's duty is owed to the court and it overrides any duty to the party who has commissioned the report. The court now has power to direct that

---

[246]Section 29, Arbitration Act 1996.

there will be only one expert witness which the court may choose in the event of a failure to agree. In such cases, both parties give instructions to the expert.

The practice direction supplementing Part 36 helpfully says what an expert's report must contain. Briefly it is as follows:

— The expert's qualifications.
— The written material relied on in making the report.
— The identity of, and qualifications of, the person carrying out tests and whether they were under the supervision of the expert.
— Summary of the range of opinion and reasons for the expert's opinion.
— Summary of conclusions.
— Statement that the expert understands and has complied with his duty to the court.
— The substance of all relevant instructions.
— It should have a statement of truth.

Generally, the expert witness is chosen to appear for one side or the other because his opinion favours them. His views, however, must be sincerely held. His principal duty is to assist the court or the arbitrator to get at the truth, and he must not attempt to conceal something which would benefit the other party. It sometimes happens that an expert witness changes his mind during the course of a hearing. In such a case he is under an obligation to notify his own party and to offer to withdraw. He is under no obligation to volunteer information which would assist the other party. To act as an expert witness is often a thankless task because he is clearly going to be subjected to a very searching cross-examination (qv) during which his reputation as an expert may be affected. The duties of an expert witness have been set out[247] broadly as follows:

— His evidence should not be influenced by the pressures of litigation.
— He should be unbiased and should never act as advocate for a party.
— Facts and assumptions supporting, and detracting from, the opinion should be stated.
— He should make clear what matters fall outside his scope.
— The expert must say if his opinion is not adequately researched and indicate if the opinion is provisional. He should make clear if he cannot say that his opinion contains the truth, the whole truth and nothing but the truth.
— If the expert changes his opinion after exchange of reports, he must inform the opposing side and the court immediately.
— All the documents referred to must be provided to the opposing side when reports are exchanged.

See also: **Arbitration; Evidence.**

Express terms Terms which are actually recorded in a written contract or which are expressed and agreed openly at the time the contract was made. Thus, JCT

---

[247]*National Justice Compania Naviera SA* v. *Prudential Assurance Company Ltd (Ikarian Reefer)* TLR 5 March 1993.

98, clause 10 clearly states that the contractor 'shall constantly keep upon the Works a competent person-in-charge'. This provision leaves no room for doubt, as far as it goes. An express term will prevail over any term which would otherwise be implied on the same subject matter.

It is the function of the court to determine what the terms of the contract are and to evaluate their comparative importance and effect. Traditionally, contract terms are either *conditions* (qv) or *warranties* (qv), the former being major terms and the latter subsidiary or minor terms. Breach of a condition entitles the innocent party to treat the contract as discharged if he so wishes. Breach of a warranty, in contrast, merely entitles him to claim damages (qv). Since the decision of the Court of Appeal in *Hong Kong Fir Shipping Co Ltd v. Kawaski Kisen Kaisha Ltd* (1962)[248] it has been recognised that this classification is not exclusive. Between conditions and warranties there is an intermediate class of 'innominate terms', the effect of whose breach depends not on classification of the term but upon the seriousness of the breach and its effects. The concept of a 'fundamental term' (qv) is also sometimes quoted.

See also: **Implied term; Interpretation of contracts.**

Expressly Definitely stated. To say that a contract expressly provides for payment to be made on the 25th of each month means that a term to that effect is written in the contract or, in the case of an oral contract, that the parties have clearly stated the same. This is in contrast to saying that such a term is implied, which means that it is not written down or clearly stated, but that it would be imported into a contract by the court, because either it is a term which goes without saying or that it is necessary to make the contract work.

See also: **Express term; Implied term.**

Extension of preliminaries The term refers to a technique of valuation carried out by the quantity surveyor under certain circumstances.

The preliminaries section of the bills of quantities (qv) is priced by the contractor at tendering stage. He may choose to do this in various ways. For example, he may price every item individually having regard to his actual costs or he may simply allow a percentage to preliminaries of the total cost of the measured work; alternatively and rarely, he may simply pluck a figure from the air to serve as a total for all the preliminaries.

When the quantity surveyor is preparing his monthly valuations prior to the issue of interim certificates (qv) he must allow a sum of money to represent a reasonable proportion of the contractor's preliminaries price. If the contractor has priced individual items (erecting offices, insurance, etc.), the quantity surveyor will look carefully at each item to arrive at his figure. If the preliminaries figure is simply a lump sum, the quantity surveyor will often merely divide the sum by the expected number of valuations to arrive at a suitable figure.

---

[248][1962] 1 All ER 474.

If it seems likely that the contract period will be extended but no financial claim is involved, the quantity surveyor will often reduce the monthly preliminaries figure so as to extend the preliminaries to the end of the contract. This practice has doubtful validity. If at all valid, it appears that the quantity surveyor should only address the time related element of the preliminaries. Reference should be made to the provisions of the Standard Method of Measurement (qv) Edition 7. The process is also known as 'adjustment of preliminaries'.

If a financial claim is made for delay, the process is rather more complex. Briefly, the monthly preliminary figure is not reduced and if the contractor's claim is valid, a calculation must be carried out to reflect the loss. However, there is no automatic claim by way of extension of the usual monthly figure for preliminaries over the prolonged period; in each and every case the claimant contractor must prove his loss, preferably by reference to records, and reimbursement is based on actual cost.

See also: **Claims; Loss and/or expense.**

Extensions of time All the standard forms of contract contain provision for the insertion of a completion date (qv) and for the employer to deduct or receive liquidated damages (qv) in the event of late completion. However, the employer would forfeit his right to liquidated damages if he were wholly or partly the cause of the delay[249]. There is no power to extend time unless the contract so provides. The standard forms provide for the architect to extend the time for completion for a variety of reasons. The grounds for extension divide into two groups:

— Those for which the employer or his agents (including his employees, etc.) are responsible.
— Those for which neither the employer nor the contractor is responsible, and which are outside the control of either party.

The first set of grounds is most important. In the absence of an express provision to extend time in the contract, the architect would be unable to extend time due to the employer's default (qv) and time would become 'at large' (see: **Time at large**)[250]. The contractor would be under no other obligation in respect of the completion date than to complete within a reasonable time and the employer would lose his right to liquidated damages. The employer could try to prove his actual loss at common law, but it would not be easy. Time will also become at large if the architect does not exercise any power he may have to extend time because of the employer's default[251] or if he fails to exercise the power properly and at the right time.

It is often said that the extension of time clauses in the standard forms are there for the benefit of the employer. That is correct as far as they concern the employer's default. However, it should not be overlooked that many of the

---

[249] *Holme* v. *Guppy* (1838) 150 ER 1195.
[250] *Astilleros Canarios SA* v. *Cape Hatteras Shipping Co Inc and Hammerton Shipping Co SA* [1982] 1 Lloyds Rep 518.
[251] *Peak Construction (Liverpool) Ltd* v. *McKinney Foundations Ltd* (1970) 1 BLR 114.

grounds, i.e. those which provide for extensions due to events outside the control of both parties, benefit the contractor. Without them, he would be obliged to stand the burden of liquidated damages. The employer may, of course, benefit indirectly by obtaining a lower tender figure than would otherwise be the case.

The provisions in respect of extensions of time are often complex[252].

Extra work Very often simply referred to as 'extras'. Work which is required by the employer, to be carried out by the contractor and which is additional to the work described in the contract documents (qv). It is usually contained in an instruction (qv) of the architect and treated as a variation (qv) of the contract to be valued by the quantity surveyor according to the rules set out in the contract.

There is no automatic right for the contractor to be paid under the contract, because he has carried out extra work. There must be a provision in the contract to allow extra work to be instructed and there must be an instruction. Extra work carried out without an instruction may amount to a breach of contract, but in any event the contractor would not be entitled to payment. The position may be different where the contractor carries out work in reliance on an oral instruction where the contract provides only for written instructions[253].

Extrinsic evidence See: **Parol evidence.**

---

[252]For a fuller discussion of the subject see: Chappell, D., *Powell-Smith and Sims' Building Contract Claims*, 3rd edn (1998), Blackwell Science.
[253]*Bowmer and Kirkland Ltd* v. *Wilson Bowden Properties Ltd* (1996) 80 BLR 131.

Facilities management The management of all the systems in a completed project. 'Systems' is to be interpreted very broadly and may include such things as the upkeep and maintenance of all the heating, lighting, cleaning and catering facilities for a specific period of time at an agreed fee. Some construction companies may offer the service as an add-on to the construction package or it may be purchased separately. There is scarcely any limit to the kinds of services which may be offered and it is possible to employ a facilities management firm to deal with everything necessary to make a building work, even to the extent of monitoring and arranging insurance cover, telephonist services and the purchase of consumables. A standard form of agreement has been produced by the Chartered Institute of Building.

Fair valuation The JCT 98 and GC/Works/1 (1998) forms of contract set out rules for the valuation (qv) of architect's or project manager's variation instructions (clauses 13 and 42 respectively). JCT 98 allows the quantity surveyor to value work at fair rates and prices if the work is not of similar character to, or not executed under similar conditions of, the work set out in the contract bills (qv). GC/Works/1 allows valuation at fair rates and prices if it is not possible to value by measurement and valuation at rates and prices deduced or extrapolated from the bills of quantities (qv) (clause 42 (5) (6)). MW 98, in clause 3.6, allows valuation on a fair and reasonable basis, using the relevant prices in the priced specification, schedules or schedule of rates, for all variations. The ACA 3 form has no equivalent provision.

Much has been written about the meaning of the word 'similar' in the context of valuations. The ordinary meaning of 'similar' would be 'almost but not precisely the same' or 'identical save for some minor particular'. When dealing with variations, however, it is not safe to consider 'similar' as anything other than 'identical' for the simple reason that even a minor difference in the description of an item in the bills of quantities may cause the contractor to considerably amend his prices[254]. The conditions under which work is carried out are those set out in the express provisions of the contract. The background information and facts against which the contract was made cannot be taken into account[255].

What is 'fair' will depend on the whole of the contractor's pricing. It has been suggested that if a contractor has priced keenly in the contract as a whole, a fair valuation will take account of the fact and vice versa. Some contractors, however, adopt a pricing strategy by which some items are keenly priced while others show a handsome profit margin. Although a fair valuation is solely the responsibility of the quantity surveyor under the standard forms, he must still

---

[254]A full discussion of the point is to be found in Chappell, D., *Powell-Smith and Sims' Building Contract Claims*, 3rd edn, 1998, pp. 110–112, Blackwell Science.
[255]*Wates Construction (South) Ltd* v. *Bredero Fleet Ltd* (1993) 63 BLR 128.

have regard to the general tenor of pricing as revealed by the bill rates. The contractor's remedy, if aggrieved, is to go to arbitration.

Fair wages clause A clause formerly found in local authority contracts which usually reproduced the terms of the House of Commons Fair Wages Resolution of 14 October 1946. This Resolution was rescinded from August 1983, and the standard form contracts have been amended accordingly.

Fees Generally a payment given or due to any professional person, public office or for entrance to museums, art galleries and the like. It is referred to in many contracts (e.g. JCT 98, clause 6.2) as a sum payable to a statutory or local authority.

Fee also refers to the quality of inherited land. The highest is *fee simple* which is, to all intents and purposes, unfettered ownership.

Fiduciary Where someone is in a position of trust in relation to another he is bound to exercise his rights and powers in good faith for the benefit of the other person and cannot make any profit or advantage from the relationship without full disclosure. A person in a fiduciary position must not put himself in a position where his duty and his interest conflict. Fiduciary relationships include trustee and beneficiary, and solicitor and client.

A number of building contracts, e.g. JCT 98, clause 30.5.1, provide that the employer's interest in the retention monies (qv) 'is fiduciary as trustee for the contractor and for any nominated sub-contractor'. The clause adds, contrary to the general law of trusts, that the employer has no obligation to invest the retention money, but the legal effect of this is doubtful because the Trustee Act 1925 and the Trustee Investments Act 1961 oblige a trustee to invest trust monies in prescribed investments. It is certain that an employer under JCT 98 or IFC 98 terms has an obligation to segregate the retention fund in a separate and identifiable bank account[256].

Final account The ACA 3 form of contract, clause 19.1 provides that the contractor shall submit within 60 working days of the expiry of the maintenance period (qv) a final account for the works. This will be a detailed summing up of the effects upon the contract sum (qv) of all additions, deductions and alterations. GC/Works/1 (1998) also refers to the final account, in clause 49 (2), but in this instance it is to be prepared by the quantity surveyor.

JCT 98, clause 30.6.1.1 provides that the contractor must provide the architect with all documents necessary for adjustment of the contract sum no later than six months after practical completion. Not later than three months after receipt, the architect or the quantity surveyor must ascertain any loss and/ or expense under clauses 26 and 34 unless it has already been done and he must then prepare a statement of any adjustments. The architect must forthwith

---

[256] *Wates Construction (London) Ltd* v. *Franthom Property Ltd* (1991) 53 BLR 23.

send the result to the contractor. However, the architect is not entitled to delay the issue of the final certificate (qv) because the final account has not been sent to the contractor[257]. Neither is there any requirement that the contractor must agree the final account. The six months period following practical completion is intended for the contractor to draw any last minute matters to the attention of the quantity surveyor who should, by that time, have developed his own final account in a gradual process from the start of the works on site.

See also: **Bill of variations.**

Final certificate The last certificate issued by the architect in connection with a contract. The effects of the final certificate vary according to the form of contract being used.

JCT 98, clause 30.8 stipulates that the final certificate must be issued not later than two months from the latest of the following events:

— The end of the defects liability period (DLP) (qv).
— The date of issue of the certificate of completion of making good of defects.
— The date the architect sent to the contractor a copy of the adjusted contract sum.

Failure to issue the final certificate in due time is a breach of contract for which the employer is liable[258]. **Figure 7** shows an example of a final certificate.

The contractor will often demand the final certificate during the earliest two months, i.e. from the end of the DLP. As it is virtually unknown for there to be no defects during the DLP, the earliest possible period will be two months from the completion of making good of defects. In practice, the period often commences with the sending of the final account to the contractor. There is, however, no excuse for undue delay in the issue of a final certificate and if it can be accomplished towards the beginning, rather than the end of the period, so much the better for all concerned. It should be noted that architects have traditionally delayed issuing a final certificate until the last possible moment for fear of the effects noted below. The final certificate must state:

— The total amounts already stated as due (note: not *paid)* in previous interim certificates (qv) plus any advance payments (qv).
— The contract sum (qv) adjusted as provided in the contract (clause 30.6.2).
— To what the amount relates and the basis of calculation of the statement.
— The difference between the above two sums shown as a balance due to the contractor from the employer or vice versa.

The balance will be a debt from one to the other as appropriate and the final date for payment (qv) will be 28 days after the date of issue of the final certificate. The conclusiveness of the final certificate has been the subject of much debate. The current position is that the final certificate is conclusive

---

[257] *Penwith District Council* v. *P. Developments Ltd* 21 May 1999, unreported.
[258] *Rees & Kirby Ltd* v. *Swansea City Council* (1983) 25 BLR 129.

Final certificate

**Final
Certificate**

JCT 98

Issued by: E W Pugin
address: Gothic Buildings
Birmingham

Employer: The Duke of Omnium
address: Trollope House
Belstead

Contractor: Gerrybuilders Ltd
address: The Crescent
Belstead

Works: Mansion
situated at: Belstead Gardens, Belstead

Contract dated: 1 February 2001

Serial no: **K 101624**

Job reference: EWP/ABC

Date of issue: 2 May 2002

Final date for payment: 30 May 2002

Copy

This Final Certificate is issued under the terms of the above-mentioned Contract.

Contract Sum adjusted as necessary £950,000.00

Total amount previously certified for payment to the Contractor
plus amount of any advance payment £948,000.00

Difference between the above stated amounts .... £ 2,000.00

I/We hereby certify the sum of (in words)   *All amounts are exclusive of VAT.*
  Two Thousand Pounds Only----

as a **balance due:**

* Delete as
appropriate

* to the Contractor from the Employer.

* ~~to the Employer from the Contractor.~~

To be signed by or for
the issuer named
above

Signed _E W Pugin._

[1] Relevant only if
clause 1A of the VAT
Agreement applies.
Delete if not
applicable.

[1] The Contractor has given notice that the rate of VAT chargeable on the supply of goods and services to which the Contract relates is 17.5 %

[1] 17.5 % of the amount certified above .... £ 350.00

[1] Total of balance due and VAT amount (for information) .... £ 2,350.00

This is not a Tax Invoice.

F852A for JCT 98   © RIBA Publications 1999

Figure 7 Final certificate.

evidence (except in the case of fraud or proceedings started as specified in clauses 30.9.2 and 30.9.3), in any adjudication, arbitration or litigation arising out of the contract, that:

— Where the contract drawings, the contract bills, the numbered documents (qv) or any instruction or other drawings expressly state that something specific about the quality of materials or the standard of workmanship is to be to the approval of the architect, then the specific attribute is to his reasonable satisfaction. The contractor still retains his obligation to have carried out the works in accordance with the contract documents (qv) and the final certificate is not conclusive in any broad sense.

— The contract sum has been adjusted as necessary in accordance with the contract provisions except in the case of a mistake when the certificate shall be conclusive as far as the other computations are concerned.

— All the due extensions of time have been given.

— All the contractor's claims relating to matters under clause 26.2 in respect of breach of contract, duty of care, statutory duty or otherwise have been fully and finally settled.

There are complex provisions to deal with the situation where one of the parties has commenced adjudication, arbitration or litigation proceedings before the final certificate becomes conclusive.

The position now set out under JCT 98 and IFC 98 in regard to the architect's satisfaction with the quality and standards of materials and workmanship is what the construction industry generally believed the position to be from about 1976 (then in relation to the JCT 63 form of contract). That was that the final certificate was only conclusive about precise things which the architect had expressly stated in the contract documents were to be to his satisfaction or approval. For example, if the architect stated that all floor finishes were to be his satisfaction, the final certificate was conclusive that he was satisfied with all floor finishes, even if he had inspected only one room or none at all.

That state of affairs was upset first by the High Court and then by the Court of Appeal in 1994[259]. They held that the final certificate was to be interpreted in a very broad way to signify the architect's satisfaction with the quality and standards of all materials and all workmanship. This was not quite the same as saying that the contractor had constructed the works in accordance with the contract, but it effectively prevented the employer from recovering against the contractor for latent defects which occurred within the limitation period (qv). Architects were understandably concerned and the Royal Institute of British Architects issued guidance together with a form of declaration and stickers which were of doubtful value. Many architects reverted to withholding the final certificate in favour of issuing an interim certificate from which a nominal sum was retained. This is obviously a breach of contract, but the idea was that if the

---

[259]*Colbart* v. *H. Kumar* (1992) 8 Const LJ 268 and *Crown Estates Commissioners* v. *John Mowlem & Co* (1994) 70 BLR 1 dealing with IFC 84 and JCT 80 respectively.

final certificate was not issued, it could not become conclusive and provided the contractor received all the money to which he was entitled, less perhaps £5, he would not press for the final certificate. It has become clear that the courts will take a dim view of this approach, where the employer is seeking to take advantage of his own breach of contract[260]. In effect, a court may treat the situation as if the certificate had been issued on the due date.

The ACA 3 form, clause 19, stipulates that the final certificate must be issued within 60 working days:
— After the end of the maintenance period (qv).
— After the architect has received the contractor's final account (qv) together with all the documents necessary for computation of the final contract sum and all documents prepared by the contractor for the work.

The final certificate must state:
— The final contract sum.
— The total amount already paid under clause 16.3.
— The difference between the above two sums shown as a balance due to the contractor from the employer or vice versa.

The balance will be a debt from one to the other as appropriate from the tenth working day after the issue of the final certificate. The fact that it is the amount already *paid* which has to be stated, immediately raises the question: What is the position if the last interim certificate issued by the architect has not been honoured by the employer by the time the architect is to issue the final certificate? In practice, the situation should never arise because the contractor will have invoked his termination powers under clause 20.2. If the contractor has not invoked his power to terminate, the architect will be obliged to certify only those sums already paid. The danger of the architect over-certifying in such a situation is avoided by clause 19.5 which empowers the architect to delete, correct or modify any sum previously certified by him.

The final certificate is specifically stated to leave the contractor's liabilities, arising out of or in connection with the contract, intact (clause 19.5). Under the ACA 3 form, therefore, the final certificate is not to be regarded as conclusive. Clauses 7.5 and 17.6 make the final certificate the only time at which the architect is empowered to adjust the contract sum in regard to matters provided for in clauses 7.2 and 17.1 when the contractor has failed to comply with the appropriate provisions.

GC/Works/1 (1998), clause 49 (4) provides that, if at the end of the maintenance period the PM has certified that the works are satisfactory and the final sum (qv) has been agreed or is to be treated as agreed under clause 49 (3) because the contractor has failed to notify his disagreement with the final account, the balance between the final sum and amounts previously paid to the contractor will be paid by the employer to the contractor or vice versa as appropriate. The contractor has three months from receipt of the final account in which to notify his disagreement and the quantity surveyor must have

---

[260] *Matthew Hall Ortech Ltd* v. *Tarmac Roadstone Ltd* (1997) 87 BLR 96.

prepared the final account within six months of certified completion of the works. Clause 50 provides for certificates to be issued. Unlike the JCT and ACA forms, GC/Works/1 is not specific in regard to what information must be included on the certificate. It would be advisable, in the interests of all parties, to indicate briefly the way in which the final amount on the certificate has been calculated. The final certificate is not stated to be conclusive and, therefore, it can be opened up and revised by the arbitrator.

MW 98, clause 4.5 lays down a time scale for the issue of the final certificate:
— The contractor has three months from the date of practical completion (qv) to supply the architect with all documentation reasonably required for the computation of the final amount to be certified.
— The architect has 28 days from the receipt of the documentation to issue the final certificate, provided that he has issued a certificate (under clause 2.5) that the contractor's obligations have been discharged. There is no suggestion that, if the architect is not in a position to issue a clause 2.5 certificate by the date of expiry of the 28 days, the 28 day period begins to run from the issue of the clause 2.5 certificate. In such circumstances, it is likely that the architect would simply hold the final certificate until the contractor had fulfilled his obligations and then issue the two certificates on consecutive days.

This form of contract makes no mention of the information to be contained in the final certificate other than that it must certify the sum remaining due to the contractor or the employer. A simple and clear calculation showing how the sum is derived would seem to be advisable. The sum certified becomes a debt from one party to the other 14 days after the date of the final certificate. The final certificate under this form is not stated to be conclusive about anything.

See also: **Certificates; Final account; Interim certificate; Conclusive evidence.**

Final sum The amount which represents the contract sum (qv) as adjusted to take into account all additions, deductions and alterations to the contract. It is the total sum payable to the contractor, inclusive of sums already paid after the issue of the final certificate (qv).

Finance Act (No 2) 1975 This measure introduced the construction industry tax deduction scheme which came into operation on 6 April 1977. From that date, the position is that all payments under a contract for what the Inland Revenue define as 'construction operations' and made by 'contractors' to 'subcontractors' are subject to a deduction by the payer on account of the payee's tax liability.

Because of the complexities of the legislation, a number of standard form contracts introduced special clauses to deal with the requirements of the Act, e.g. JCT 80, clause 31 and ACA 3, clause 24.

Minor amendments to the scheme were made by the Finance Act 1980. The scheme has been replaced by the Construction Industry Scheme (CIS) (qv).

Finance charges The financial burden to the contractor who receives money later than they should have received it under the terms of the contract. It is settled that such charges are a constituent part of 'direct loss and/or expense' (qv) under JCT contracts[261]. Financing charges may also be recoverable as a head of special damages (qv) in appropriate cases[262]. GC/Works/1 (1998) deals specifically with finance charges in clause 47. Clause 47 (6) precludes finance charges being claimable at all. Clause 47 (1) provides for automatic reimbursement of finance charges (as defined) to the contractor in very limited circumstances.

See also: **Expense; Direct loss and/or expense; Interest on money.**

Firm price contract A contract in which the price of labour and materials is not subject to fluctuations; sometimes referred to as a fixed price contract (qv).

Fit and ready A term used only in the ACA 3 contract in conjunction with 'taking-over'. It is not defined in the contract, but it is clear that it is not the same as 'practical completion' (qv) in the JCT forms because clause 12.1 gives the architect the option of issuing his certificate that the works are fit and ready for taking-over (qv) provided that the contractor gives a written assurance to complete with all due diligence (qv) items contained on the architect's or the contractor's list. The architect can wait until the items are completed before issuing his certificate if he so desires.

Fitness for purpose Under the Sale of Goods Act 1979, s. 14, there is an implied condition (qv) that the goods (qv) are reasonably fit for the purpose required, if this has been made known to the seller, expressly or by implication. In business dealings – as opposed to consumer transactions – it is possible to contract out of this to a limited extent, provided the exemption clause (qv) is 'fair and reasonable'. This applies to goods supplied to a contractor by a merchant, and the seller is liable even if he has taken every care or did not know of the defect.

A similar term of reasonable fitness for purpose will be implied at common law where the contractor undertakes to carry out both design and build as regards the completed structure in the absence of an express term to the contrary[263]. In a significant judgment in the House of Lords, it was said:

> 'It is now well recognised that in a building contract for work and materials a term is normally implied that the main contractor will accept responsibility to his employer for materials provided by nominated sub-contractors. The reason for the presumption is the practical convenience of having a chain of contractual liability from the employer to the main contractor and from the main contractor to the sub-contractor – *Young & Marten Ltd* v. *McManus Childs Ltd* (1969).'[264]

---

[261] *F. G. Minter* v. *Welsh Health Technical Services Organisation* (1980) 13 BLR 1; *Rees & Kirby Ltd* v. *Swansea City Council* (1983) 25 BLR 129.

[262] *Holbeach Plant Hire Ltd* v. *Anglian Water Authority* (1988) 14 Con LR 101.

[263] *Viking Grain Storage Ltd* v. *T. H. White Installations Ltd* (1985) 3 Con LR 52.

[264] *Independent Broadcasting Authority* v. *EMI Electronics Ltd and BICC Construction Ltd* (1980) 14 BLR 1 per Lord Fraser at 44.

The Supply of Goods and Services Act 1982 (qv), which applies to building contracts, is also relevant.

Fixed price contract A contract in which the price of labour and materials is not subject to fluctuations; sometimes referred to as a firm price contract (qv). See also: **Lump sum contract.**

Fixtures and fittings Fixtures are goods which have become so affixed to land as to have become in law part of the land. They are contrasted with fittings which are goods which retain their character as personal property (qv). The general rule is that fixtures installed by a tenant become the property of the landlord and may not be removed by the tenant when his tenancy comes to an end, but three groups of 'tenant's fixtures' can be removed:
— Ornamental and domestic fixtures which can be removed provided that no serious damage is caused to the fabric of the premises by the removal.
— Trade fixtures, e.g. fittings of a public house, including the beer pumps.
— Agricultural fixtures.

It is often difficult to decide whether a thing is a fixture or not. The word implies something fixed to the soil or attached in a substantial way. The tests applied by the courts are the degree of annexation to the land and the purpose for such annexation[265]. A useful example is the distinction between a stock of stone (a chattel) and the same stones constructed as a dry stone wall (a fixture)[266]. Whether an item is a fixture is a mixed question of law and fact to be determined by the judge in all the circumstances[267].

The rule relating to fixtures is largely important in building contracts in that once the contractor has affixed materials to the building, the property in them passes from him to the employer:

'Materials worked by one into the property of another become part of that property. This is equally true whether it be fixed or moveable property. Bricks built into a wall become part of the house, thread stitched into a coat which is under repair, or planks and nails and pitch worked into a ship ... become part of the coat or the ship.'[268]

Float A term used in programming, particularly in connection with network analysis and precedent diagrams. It is the time difference between the time required to perform a task and the time available in which to do it. If there is a three days float to a particular activity, it means that the activity can be delayed by up to three days before it becomes critical and affects the date for completion of the whole project. Critical activities have no float, because if they are delayed at all, the completion date will be affected.

---

[265]*Elitestone Ltd* v. *Morris* [1997] 1 WLR 687 per Lord Lloyd of Berwick at 692–693.
[266]*Holland* v. *Hodgson* (1872) LR 7 CP 328 per Blackburn J at 335 approved in *Elitestone Ltd* v. *Morris* [1997] 1 WLR 687.
[267]*Holland* v. *Hodgson* (1872) LR 7 CP 328.
[268]*Appleby* v. *Myers* (1867) LR 2 CP 651 per Blackburn J at 660.

It is common for contractors to claim that they 'own' the float in their programmes so that they are entitled to an extension of time if even a non-critical activity is delayed. No one owns the float[269]. The argument is often extended to say that if a contractor programmes to complete an eight weeks contract in seven weeks, the extra week is his float and if the project is delayed at all, he is entitled to an extension of time, even though he will still finish before the completion date. That contention has no basis. The more likely analysis, we consider, is that no one owns the float. For example, if activity A has a float of five days and this float is used up as a result of the architect giving late information, there is no entitlement to an extension of time, such that if critical activity B is delayed by three days by the contractor's own default, he will be liable for three days' liquidated damages (qv), even if activity B became critical because of the delay to activity A. Conversely, if activity A was delayed by the contractor and activity B by the architect, there would be an entitlement to a three day extension of time. This does not mean the contractor in the first example could not have any claim against the employer.

Fluctuating price contract A contract in which adjustment is allowed for fluctuations in the prices of labour, materials, etc. Various degrees of fluctuation are allowed under the provisions of the standard forms. The extent to which fluctuations are allowed will have a significant effect upon the contractor's tender figure.
See also: **Firm price contract; Fixed price contract; Fluctuations.**

Fluctuations The cost to the contractor of labour and materials etc. used in the works will alter during the contract period. It may fall but, more usually, it will rise. In the absence of any provision in the contract, the contractor would have to take the risk. In order to cover himself, he would probably make an estimate of the likely rise in costs before inserting his prices in his tender (qv); higher tender figures result. It is often thought to be of overall advantage to the employer, as well as giving the contractor some guarantee of recovering his costs, to insert a clause in the contract to recover some or all of the increases if and when they occur; rather than price the risk. Most standard forms allow for this to be done by providing clauses which may be included or deleted as the parties agree. JCT 98, for example, has a selection of three clauses:
- Clause 38, which allows contribution, levy and tax fluctuations – a bare minimum provision to take account of statutory adjustment to items such as national insurance contributions.
- Clause 39, which allows labour, materials cost and tax fluctuations. The contractor can recover full fluctuations on the construction work, but not his preliminaries. This is calculated by reference to awards by the National Joint Council for the Building Industry, in the case of labour costs, and to the contractor's basic prices (qv) in respect of materials.

---

[269]*How Engineering Services* v. *Lindner Ceilings Partitions*, (first judgment) 17 May 1995, unreported; *Ascon Contracting Ltd* v. *Alfred McAlpine Construction Isle of Man Ltd* (1999) CILL 1583.

— Clause 40, which allows fluctuations in accordance with price adjustment formulae rules issued by the Joint Contract Tribunal. Details of price changes are issued monthly. There is usually provision for making part of the contract sum not subject to this formula (the non-adjustable element). With this exception, full fluctuations are recovered.

Force majeure A French law term, found in many standard contracts as a ground for granting extension of time (qv). It is used 'with reference to all circumstances independent of the will of man, and which it is not in his power to control'[270]. It is wider in its meaning than Act of God (qv) or *vis major* (qv) but in building contracts it generally has a limited and restricted meaning because such matters as war (qv), strikes (qv), fire and weather conditions are dealt with expressly.

A strike, a breakdown of machinery, supply shortages as a consequence of war, refusal of an export licence and fire caused by lightning have all been held to be within the definition of *force majeure* in varying types of contract, but not delay due to bad weather, football matches or a funeral[271].

Force majeure is referred to in JCT 98, clause 25.4.1 and 28A.1.1.1, IFC 98, clause 2.4.1 and 7.13.1 (a) and ACA 3, clause 11.5 (a) and 21 (a). It is not referred to in GC/Works/1 (1998).

Forecast tender price This is the term used in the BPF System (qv) to describe the forecast by the design leader (qv) and agreed by the client's representative (qv) of the likely cost of constructing a project. It forms part of the master cost plan (qv).

Foreseeability 'Reasonable foreseeability' is the standard generally used by the law to determine whether a defendant is liable for his actions in tort (qv), and a somewhat similar test is applied in respect of remoteness of damage (qv) in contract[272].

> 'You must take reasonable care to avoid acts or omissions which you can reasonably foresee would be likely to injure your neighbour ... persons who are so closely and directly affected by my act that I ought reasonably to have them in contemplation as being so affected when I am directing my mind to the acts or omissions which are called in question.'[273]

It is this principle on which the tort of negligence (qv) is based, but the rule is not, it seems, of universal application, e.g. in tort a person takes a victim as he is found, so that if a person injures another who subsequently dies because he reacted abnormally to the injury, the person will be liable for his death[274].

---

[270] *Lebaupin* v. *Crispin & Co* [1920] 2 KB 714 per McCardie J at 718.
[271] *Matsoukis* v. *Priestman & Co* [1915] 1 KB 681.
[272] *Hadley* v. *Baxendale* (1854) 9 Ex 341.
[273] *Donaghue* v. *Stevenson* [1932] AC 562 per Lord Atkin at 580.
[274] *Smith* v. *Leech Brain & Co Ltd* [1961] 3 All ER 1159.

In general, however, the defendant is liable only for the consequences of his act which a reasonable man could have foreseen.

In claims for breach of contract or for loss and/or expense under the standard contract forms (e.g. JCT 98, clause 26) the damages or amount recoverable are subject to the test of foreseeability set out in *Hadley* v. *Baxendale* (1854)[275] as explained in *Victoria Laundry (Windsor) Ltd* v. *Newman Industries Ltd* (1949)[276] and in *The Heron II* (1967)[277], i.e. damages are recoverable in respect of losses which the contracting parties might reasonably contemplate at the time the contract was made, as a not unlikely consequence of the breach or event relied on.

See also: **Injury; Negligence; Remoteness of damage.**

Forfeiture The loss of some right or property as a result of specified conduct, but in building contracts usually referring to the employer's right to determine the contractor's employment or seize plant and materials etc.

See: **Forfeiture clause.**

Forfeiture clause A clause in a building contract which gives one party, usually the employer, the right to determine the contractor's employment, turn the contractor off site etc. In standard form building contracts it is usually referred to as 'determination of employment' or 'termination'. In this sense JCT 98, clause 27, ACA 3, clause 21 and GC/Works/1 (1998), clause 56 are forfeiture clauses. Forfeiture clauses are strictly interpreted by the courts and any prescribed procedure must be followed. Wrongful forfeiture or determination will normally amount to a repudiation of the contract by the employer, but it has been held not to amount to repudiation if a party honestly, albeit mistakenly, relies on a contract provision[278].

See also: **Determination.**

Formal contract An alternative description of a contract made by deed or specialty (qv). Sometimes the expression is used to describe simple contracts (qv) which are entered into in a formal way, e.g. in a standard printed form, duly signed by the parties.

Formalities of contract In general, there are no formalities attached to the making of a contract. A contract (qv) may be made orally, in writing, or even implied from conduct. In some cases, however, the law requires the presence of additional formalities before a contract can be enforced. Some contracts must be made by deed (qv); others must be in writing and in a few cases there must be written evidence of the contract. If these formalities are not complied with the contract is unenforceable by legal action. An assignment of copyright (qv)

---

[275](1854) 9 Ex 341.
[276][1949] 1 All ER 997.
[277]*Koufos* v. *C. Czarnikow Ltd* [1969] 1 AC 350.
[278]*Woodar Investment Development Ltd* v. *Wimpey Construction UK Ltd* [1980] 1 All ER 571.

must be in writing, otherwise it is void, as must a bill of exchange, e.g. a cheque. Contracts of guarantee (qv) must also be in writing, in contrast to contracts of indemnity (qv) which need not. Hire-purchase contracts must also be in writing. It should be noted in particular that if Part II of the Housing Grants, Construction and Regeneration Act 1996 (qv) is to apply to a construction contract (qv) there must be an 'agreement in writing' under s. 107. Section 107 (6) makes clear that references to anything being written or in writing include 'its being recorded by any means'. In the Act, the term is given its widest possible meaning.

Formerly contracts relating to the sale or disposition of any interest in land were enforceable at law only to the extent that they were in or evidenced in writing[279]. In equity it was possible to enforce such a contract where there had been part-performance[280]. However, since 26 September 1989, such contracts can only be made in writing and only by incorporating all the terms of the contract[281]. It appears, therefore, the equitable doctrine of part-performance no longer exists[282], although it may be possible to enforce an otherwise void contract by means of a constructive trust[283].

Formula price adjustment See: **Fluctuations**.

Formulae When making application for the head office overheads (qv) part of loss and/or expense under standard form building contracts, contractors often base their claim on a formula. The courts have never given global approval to the use of formulae in this way although they have accepted the use of formulae in certain cases which generally were decided on their own facts. Indeed, the courts have tended to disapprove formulae unless as a last resort or the parties have agreed their use[284]. Actual costs are normally required.

Claims for fixed head office overheads are essentially claims for lost opportunity to contribute to those overheads, because the overheads do not actually change or, if they do, the amount of any extra overheads directly resulting from the delay can be claimed separately. Formulae assume a healthy construction industry and a contractor with finite resources with the result that if he is delayed on a project, he will be deprived of the chance to undertake other work. Where the industry is sluggish or where the contractor is so large that turning away work just does not arise, the contractor will face difficult problems in showing the lost opportunity[285]. There are several formulae in common use.

See also: **Emden formula; Eichleay formula; Hudson formula.**

---

[279]Section 40 of the Law of Property Act 1925.
[280]This equitable doctrine was preserved by ss. 40 (2) and 55 (d) of the Law of Property Act 1925.
[281]Section 2 of the Law of Property (Miscellaneous Provisions) Act 1989.
[282]*Yaxley* v. *Gotts* [2000] Ch 162 per Robert Walker LJ at 172.
[283]Under s. 2 (5) of the Law of Property (Miscellaneous Provisions) Act 1989: *Yaxley* v. *Gotts* [2000] Ch 162.
[284]*Alfred McAlpine Homes North Ltd* v. *Property and Land Contractors Ltd* (1995) 76 BLR 65.
[285]*AMEC Building Ltd* v. *Cadmus Investments Co Ltd* (1997) 13 Const LJ 50.

Forthwith As soon as is reasonable[286] or 'without delay or loss of time'[287]. The word is used in most forms of building contract to convey the fact that the action required must not be delayed. For example, in JCT 98, clause 4.1.1: 'The Contractor shall forthwith comply with all instructions ...'; ACA 3, clause 20.1: '... the Employer may by further notice ... forthwith terminate ...'; MW 98, clause 3.5: '... the Contractor shall forthwith carry out ...'. It is sometimes mistakenly taken to mean 'immediately' (qv) or 'instantly'.

Fossils 'A relic or representation of a plant or animal that existed in a past geological age, occurring in the form of mineralized bones, shells, etc': *The New Collins Concise English Dictionary*. In the absence of a special clause in the building contract the employer is entitled to fossils found under or fixed in any way to his land, but the legal position is unclear as to who has the right to fossils found lying on the surface.

The standard form building contracts usually contain an express clause covering the position. JCT 98, clause 34, provides that 'all fossils, antiquities and other objects of interest or value' found on the site or during excavation are the property of the employer. The contractor must try not to disturb the object, ceasing work if necessary, and inform the architect or clerk of works. The contractor may claim any direct loss and/or expense caused to him by compliance with this provision and may also be entitled to an extension of time.

GC/Works/1, clause 32 (3) provides to similar effect and any instruction issued by the project manager under clause 32 (4) could give rise to similar claims. ACA 3, clause 14 is to the same effect.

See also: **Antiquities.**

Foundations Broadly, anything which supports something else. In building work, the term is generally used to describe the lowest artificial works placed in contact with the natural ground to support a structure, e.g. piles, concrete rafts, concrete strip footings etc. More rarely, it is applied to the ground itself. In *Worlock* v. *SAWS & Rushmoor Borough Council* (1982)[288], the question whether a floor slab which supported internal partition walls of a building was a foundation for the purposes of the then current building regulations was considered. The court held that it was. A foundation is 'an object which is placed in position on or in the ground in the course of constructing a building, or for the purposes of a building which is to be constructed, the function of which is to provide support for that building so that in fact it transmits load to the material beneath ...'. This definition has been approved in several subsequent cases.

GC/Works/1 (1998), clause 16 stipulates that the contractor must not lay foundations until the PM has examined and approved the excavations. This clause simply clarifies what is normal practice on most building contracts.

---

[286]*London Borough of Hillingdon* v. *Cutler* [1967] 2 All ER 361.
[287]*Roberts* v. *Brett* (1865) 11 HLC 337.
[288](1981) 20 BLR 94 per Woolf J at 112 (the first instance decision upheld on this point).

Fraud Fraud is deliberate deception and is a type of tort known as deceit. It is one of the torts affecting business relationships. Usually it takes the form of fraudulent misrepresentation (qv) which was defined as a 'false statement, or one which (the maker) did not believe to be true, or was recklessly careless whether what he stated was true or false'[289]. The motive behind the fraud in this context, whether dishonest or not, is irrelevant. Someone who is induced to enter into a contract by a fraudulent misrepresentation may repudiate their obligations under the contract and also recover damages (qv). Alternatively, they can affirm the contract, and still recover damages for deceit. It should be noted that the fraudulent misrepresentation must be one of the inducing causes of the contract. It is not possible to contract out of liability for fraudulent misrepresentation[290].

Section 32 of the Limitation Act 1980 (formerly Limitation Act 1939, s. 26), which is concerned with 'deliberate concealment' of, *inter alia*, defective construction work so as to postpone the start of the limitation period, does not require fraud in the sense of moral turpitude. All that the claimant needs to show is that the contractor has knowingly done bad work which is not of a trivial kind and which he has covered up, so that the bad work is not likely to be detected[291].

See also: **Rescission**.

Fraudulent misrepresentation A false statement of fact which the maker knows to be false or is reckless as to the truth of it. The absence of 'honest belief' is essential. If a fraudulent misrepresentation induces one party to enter into a contract, on discovering the fraud he can void the contract and treat it as at an end. The fraudulent misrepresentation must, of course, have affected his initial decision to enter into this contract[292]. Alternatively, he can affirm the contract and go ahead. In either case he can recover damages for the tort of deceit. A contracting party cannot escape liability for fraudulent statements made by him or on his behalf by putting an exclusion clause in the contract[293].

See also: **Fraud; Misrepresentation**.

Freezing injunction A freezing injunction[294] was formerly called a *Mareva* injunction. It is an order of the court whereby a party is prohibited from disposing of his assets within the jurisdiction (qv). Exceptionally, the court may make an order prohibiting disposal of assets outside of the jurisdiction[295]. The applicant must have a 'good arguable case' on the substantive merits of his case and have cogent evidence that there is a real risk of disposal of assets. Usually, such orders are made against specified assets within the jurisdiction, e.g. bank

---

[289] *Derry* v. *Peek* (1889) 14 App Cas 337.
[290] *S. Pearson & Son Ltd* v. *Dublin Corporation* [1907] AC 351.
[291] *Kijowski* v. *New Capital Properties Ltd* (1989) 15 Con LR 1.
[292] *Convent Hospital Ltd* v. *Eberlin & Partners* (1988) 14 Con LR 1.
[293] *S. Pearson & Son* v. *Dublin Corporation* [1907] AC 351.
[294] CPR Rule 25.1 (1) (f).
[295] *Derby* v. *Weldon (No. 2)* [1989] 1 All ER 1002.

accounts, but the court may order disclosure of details of assets, where appropriate.

Frontager Someone who owns or occupies land which abuts a highway (qv), river or seashore. The Highways Act 1980 contains procedures whereby private streets, as defined in the Act, can be made-up at the expense of the frontagers and formally adopted by the highway authority so that for the future the highway (qv) becomes maintainable at the public expense.
See also: **Boundaries.**

Frost damage Most standard forms limit the contractor's liability to make good damage caused by frost. He is not required to make good such damage if it was caused by frost occurring after practical completion (qv). JCT 98 limits the contractor's liability in clauses 17.2 and 17.3. Clause 17.5 states emphatically that the contractor may only be required to make good frost damage which may appear after practical completion if the architect specifically certifies that such damage was due to injury which took place before practical completion.

ACA 3 does not mention frost damage specifically. In clause 12.2, it refers only to 'defects shrinkages or other faults which may appear during the Maintenance Period'. But 'other faults' must be interpreted *ejusdem generis* (qv) to mean other faults like defects or shrinkages (qv). Damage by frost before the works are fit and ready for taking-over will clearly create a defect because the contractor has an obligation to protect the works under clause 1.2. His obligation in this respect must cease after taking-over when he is no longer in possession of the works. Frost damage after taking-over becomes the responsibility of the employer.

GC/Works/1 (1998) provides (clause 40 (2) (g)) for the PM to instruct the contractor to suspend the execution of the works and he might well do this on the ground that frost damage may result from continuation. The contractor retains his general obligations in respect of the works during and after such suspension. Clause 21 does not expressly limit the contractor's liability to make good frost damage during the maintenance period, but clause 21 (2) provides for him to be reimbursed by the employer for the costs he has incurred in remedying defects to the extent that the employer is satisfied that any defects were not caused by his neglect or default or by circumstances within his control. Hence, if the contractor is instructed to remedy defects caused by frost he would in appropriate circumstances be entitled to payment.

MW 98 provides, in clause 2.5, that the contractor is liable to make good defects etc. caused by frost occurring before practical completion. He is not liable for the effects of frost after practical completion.

Frustration The release from contractual obligations of the parties to a contract which, as a result of events completely outside the control of the contracting parties, is rendered fundamentally different from that contemplated by the parties at the time the contract was made. It is not sufficient that the contract has turned out more difficult and expensive for one party to perform than he expected[296]. There are very few cases in which a building contract has been held to be frustrated, although it is sometimes put forward as an excuse for non-completion. The position was aptly summarised as follows:

'Frustration occurs whenever the law recognises that without default of either party a contractual obligation has become incapable of being performed because the circumstances in which performance is called for would render it a thing radically different from that which was undertaken by the contract.'[297]

This is a question of law which must depend not only on the event relied on but also on the precise terms of the contract.

In *Wong Lai Ying* v. *Chinachem Investment Co Ltd* (1979)[298] a massive landslip took with it a thirteen-storey block of flats, the debris from which, together with many tons of earth, landed on a building site. The landslip was held to be a frustrating event as it made further performance uncertain. The character and duration of any further performance would be radically different from that contemplated by the original contract. The landslip was an unforeseen natural disaster and a clause in the contract referring in general terms to what was to happen 'should any unforeseen circumstances beyond the vendor's control arise' could not be interpreted so as to cover the landslip.

A building contract may be frustrated if Government order prohibits or restricts the work[299] and the total destruction of premises by fire has been held to frustrate an installation contract[300]. Extreme delay through circumstances outside the control of the parties may frustrate a building contract, but only if the delay is of a character entirely different from anything contemplated by the contract.

Where a contract is discharged by frustration, both parties are excused from further performance and the position is governed by the Law Reform (Frustrated Contracts) Act 1943. Money paid under the contract is recoverable, but if the party to whom sums were paid or payable has incurred expenses, or has acquired a valuable benefit, the court has a discretion as to what should be paid or be recoverable. The various standard form contracts often make provision for what is to happen should certain events occur, and in principle those express provisions prevail. JCT 98, clause 28 entitles the contractor to determine employment under the contract for certain matters, some of which would be capable of being frustrating events, provided the works are suspended for a stated period as a result. ACA 3, clause 2.1 is a special clause dealing with termination resulting from causes outside the control of the parties, and includes frustrating events such as *force majeure*, war and allied events.

See also: **Discharge of contract; Illness.**

---

[296]*Davis Contractors Ltd* v. *Fareham UDC* [1956] 2 All ER 148.
[297]*Davis Contractors Ltd* v. *Fareham UDC* [1956] 2 All ER 148 per Lord Radcliffe at 729.
[298](1979) 13 BLR 81.
[299]*Metropolitan Water Board* v. *Dick, Kerr & Co Ltd* [1918] AC 119.
[300]*Appleby* v. *Myers* (1867) LR 2 CP 651.

Functus officio Having discharged his duty or performed his function. The term
is used of an architect who has discharged his duties under a building contract
and has exhausted his authority. In *H. Fairweather Ltd* v. *Asden Securities Ltd*
(1979)[301] it was held, under JCT 63 terms, that once the architect had issued the
final certificate (qv) under the contract then, if no notice of arbitration had
been given under the contract conditions, the architect was thereupon *functus
officio*, with the result that he could not thereafter issue any valid certificate
under the contract. An architect who attempted to issue a certificate of non-
completion after issuing the final certificate was also held to have no power to
do so[302].

The phrase is also used of an arbitrator who makes a valid award. His
authority as arbitrator then comes to an end and with it his powers and duties.
See also: **Arbitrator; Certificates.**

Fundamental term An expression used to describe a term in a contract, breach
of which entitles the innocent party to treat their obligations under the contract
as discharged. It is a vitally important term going to the very basis of the
contract. The expression is sometimes used in respect of a contract term,
breach of which cannot be avoided by an exemption clause (qv).

The phrase 'fundamental breach of contract' is sometimes used to mean the
same as 'breach of a fundamental term'. It has two different senses:
— A breach of contract so serious that the other party may treat his future
  obligations under the contract as at an end, i.e. there has been a
  repudiation of the contract.
— A so-called principle of law that some breaches of contract are so
  destructive of the parties' obligations that liability for such a breach
  cannot be limited by an exemption clause. Case law states that there is
  no such principle of law; it is merely a rule of interpretation based on
  the presumed intention of the contracting parties[303].
See also: **Condition; Express term; Implied term; Repudiation.**

---

[301](1979) 12 BLR 40.
[302]*A. Bell & Son (Paddington) Ltd* v. *CBF Residential Care and Housing Association* (1990) 46 BLR 102.
[303]*UGS Finance Ltd* v. *National Mortgage Bank of Greece* [1964] 1 Lloyds Rep 438; *Suisse Atlantique
Société d'Armement Maritime SA* v. *NV Rotterdamsche Kolen Centrale* [1966] 2 All ER 61; *Photo
Production Ltd* v. *Securicor Transport Ltd* [1980] AC 827.

Garnishee order Where a creditor obtains a judgment or order for the payment of a sum (presently being in excess of £50) from a debtor ('the judgment debtor'), any person within England and Wales who owes money to the judgment debtor may, subject to certain provisions of the Civil Procedure Rules (qv), be ordered by the court to pay to the creditor, instead, enough of that amount owed to the debtor as would be sufficient to satisfy the judgment made in favour of the creditor[304]. Commonly, a garnishee order will be made in respect of a credit balance at the judgment debtor's bank[305].

By way of example: A has obtained a judgment in the sum of £5000 against B. B will not pay although B has £10 000 standing to his credit at the bank. A garnishee order can be made requiring the bank to pay £5000 from B's account directly to A. Alternatively, the garnishee order could be made requiring any third party, owing £5000 to B, to pay it directly to A.

A garnishee order cannot be made unless there is a legal debt currently owing to the judgment debtor. Under the JCT and most other standard form building contracts, payments to the contractor are not existing debts until the architect's certificate has been issued (see: JCT 98, clause 30.1.1). A garnishee order made before the issue of the architect's certificate would be invalid, because there would be no debt to be garnisheed[306].

GC/Works/Contracts The abbreviated reference given to a suite of contracts prepared by the Property Advisors to the Civil Estates (PACE) and published under the full titles of General Conditions of Contract for:
— Building and Civil Engineering Major Works; abbreviated to GC/Works/1.
— Building and Civil Engineering Minor Works; abbreviated to GC/Works/2.
— Mechanical and Electrical Works (of any value); abbreviated to GC/Works/3.
— Building and Civil Engineering, Mechanical and Electrical Small Works; abbreviated to GC/Works/4.
— Appointment of Consultants; abbreviated to GC/Works/5.
— Standard Form of Daywork Contract; abbreviated to GC/Works/6.
— Measured Term Contracts Based on Schedules of Rates; abbreviated to GC/Works/7.
— Specialist Term Contracts for use where specified maintenance of equipment is required and is capable of being priced on a task by task basis; abbreviated to GC/Works/8.

---

[304]Civil Procedure Rules Schedules 1 and 2 and RSC Order 49 Rule 1 (1) and CCR Order 30 Rule 1 (1) .
[305]*Rogers* v. *Whitely* [1892] AC 118.
[306]*Dunlop & Ranken Ltd* v. *Hendall Steel Structures Ltd* (1957) AC 79.

— Lump Sum Maintenance Contracts for Maintenance and Repair of Fixed Mechanical and Electrical Plant, Equipment and Installations; abbreviated to GC/Works/9.
— Standard Form of Facilities Management Contract; GC/Works/10.
— GC/Works/11 (2000) was published on 31 January 2001 and is the Minor Works Term Contract, General Conditions, Model Forms and Commentary.

Both the Major Works (GC/Works/1) and Consultants Appointment (GC/Works/5) Editions are further sub-divided into 'Parts' so that, in the case of GC/Works/1, versions are provided for use:

— With Bills of Quantities (GC/Works/1 Part 1).
— Without Bills of Quantities (GC/Works/1 Part 2).
— Single Stage Design and Build (GC/Works/1 Part 3).
— Two Stage Design and Build (GC/Works/1 Part 5).

A fourth volume (GC/Works/1 Part 4), drafted as a companion to the General Conditions, GC/Works/1, provides a brief comparative analysis of the material changes made between this (1998) With Quantities Edition and the previous GC/Works/1 (Edition 3), a useful clause-by-clause commentary on the Conditions and a variety of Model Forms designed for use in compiling and administering whichever version of the General Conditions has been adopted by the parties. Model documents are available for use in relation to:

Documents collateral to the contract provisions for:
— Insurance documents
— Performance bond
— Parent company performance guarantee
— Retention payment bond
— Mobilisation payment bond
— Sub-contractor's collateral warranty
— Parent company guarantee of sub-contractor's collateral warranty
— Adjudicator's appointment;

Administrative documents providing pro forma:
— Notice of delegation
— Notice of possession
— Certificate of completion
— Maintenance certificate
— Project manager's instruction
— Interim payment certificate
— Final payment certificate
— Notice of intention to withhold payment
— Notice of non-compliance with instruction
— Employer's warning notice
— Employer's notice of intention to refer to adjudication
— Employer's notice of referral to adjudication;

Stage payment chart, etc.:
— Example stage payment chart
— Chart banding calculation sheet.

An important note to the commentary reminds prospective users of the form(s) that 'Legal Advice should be taken if it is proposed to amend any of the General Conditions'.

Consultants Appointment (GC/Works/5) Edition is further sub-divided into 'Parts' so that versions are provided for:

— Appointment of Consultants (GC/Works/5 Part 1).

— Appointment of Consultants — Term Contracts (GC/Works/5 Part 2).

A further wide range of sub-contracts for use with the more commonly used main contract forms is also available.

General damages Monetary compensation payable to a claimant (qv) by a defendant (qv) as a consequence of the defendant having infringed a legal right of the claimant. Distinguished from special damages (qv) mainly for procedural purposes; general damages need not be specifically pleaded. They are awarded to compensate a claimant for such injuries, losses, costs, expenses and/or other damages as the law presumes to result from the breach of right or duty.

As a rule the measure of general damages for breach of any kind of contract is that the aggrieved party should recover such part of the damage actually caused by the breach as the defaulting party should reasonably have contemplated would flow from the breach[307]. In the circumstances, a claimant must prove damage, but need not quantify precise items within it.

Damages for breach of contract must reflect, as accurately as the circumstances allow, the loss which the claimant has sustained so that, so far as possible and so far as money is capable of doing so, the injured party is placed in as good a position as he would have been had the contract been performed[308]. Their purpose is not that of a punishment. In the contractual context, and in particular in the context of building contracts, the courts will generally assess the measure of loss or damage in purely economic terms. In cases of defective building work the normal approach will therefore be to assess the measure of damages by reference to the cost of reinstatement, provided that reinstatement is not only necessary to make the works conform but also that to undertake such reinstatement is, in all the circumstances, a reasonable action to take. In appropriate circumstances, where, for example, the objective of the building works involves a particular element of personal preference and where the cost of reinstatement would be entirely unreasonable, the courts may, instead (or in addition)[309], measure the claimant's loss by assessing and awarding a sum based on such matters as, diminution in value, loss of amenity, inconvenience, aesthetic dissatisfaction, unhappiness, frustration, disappointment and the like[310].

Where the extent of the injury or loss cannot be precisely ascertained the claimant will nevertheless be entitled to an amount which the court (or other tribunal) will assess.

See also: **Damages; Nominal; Special damage(s); Liquidated damages.**

---

[307] *Koufos* v. *Czarnikow Ltd* [1969] 1 AC 350.
[308] *Robinson* v. *Harman* (1848) 1 Ex 850.
[309] *Hannant & Curran* v. *Harrison* (t/a Grafton Builders & Roofers) [1995] 2 CL 157.
[310] *Ruxley Electronics and Construction Limited* v. *Forsyth* (1995) 73 BLR 1.

Gift A voluntary and gratuitous transfer of ownership of any property from one person who is in lawful possession of that property to another and where the giver of the property intends that the transfer of ownership should be permanent[311].

Most building contracts whether Local Authority or Private Editions now make express provision entitling the employer to determine the contractor's employment under that, and in the case of the JCT contracts any other contract where the contractor is found to have offered, given or agreed to give any person, '*any*' gift. Under MW 98, clause 5.5 the employer is entitled to 'cancel' the contract. In practice, it probably amounts to the same thing. Additional to any civil right or proceedings that may arise, the giving or offering of gifts etc. for illegal purposes or to promote some illegal act may also give rise to separate criminal proceedings[312].

Giving notice Express contractual and/or statutory provisions (such as those found in JCT 98 clauses 1.7 and 30.9.1; MW 98 clauses 1.5, 1.6 and 7.2; ICE 6th Edition Amended 1998, clauses 1 (6), 60 and 66; and s. 115 of the Housing Grants, Construction and Regeneration Act 1996) specify the method and timing by which parties must make known their intention to claim for additional payment, rights of set-off or extensions of time or to initiate dispute resolution procedures and/or to exercise some other right or fulfil some express obligation under the contract. The giving of such notice is often made a condition precedent (qv) to the further entitlement and it is well established that where the contract or statute concerned specifies precise methods and timing for such notices a failure to meet all of those requirements precisely is likely to be fatal to the right or claim to which the notice relates[313].

See also: **Service of notices, etc.**

Good faith A concept whereby a contracting party has imposed on him an implied obligation not to act intentionally in a manner likely to cause the other to be deprived of a benefit or benefits that the other party would otherwise have obtained under the contract. The doctrine applies most commonly in contracts of special relationship such as contracts of employment, contracts of insurance (i.e. contracts '*uberrimae fidei*' (qv)), professional engagements (e.g. solicitor and client, principal and agent and the like), where an increased burden of confidentiality or disclosure would normally exist[314].

In employment contracts an act which brings an employee's own interest into conflict with his duty to his employer will amount to a breach of a duty of good faith[314] so that, for example, an architect employee who uses his position

---

[311] Halsbury's *Laws of England*, 4th edn, vol. 20, Butterworths.
[312] *Rashleigh Phipps Electrical Ltd* v. *London Regional Transport Executive* (1985) 11 Con LR 66.
[313] *Pearce and High* v. *Mr & Mrs Baxter* (1999) BLR 101, CA (effect of notification provisions during defects liability periods in JCT Minor Works) and *Cambs Construction Ltd* v. *Nottingham Consultants* (1996) 13-CLD-03-19.
[314] *Boston Deep Sea Fishing & Ice Co* v. *Ansell* (1888) Ch D 339.

with his employer's company to obtain and carry out private work on his own account for clients who would otherwise be engaging his services through his employers and who by doing so is effectively competing with his employer[315] will breach his duty to faithfully serve (i.e. his duty of good faith to) his employer. Similarly, an employee, agent or other person bound to act in good faith will breach that duty where he/she misuses confidential information or impedes the work of his employer.

The doctrine of good faith should be distinguished from the implication of terms requiring parties to co-operate where co-operation by one party is essential to facilitate performance by the other. Otherwise than in contracts giving rise to special relationships such as those described above or where the duty of good faith is either expressly made a term of the contract or implied into the contract (such as in consumer contracts to which the Unfair Terms in Consumer Contracts Regulations 1994[316] refer), English law has yet to accept the implication of this concept generally into commercial contracts[317].

See also: **Confidentiality; Misrepresentation; Uberrimae fidei.**

Goods A word used in many standard form contracts (as, for example, in JCT 98, clause 8.1, where 'materials, goods ... shall be of the respective kinds and standards specified ...'), to expand references to materials to ensure the clause concerned encompasses all tangible, movable, property (excluding land and money). 'Goods' may be taken in the wider sense than materials and connotes composite and/or manufactured items comprising a variety of 'materials' (qv). Although not appearing in earlier editions of ICE contracts, the word 'goods' was added with publication of the 5th edition of that contract in 1973 and has been included both in subsequent reprints of that edition and in the later 6th edition, where it was expressly incorporated to cover all tangible movable property.

In the context of the Sale of Goods Act 1979 and Supply of Goods and Services Acts 1982 all items for incorporation into a building project may be classified as goods. Despite a tendency for numerous construction-related European Directives to use terms such as 'construction products' and 'construction materials' such 'products' and 'materials' will no doubt likewise fall under the wider general category of 'goods'.

Government action Power exercised by the UK Government acting through one or more of its authorised departments and acting in pursuance of an express statutory power. Where the exercise of such powers affects a contractor's ability to meet the completion date for building works carried out under some of the more commonly used standard form contracts (such as the JCT and

---

[315]*Hivac Ltd* v. *Park Royal Scientific Instruments Ltd* (1946) 1 All ER 350, CA.
[316]SI 1994 No 3159.
[317]For a useful overview of the topic see the judgment of Vinelot J in *London Borough of Merton* v. *Stanley Hugh Leach* (1985) 32 BLR 51.

ACA forms), then the contract may expressly provide for the granting of an appropriate extension of time for completion.

Any such entitlement will, however, be limited to and will depend on the express terms of the contract so that, for example, under the JCT 98 standard forms (clause 25.4.9), any delay caused must be the consequence of the positive exercise of the relevant power and not some failure or delay in exercising such power. Under JCT contracts the entitlement will also be dependant on the exercise of the power:

— occurring after the 'Base Date' (qv);
— being the direct cause of restrictions on the 'availability' (qv) or 'use' of 'essential' (qv) labour; and/or
— being the direct cause of delaying or entirely precluding the contractor's ability to 'secure' essential 'goods' (qv), 'materials' (qv), 'fuel' and/or 'energy.' affecting execution of the 'Works' (qv).

Where contracts containing such express provisions are used outside the UK, express references to the UK Government should be suitably amended. It also remains to be seen whether, and if so to what extent, the exercise of powers by any devolved Scottish or Welsh Parliament or any EC Regulation or Directive may automatically fall within the scope of such clauses without express amendment.

Without the benefit of express entitlement under the contract, government action probably in any event falls under *force majeure* (qv).

Government contracts See: **GC/Works/Contracts.**

Gross valuation Subject to any contrary express agreement between the parties or to provisions requiring interim payments to be made by reference to estimates of the value of work done, by way of stage payments or by reference to a 'Priced Activity Schedule' (qv), the gross valuation for purposes of interim and final payment(s) under building contracts will generally comprise the total value – before retention and/or discounts – of any amount which the contract expressly provides shall be ascertained for the purposes of addition to, deduction from or adjustment of the contract sum (qv). Under JCT contracts such gross valuations will include sums in respect of:

— All properly executed work, including that of nominated sub-contractors, and work resulting from variations (qv).
— The total value (qv) of all adequately protected, permanent, but as yet unincorporated, materials and goods (qv), properly brought to site.
— If so agreed by the employer at the time the contract is made and subject to the contractor satisfying any other express conditions of such payment, the total value of any 'Listed items' as have been pre-fabricated but not yet delivered to the site.
— Amounts due in respect of loss and expense, as and when ascertained.

JCT contracts expressly provide that payment for unincorporated materials is dependent upon them not being delivered to site prematurely. Not all contracts make such express qualifications and without such a provision it is open to a contractor to argue an entitlement to be paid for any and all materials brought to site irrespective of timing.

Notably, under the JCT 98 contract the gross valuation for interim purposes makes no express provision for inclusion in the gross valuation of sums in respect of variations for which the contractor has given and the architect has confirmed acceptance of a quotation under clause 13A (contrast JCT 98, clauses 30.2.1.1 and 30.6.2). However, this omission will probably be of no real effect given the terms of clause 3.

Ground investigation
Examination (e.g. by means of visual inspection, trial pits, rotary drilled or light cable percussion boreholes, soils sampling and laboratory analysis and/or instrument testing), for the purposes of determining the nature of the ground, its bearing capacity, the extent, if any, of existing contamination, gross instability, incidence of old mine workings, underground obstructions or past or future likelihood of subsidence etc. It is fundamental to the successful and safe design and construction of all building and civil engineering works that proper examination of the site is done. The extent of any initial ground investigation that must be undertaken will depend on many factors, including the nature of the particular project being undertaken, the obvious likelihood of any prior contamination of the land from previous use, any peculiar geological features, any likelihood of interference from underground services or natural underground phenomena etc. In the context of projects such as opencast mining, major civil engineering or tunnelling works and the like, ground investigations will clearly need to be highly sophisticated and undertaken by specialists.

Where the contract documents include bills of quantities and the contractor is required under the contract to carry out ground investigations the nature and extent of the investigations to be undertaken will generally be described according to the rules laid down in the relevant standard method of measurement applicable to those bills. Under the JCT contracts the General Rules (clause 11) of Standard Method of Measurement of Building Works (SMM) 7th edition will apply.

Where the investigation works are likely to be significant and of a particularly specialist nature other specialist standard forms of contract and specifications are published specifically for use in connection with ground investigation works (e.g. the ICE Conditions of Contract for Ground Investigation 1993 Edition with Amendments and the Specification for Ground Investigation 1993 Edition).

Generally, if an architect or engineer fails to exercise reasonable precautions in carrying out, either by himself or through the main contractor or other independent specialist contractors, an examination of the site of the works, he may be liable to the employer for losses incurred if, as a consequence, the resulting design is impractical or unsafe. A positive obligation to make a reasonable investigation of existing ground conditions will also arise under the Construction (Design and Management) Regulations 1994 (qv). In drawing up the health and safety plan (qv) for the purposes of Regulation 15 (1)–(3), it will be essential to know and record details of the existing environment on, under and surrounding the site of the works. That will almost invariably involve the employer and/or his designers in making reasonable and appropriate investigations of the ground conditions. So, too, will Regulation 13 which imposes on every designer of a qualifying project an obligation to take all reasonably practicable steps in his design to:

— Avoid foreseeable risks to the health and safety of construction workers on the project.

— Combat, at source, risks to the health and safety of construction workers on the project.

— Ensure that the design includes adequate information about any aspect of the project or structure or materials (including substances) which might affect the health or safety of construction workers.

Ground water Water standing at or below the level at which saturation point of the soil or rocks is reached. Ground water level may not necessarily remain constant and may be periodically affected by, for example, tidal or similar effects. Under building contracts where bills of quantities are made a contract document, unless otherwise expressly stated, particulars of the ground water level (the 'pre-contract water level') and the date when it was established should be given. When excavations are carried out the 'post-contract level' should then be re-established. Ground water levels subject to known periodic change caused by tidal or similar influences should also be specifically given, as should details of the mean high and low levels to be expected.

No such express requirements apply where the measured works are of a civil engineering nature, carried out under the ICE Conditions of Contract.

Guarantee There is some overlap in the use of the words 'guarantee' and 'warranty' (qv), e.g. in relation to warranties given on new vehicles. There is an increasing distinction in the use of these terms.

A guarantee is a written[318] contract in which the surety takes on a secondary liability to the beneficiary in respect of the primary obligations of another party[319]. It is not an indemnity[320].

In a building context, the employer is often the beneficiary of a parent or other company guarantee of the contractor performing its obligations. The primary obligations remain in force so that the contractor remains liable to the employer, irrespective of the existence of the guarantee[321], although a surety is generally entitled to be indemnified by the defaulting party, unless no demand has been made of the defaulting party and the surety had no necessity to discharge the debt[322].

---

[318]Section 4 of the Statute of Frauds 1677.
[319]*Lakeman* v. *Mountstephen* (1874) LR 7 HL 17.
[320]For an analysis of the difference between guarantees and indemnities, see *General Surety and Guarantee Co Ltd* v. *Francis Parker Ltd* (1977) 6 BLR 16 per Donaldson J at 21.
[321]*Yeoman Credit Ltd* v. *Latter* [1961] 2 All ER 294, 296.
[322]*Owen* v. *Tate* [1976] 1 QB 402, CA.

Where a guarantee underwrites performance of a contract which itself contains an arbitration clause, the surety and beneficiary should refer their disputes to arbitration so long as there is sufficient incorporation of the substantive arbitration clause by reference to that underlying contract[323].
See: **Arbitration; Bonds; Indemnity contract.**

Guaranteed cost contract A contract under which the employer agrees to pay the contractor his costs of labour, materials and overheads plus an additional sum of money which may be calculated in various ways. An example of such a contract is the JCT Standard Form of Prime Cost Contract, first published in 1967, substantially revised in 1992 and again in 1998. It provides for the employer to audit and pay the contractor's prime cost of executing the works plus a fee which may be fixed or on a percentage basis at the employer's choice. Practice and guidance notes on the form are also published.
See: **Cost reimbursement contract.**

Guaranteed maximum price (GMP) contract A generic term used to describe contracts aimed at placing the onus on the contractor to seek cost saving solutions and encouraging a value engineering approach to the construction process. Like traditional design and build contracts, such as WCD 98, GMP contracts offer the employer considerable certainty over what will be the maximum final cost of the project, with the contractor's entitlement to payment being capped at the maximum guaranteed (or target) price, subject only to any legitimate increases resulting from authorised changes to the employer's outline requirements or basic design criteria that the contract provides may be instructed and paid for under the contract by the employer. GMP contracts also generally make provision for the distribution of any additional profit achieved through the contractor's cost saving measures, the benefit of that additional profit often being shared in a pre-agreed ratio between the employer and contractor and also, perhaps, with the design team.
It is important to note that, despite the rather misleading title, the price under such contracts is neither guaranteed nor maximum. On the one hand, if used effectively the objective is for the contractor to create savings so that the cost to the employer falls short of the ceiling price, with the contractor and employer then sharing the benefit of the saving whilst, on the other hand, the employer must ensure that the contract is capable of upward adjustment to reflect any extra work that the contractor is properly instructed to undertake. Without retaining such provisions for additional payment any request for extra work may otherwise have the effect of setting up separate obligations and entitlement to payment.
The key to a successful and dispute-free GMP contract depends largely on the skill of those responsible for drawing up the original specification or outline design brief. A fine balance must be struck between providing the contractor

---

[323] *Roche Products Ltd* v. *Freeman Process Systems*; *Black Country Development Corpn* v. *Keir Construction Ltd* (1996) CILL 1171.

with design parameters that are sufficiently tightly drawn to ensure the employer's requirements are met without later argument over what is expected to be part of the intended scope of his work, whilst at the same time leaving the contractor with a proper degree of risk and responsibility for completing the work within the ceiling price and where possible achieving the maximum possible cost savings.

Handover A commonly misused term denoting the stage at which work is complete and the contractor hands over to the employer the keys and any useful documentation and the employer takes over the building. In the standard forms, the stage is variously described as 'practical completion', 'taking-over' and 'completion'. 'Handover' is slightly misleading, suggesting that the contractor hands over the building when he considers that it is ready, but the term continues in popular use probably because it is easier to refer to a 'handover meeting' than a 'meeting to carry out an inspection prior to the issue of a certificate of practical completion'.
See also: **Completion; Practical completion; Taking-over.**

Head contract A term commonly used to describe a contract under which the main contractor or perhaps a developer undertakes the entire obligation for construction of the project for which he then engages others under other sub-contracts to carry certain aspects of the construction process. Such sub-contracts will then be let either on terms of a standard form specifically designed for use in conjunction with the head contract (e.g. DOM/1 and DOM/2 (qvv)) standard forms for use with JCT 98 and WCD 98 contracts) or will be specifically drafted by the parties in terms that ensure the sub-contracts and head contract will operate 'back to back'.

Health and safety file Documented information required by the Construction (Design and Management) (CDM) Regulations (qv)[324] to be brought together in a file or series of files and given to the employer on completion of the construction phase of the project. The primary purpose of the file is to provide assistance to anyone directly or indirectly involved in carrying out other construction works on or associated with the project for which the file has been prepared.

There is combined responsibility for producing the information comprising the file as between designer(s), contractor and his sub-contractors, principal contractor and planning supervisor. However, ultimate responsibility rests with the planning supervisor to ensure, so far as practicable, that a file is prepared for each structure making up the project. In respect of each structure the file must contain adequate design information to identify potential health and safety risks inherent in the project, structure, materials, articles and/or substances used which affect or are likely to affect anyone who may at any time be directly or indirectly exposed to those risks during other construction or cleaning works carried out on the structure. Other information which is reasonably foreseeable as having potential for risk to the health, safety and welfare of such people should also be included.

---

[324]In particular, Regulations 12, 13, 14 and 16 of CDM Regulations 1994.

A health and safety file will typically include:
— As built details and drawings.
— Details of any particular emergency procedures necessary for or associated with the use or operation of the structure or any parts of it.
— Information on the methods of construction adopted.
— Information on the types of materials used in the construction.
— Particulars of any specific maintenance requirements for the whole or any individual aspects of the structure along with any useful testing, commissioning or maintenance manuals associated with any specialist plant, equipment or the like installed in the structure.
— Particulars of the type and location of any mains or other services.

Health and safety plan Projects which are subject to compliance with all of the Construction (Design and Management) (CDM) Regulations (qv) require such a plan. To conform with Regulation 15 the plan must be prepared and developed in two stages. The first stage takes place before the construction process begins and at this time its preparation is the responsibility of the planning supervisor (qv). Thereafter, it is maintained and developed until the completion of the construction phase of the project by the principal contractor (qv).

It is the duty of the planning supervisor, in the first instance, to ensure that an appropriate document is prepared and completed so that it may be provided to the contractor even before arrangements are made for the construction and/or management of the project. In addition to providing a general description of the project works and details of the time(s) within which the whole or specified parts of the project are to be completed, the contractor must also have details of:
— Any known and/or reasonably foreseeable health and safety risks likely to affect those carrying out the works.
— Any other information known or reasonably discoverable by the planning supervisor which would reasonably assist a contractor to demonstrate that he has or will employ adequate resources and has the ability to perform the works and/or the management of them without contravention of any relevant statutory prohibition.
— Such other information as the planning supervisor knows or could reasonably know of and can expect will be needed to enable the principal contractor to subsequently carry out his duty to maintain and develop the health and safety plan until completion of the construction phase of the project.
— Such other information as the planning supervisor knows or could reasonably establish for the purposes of assisting the contractor to understand how to comply with any statutory requirements in relation to welfare.

Once prepared by the planning supervisor and passed to the contractor it then becomes the responsibility of the principal contractor (qv) to take all reasonable measures until completion of the construction stage of the works to ensure that the plan is maintained and developed to contain adequate information concerning:

— The arrangements for construction and/or management that are necessary to monitor compliance with relevant legislation and, having regard to the risks involved, to reasonably and practicably safeguard the health and safety of those engaged on and/or affected by the construction work.

— Such other information as is sufficient to enable the contractor to understand how he can comply with his statutory obligations in relation to welfare.

It is a most important requirement of the Regulations that the plan be prepared and provided to the contractor before arrangements are made to carry out and/or manage the project. Hence, where a letter of intent is used before a formal contract is in place, careful thought must be given not only to ensuring that this requirement is fulfilled but also to making express provisions, avoiding later argument over issues such as how the acceptability of the initial plan is to be measured (e.g. whose judgement prevails) and/or the position regarding extensions of time and additional reimbursement where delay is experienced in issuing the initial plan.

Where the project is subject to all of the CDM Regulations and a health and safety plan is necessary, neglect to draw up and implement a plan before the 'construction phase' (qv) begins[325], or to produce one without proper attention to detail may lead to prosecution. Prosecution may also occur where there is a failure to update and develop the plan so that it contains all of the features required during the life of the project. Under JCT contracts, where the contractor and principal contractor are one and the same entity any failure to update and/or develop the plan or to notify the employer of any amendments made to it will also amount to a breach of contract (JCT 98, clause 6A.2).

CDM computer software programs to deal with the demands of the legislation are available to assist in preparation of the health and safety plan. See also: **Construction Sites Directive.**

Hearing A general term referring to an occasion when disputing parties may bring (i.e. 'adduce') oral evidence or arguments before the court, arbitrator or other tribunal charged with deciding the issues.

In arbitration, unless the parties agree otherwise, the question whether and if so to what extent there should be a hearing of oral evidence or submissions is a matter to be decided by the tribunal[326]. The form which the hearing will take and/or the length of time afforded to each party to be heard may also be the subject of agreement between the parties (e.g. Rule 7 of the Construction Industry Model Arbitration Rules (CIMAR)), or may be imposed by the court, arbitrator or other tribunal concerned. Certain kinds of hearing have technical names such as trial, appeal, etc.

---

[325]Regulation 10 of CDM Regulations 1994.
[326]Section 34-(1)(h) of the Arbitration Act 1996.

In litigation, the court has an obligation under the Civil Procedure Rules (CPR) (qv) to manage proceedings. To that extent, the court may direct how a 'hearing' is to be conducted, e.g. by telephone[327], to limit the time which the parties may take[328].

See also: **Appeal; Reference; Adjudication.**

Hearsay A statement given as evidence of the truth of some fact or matter will be hearsay where the giver of the statement does not make it based on his own direct knowledge or observation of the fact or matter concerned but based, instead, on what someone else told him or indicated to him was true. In this context a 'statement' may be written or oral and documentary evidence may also, in appropriate circumstances, be considered as hearsay. As a general rule, hearsay evidence is non-admissible in civil proceedings[329].

A subtle yet important distinction must be made between the purposes for which such statements may be given, because evidence of a statement made to a witness by a person who is not himself called as a witness may or may not be hearsay. It is hearsay when the object of the evidence is to establish the truth of what is contained in the statement. It is not hearsay when it is proposed to establish by the evidence, not the truth of the statement, but the fact that it was made[330].

For example, B can give evidence that X said that he saw Y at the bus stop, to the extent that B is trying to prove that X said something (i.e. he was not mute), but to use it to prove Y was in fact at the bus stop would be hearsay.

Under the Civil Evidence Act 1995, the position is that hearsay evidence will not be excluded on that ground alone, although the weight (if any) attached to such evidence by the court depends upon the circumstances.

It should be noted that, in arbitration proceedings, the strict rules of evidence are not in any event necessarily complied with[331].

See also: **Admissibility of evidence.**

Highway A public right of way for vehicular or other traffic, including a way for pedestrians only. A comprehensive definition is contained in the Highways Act 1980, which is a consolidating Act drawing together earlier enactments. Local authorities have wide powers and duties in relation to highways, and over and above those generally wide powers and duties, highways in London are also subject to certain other unique provisions.

At common law highways may be further classified into carriageways, bridleways and footpaths, according to limitations on the type of traffic that may be used on them. Under the common law, the owner of property adjoining

---

[327]CPR Rule 3.1 (2) (d).
[328]For example, CPR Rule 28.6 (1) (b) (fast track trial timetable); Rule 29.8 (c) (i) (multi-track trial timetable); Rule 32.1 (Court power to restrict evidence and cross-examination) (qvv).
[329]Section 1 of the Civil Evidence Act 1995.
[330]*Subramaniam* v. *Public Prosecutor* [1956] 1 WLR 965.
[331]For a most enlightening explanation of what must be the earliest ever example of hearsay evidence see footnote 45 at page 137 of the 1984 reprint of *A Practical Approach to Evidence* by P. Murphy, published by Financial Training.

a highway is entitled to access to it at any point, but there are many statutory modifications of this right, e.g. the formation or laying-out of a means of access to a highway is development for which planning permission is required.

Hindrance or prevention At common law, almost without exception[332] it will be an implied term of every building contract that the employer will not himself, or through his employees or agents, hinder or prevent the contractor from performing the contract. A similar implication will almost invariably apply to the relationship as between contractor and sub-contractor under building sub-contracts[333].

If there are acts of hindrance or prevention which cause delay then, without express terms of the contract enabling the employer to make an appropriate extension to the period for completion, the employer cannot enforce any liquidated damages (qv) clause and time will be at large. Even where a right to extend completion does exist, if the delay caused by the act of hindrance is unreasonable the contractor may have a claim for damages against the employer[334]. In extreme cases, the contractor may even treat the contract as repudiated by the employer[335]. Generally, standard form contracts allow acts of hindrance and prevention by the employer or others for whom he is responsible as grounds for both extension of time and for money (e.g. JCT 98, clauses 25 and 26), but those provisions are not necessarily exhaustive.
See also: **Claims; Extension of time.**

Hire A type of bailment (qv) whereby an agreement is made under which a person, called the hirer, obtains possession and generally uninterrupted use of chattels (qv) for a specified or indeterminate period in return for payment. It is a type of contract and the rights and liabilities of the parties will be governed by the express and implied terms of the contract. Legislation now affects the position, e.g. the Unfair Contract Terms Act 1977 (qv) and the Supply of Goods and Services Act 1982 (qv), which implies into such hire contracts terms broadly similar to those implied into sale of goods contracts relating to fitness and quality. Depending upon the nature and terms of the hire contract it may be a 'consumer hire agreement', in which case it will also be regulated by the Consumer Credit Act 1974. Because most construction industry plant is hired, there are important implications should the contractor's employment under the building contract be determined by the employer who will then have no rights in the hired plant, irrespective of any contrary provisions in the building contract.

Hire-purchase A hire-purchase agreement is one 'under which an owner lets chattels (qv) of any description out on hire and further agrees that the hirer may either return the goods and terminate the hiring or elect to purchase the

---

[332]*London Borough of Merton* v. *Stanley Hugh Leach* (1985) 32 BLR 51 per Vinelot J at pp. 79–80.
[333]*Jardine Engineering* v. *Shiizu* (High Court of Hong Kong) (1992) 63 BLR at p. 98.
[334]*Lawson* v. *Wallsey Local Board* (1882) 47 LT 625.
[335]*Holme* v. *Guppy* [1838] 150 ER 1195.

goods when the payments for hire have reached a sum equal to the amount of the purchase price stated in the agreement or upon payment of a stated sum'[336]. It is, in effect, a means of buying goods on long-term credit which is regulated by complex legislation designed to protect private individuals. Goods which are subject to a hire-purchase agreement do not belong to the purchaser (hirer) until he has exercised his right to purchase.

At common law, subject to any qualification made by virtue of any agreed express terms, there will be an implied term in the hire purchase agreement to the effect that, if and when the time comes that the hirer wishes to exercise his option to purchase, the owner will still have the right to sell the goods on to the hirer. If it transpires that he does not, then the hirer may be entitled to repayment of all sums paid during the period when he hired and had the use of the goods concerned[337].

The fact that the goods do not belong to the hirer until he exercises his right to buy them has implications where a building contract contains, for example, a vesting clause (qv) or forfeiture clause (qv). Such clauses are ineffective as regards third parties, including the owner of the goods let on hire-purchase. See also: **Bailment.**

Hoarding Erection of a hoarding or fence separating proposed building, demolition, repair, maintenance or other such works from the street, highway (qv), road, lane, path or the like is governed by the Highways Act 1980. Among other things, the appropriate local authority will generally require a close boarded hoarding or fence which is to its satisfaction and which, if and when so required by the local authority, must include:
— A convenient covered platform and handrail outside the hoarding for the benefit of pedestrians.
— Sufficient lighting during hours of darkness.
— Maintenance.
— Removal when directed.

Erection of suitable hoarding is the contractor's responsibility and is covered under most forms of contract, e.g. JCT 98, clause 6.1, in which case failure to erect any, or any suitable hoarding will be a breach of the contract. Where bills of quantities are a contract document and where those bills are measured in accordance with the Standard Method of Measurement of Building Works, appropriate provision should be made for the contractor to price the work associated with erecting, maintaining and dismantling such hoarding.

A failure by the contractor to erect a suitable hoarding before the works proceed may also expose the employer to civil and statutory liabilities.

Hostilities A state of armed conflict between two or more states during which a declaration of war (qv) may or may not have been made by the Prime Minister in the House of Commons. The UK's participation against Iraq following the

---

[336]*Chitty on Contracts* (1999) 28th edn, at Chapter 38–267, Sweet & Maxwell.
[337]*Warman* v. *Southern Counties Car Finance Corporation Ltd* [1949] 1 All ER 711.

United Nations Security Council Resolutions against that country in 1991 illustrates hostilities falling short of a UK declaration of war. It is thought that, as a matter of public policy, contracts with, involving, or somehow likely to aid a state with whom hostilities have broken out will generally be dissolved even where no declaration of war is made. In the event of hostilities involving the UK, the mere suspension of the contract during hostilities, for whatever period, would probably never arise and so, albeit that JCT 98, clause 28A expressly provides rights for either party to determine the contractor's employment in the event of hostilities involving the UK causing excessive suspension of all or most of the works, it is thought unlikely that such a clause would, in practice, be used.

In contrast, provisions such as those in ACA 3, clause 11.5 which provide for granting an extension of time in the event of delay due to hostilities, and clause 21 (c) which gives the parties rights of termination in the event of excessive suspension due to hostilities, war etc., may be a most useful addition since, unlike the JCT contract, the hostilities that are at the root of the suspension or delay referred to under the ACA 3 form are not limited to those 'involving the United Kingdom'.

See also: **Alien enemy; Force majeure; Frustration; Vis major.**

Housing Grants, Construction and Regeneration Act 1996 An Act of Parliament in five parts:

Part 1 makes provision relating to the application for and payment of various grants, etc. for renewal of private sector housing.

Part 2 makes provision for the mandatory introduction of certain payment and dispute resolution procedures within specific categories of construction contracts (as to which see further below).

Part 3 makes provision concerning the constitution of and registration with the Architects' Registration Board (qv). Associated matters of education, code of practice and discipline of those registered with the Board are also covered as is the offence of practising under the title of 'Architect' while not registered.

Part 4 makes provision for grants and other financial assistance for regeneration, development and relocation.

Part 5 makes miscellaneous and general provisions governing existing housing grants, home energy efficiency schemes and certain other provisions relating to dissolution of urban development corporations, housing action trusts and commissions for new towns.

Of greatest impact on the construction industry generally has been the introduction by this Act of the legislation set out in ss. 104–117 in Part 2. Those sections provide for mandatory incorporation of certain terms and conditions within what the Act refers to and defines as construction contracts (qv), where parties to a construction contract neglect to set out in their agreement any appropriate terms, or those that they have incorporated do not conform to the requirements of the Act in relation to:

— The right for either party to refer any contractual disputes or differences to adjudication (qv).

— The right to receive interim payment(s) at stipulated times throughout the course of the work.

— The right to receive advance notice of the amount(s) which are due to be paid.

— The right to receive advance notice of any intention to withhold any part of the forthcoming payment.

— The right to suspend continued performance of the works in the event that payment is not, and until such times as it is, made properly in accordance with the contract.

— The outlawing of pay when paid provisions other than in very particular circumstances of insolvency.

In that case, the Act provides that conforming terms and conditions will, instead, be implied into, and will be deemed to be incorporated into, the contract. The precise wording of those implied terms is set out in other associated, secondary, legislation, called the Scheme for Construction Contracts (England and Wales) Regulations (1998).

Under the Act, 'construction contracts' (qv) are given an extremely wide definition, covering not only traditional building and civil engineering contracts and sub-contracts but extending, too, to many other types of agreement not normally thought of as falling within the general description of a construction contract[338] (e.g. architectural, design or surveying services and contracts for the provision of other advice on building, engineering, decoration and landscaping).

Following the imposition of those statutory rights of payment and adjudication in all defined construction contracts entered into after 1 May 1998, all of the standard form contracts commonly in use within the industry have been amended or reprinted to incorporate terms conforming to the requirements of the Act, thus avoiding the need for implication of those terms set out in the Scheme. (See, for example, JCT 98, clauses 30.1.1.3, 30.1.1.4, 30.1.1.5 and 41A; MW 98, clauses 4.4.1, 4.4.2, 4.4.3 and Supplementary Condition D; IFC 98, clauses 4.2, 4.2.3, 4.4A, 4.6.1 and 9A.)

In the early days of the Act, there has been extensive litigation by parties seeking to test the force and effect of the adjudication provisions[339]. However, it now appears well settled that except in those cases where the adjudicator has acted outside his jurisdiction, the courts are most unlikely to overturn or interfere in any way with an adjudicator's decision. The adjudicator does not have jurisdiction to decide his jurisdiction[340]. Where jurisdiction is or is likely to be an issue, care must be taken to preserve any right of challenge. A party may submit to an adjudicator's jurisdiction thereby losing any right to object[341]. The radical introduction of a statutory process of adjudication as an

---

[338]See, in particular, ss. 104–107 of the Act.
[339]See, for example, *Macob Civil Engineering* v. *Morrison Construction* (1999) 3 BLR 93.
[340]*Homer Burgess Ltd* v. *Chinx (Annan) Ltd* [2000] BLR 124.
[341]*Grovedeck Ltd* v. *Capital Demolition Ltd* [2000] BLR 181.

inexpensive, quick and relatively informal means of temporary dispute resolution has, therefore, to that extent proved successful and has met with general approval in the industry.

Hudson formula A proposed calculation of the 'head office overheads and profit' element in a contractor's claim for direct loss and/or expense arising under standard forms of contract, and in particular under JCT 98, clause 26.

It is one of a number of different methods adopted 'to produce a reasoned estimate of the sort of profit and fixed overhead recovery combined which a delayed contract organisation, viewed as a profit earning entity, might be expected to earn in the market had it been free to demobilise and leave the project on time'[342]. The formula is based upon information abstracted from a contractor's annual accounts or from such other information as provides a means of calculating the combined profit and fixed overheads which the delayed contract organisation might have expected to earn elsewhere but for being required to keep all or some of its resources on the delayed project.

First so called because it appears in the 10th edition of *Hudson's Building and Engineering Contracts* (p. 599), the formula has since been much criticised Despite initial widespread use of the formula in the preparation and presentation of contractor's claims, unlike the Eichleay formula (qv), which is of trans-Atlantic origin, Hudson's approach to ascertainment of this aspect of contractors' loss and expense does not appear to have received 'judicial approval' in any reported case in this country[343].

As put forward in the 10th edition of the work the formula was expressed as:

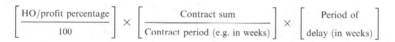

$$\left[\frac{\text{HO/profit percentage}}{100}\right] \times \left[\frac{\text{Contract sum}}{\text{Contract period (e.g. in weeks)}}\right] \times \left[\begin{array}{c}\text{Period of}\\\text{delay (in weeks)}\end{array}\right]$$

When first proposed, the authors even then identified a number of constraints and shortcomings which require careful consideration and in the 11th edition Hudson now further acknowledges the rather simplistic, outdated and relatively arbitrary nature of the formula. Although not altogether abandoning Hudson's approach in appropriate circumstances, the 11th edition tends to suggest a preference (albeit still heavily qualified) for use of the Eichleay formula (qv).

In recent years, some of the more fundamental criticisms levelled at the Hudson formula have been that:

— It assumes the profit budgeted for by the contractor in his prices was in fact capable of being earned by him elsewhere had he been free to leave the delayed contract at the proper time.

---

[342]Wallace (1995) *Hudson's Building and Engineering Contracts* 11th edn, Sweet & Maxwell at pp. 8–182.
[343]But, in contrast, see the judgments given on this point in *Shore and Horwitz Construction Co Ltd* v. *Franki of Canada Ltd* [1964] SCR 589 and *Ellis-Don Ltd* v. *Parking Authority of Toronto* (1978) 28 BLR 98: note, in particular, the bracketed caveat at 28 BLR p. 126.

— At best it requires adjustment to be made for the various factors for which recovery is not permitted, e.g. the contractor's own inefficiency.

— It ignores the contractor's duty to make realistic attempts to deploy his resources elsewhere during any period of delay[344].

— The value of the final account may well exceed the contract sum. Thus, any proper valuation for variations must include an element of reimbursement for overheads and profit and hence there is risk of duplicating recovery.

— The use of the formula as it stands results in profit being added to the profit already in the contract sum so that, at the very least, the Hudson formula as first set out should read 'Contract sum less overheads and profit' rather than 'Contract sum'.

— The formula can also produce under-recovery where inflation during the period of delay increases the overhead costs envisaged at the time of tender.

It is sometimes said that Hudson's formula has received judicial approval and the judgments in *J. F. Finnegan & Son Ltd* v. *Sheffield City Council* (1989)[345] and *Whittal Builders Co Ltd* v. *Chester le Street District Council* (1985)[346] are often cited in support of that contention. On a close reading of those and similar reported English cases, it appears that, strictly speaking, reference was not in fact made to Hudson's formula. In *Finnegan*, having referred to Hudson's formula the court then went on to apply another (the Emden) formula (qv) which, unlike Hudson's approach, is based on a percentage taken from the contractor's organisation as a whole. In *Whittal* the court adopted a formula based on average and essentially notional figures.

Whether it be Hudson, Eichleay, Emden or any other such formula, that approach to *ex contractu* claims (qv) should, if at all possible, be avoided.

Human Rights Act 1998 An Act of Parliament implemented on 2 October 2000 to give effect in the English courts to rights and freedoms guaranteed under the European Convention of Human Rights. It is a complex piece of legislation designed to provide statutory safeguards over existing human rights, freedom of speech, thought, conscience and religion. Although not directed specifically at the construction industry, the Act nevertheless has relevance to the conduct of construction dispute resolution procedures in that it underpins parties' rights to have their disputes, differences or grievance fairly and publicly heard within a reasonable time, by an independent and impartial tribunal established by law[347]. It thus adds weight to the general principles of natural justice, impartiality and good faith which already exist to provide safeguards

---

[344]*Peak Construction (Liverpool) Ltd* v. *McKinney Foundations Ltd* (1970) 1 BLR 114.
[345](1989) 43 BLR 124.
[346][1996] 12 Const LJ 356.
[347]Article 6 (1) of the Convention.

against potential maladministration of the dispute resolution procedures provided for under the various standard form building contracts. However, whether or not the Act will have any tangible effect in overcoming the potential for 'rough justice' that is a recognised feature of the speedy construction adjudication process is questionable[348].

---

[348]Notwithstanding the observations of Dyson J at first instance in *Bouygues* v. *Dahl-Jensen* [2000] BLR 49; see for example the tests laid down in *Locabail (UK) Limited* v. *Bayfield Properties Limited* [2000] QB 451 (CA).

Identified terms Terms, such as 'employer', 'contractor', 'project manager', 'works', etc. which are used regularly throughout a standard form contract but which, for the purposes of each individual contract, must be specifically identified in an appendix, schedule or some other part of that contract. To a greater or lesser extent the use of identified terms is essential and is common place in all popular standard form contracts such as the JCT 98, IFC 98, etc. Most contracts, by convention, capitalise the initial letter of such terms to avoid confusion. This is specially important to clearly differentiate between such words as 'Works' and 'work'. It is a particular feature of the New Engineering Contract (NEC) (qv) where provision is made in the contract data (qv) section (at Part One) of the contract for identification of terms such as; 'employer', 'currency of the contract,' 'starting date,' 'possession dates,' 'weather measurements', 'boundaries of the site,' 'interest rate', etc. Terms used in the contract data are identified by being italicised throughout the contract. See also: **Defined terms.**

Illegal contract A contract which, when it was formed, contravenes statute or common law. Such a contract will be unenforceable by either party and will be void (qv). Thus, for example, a party who is paid money on the basis of an illegal contract but then refuses to carry out his side of the bargain cannot as a general rule be made to refund the money. The money can only be recovered if the other party can show that some fraud, duress, misrepresentation or mistake induced him to enter into the contract. A building contract which had as its primary objective the contravention of the planning law would be an illegal contract. Such a case might arise if the parties made an agreement to build in a green belt area. The courts sometimes extend the concept to embrace contracts which are considered to be against public policy, e.g. a contract which has effect to deprive a person of their livelihood[349] or where the effect is restraint of trade.

For the purposes of determining whether, and, if so, to what extent the contract is void and/or unenforceable, it is important to distinguish those contracts which are illegal at their inception from those made for legitimate purposes but which one or both parties knew or intended should be performed in an illegal manner[350]. Contracts may also become illegal because, following formation or part performance, some subsequent enactment or change in the law makes further performance illegal in which case it is discharged by frustration. See also: **Contract; Frustration.**

Illness Illness may result in frustration of contracts for personal services such as contracts of employment[351]. However, a great many factors must be taken into

---

[349]*King* v. *Michael Farraday & Partners Ltd* [1939] 2 All ER 478.
[350]*Archbolds (Freightage) Ltd* v. *S. Spanglett Ltd* [1961] 1 All ER 417 per Devlin LJ at 424 and 425.
[351]*Marshall* v. *Harland & Wolff Ltd* [1972] 2 All ER 715.

account including the terms of the contract, the nature of the employment, the nature and duration of the illness, and the prospects of recovery. Where a contract is 'personal' in character, e.g. a well-known sculptor producing a work of art, grave and lengthy illness may also frustrate the contract. As the personality of an architect is generally of vital importance to the employer, the same principle will apply. Conceivably, if a builder is an individual and his personality is of importance to the completed work, serious illness could also result in frustration, but there appear to be no reported cases on the point.
See also: **Frustration.**

Immediately – Payment/Notice etc. Where, under a building contract, it is stipulated that action shall be taken or notice given 'immediately', the act or notice will generally be required to be instigated with all reasonable speed. Whether or not the speed taken has or has not been 'reasonable' will depend upon the particular circumstances surrounding the notice or action concerned. But, a requirement for immediacy will in any event connote greater urgency than would normally be called for where, for example, the action is described to be taken simply within a reasonable time[352].
See also: **As soon as possible; Forthwith; Directly.**

Implied contract A contract which is implied from the actions or conduct of the parties in contrast to the terms of the agreement being expressed by the parties in words, e.g. by the contractor starting work on receipt of an order[353].
See also: **Contract; Simple contract.**

Implied term A term of a contract which the parties to that contract did not expressly agree either in writing or orally and which is not negatived[354] by or inconsistent with some express term. If held to be a part of the bargain it will bind the parties as if it were expressly incorporated into the contract. Terms may be implied in various ways, e.g.
— By statute, e.g. by s. 114-(4) Part II of the Housing Grants, Construction and Regeneration Act 1996, in respect of any construction contract: 'where any provisions of the Scheme for Construction Contracts apply by virtue of this Part in default of contractual provisions agreed by the parties, they have effect as implied terms of the contract concerned', and e.g. under the Supply of Goods and Services Act 1982 and Sale of Goods Act 1979, in appropriate circumstances important terms as to fitness, quality, price and the like may be implied into a contract.
— At common law where, for example, subject to any exclusion or express term to the contrary, certain warranties will be implied, e.g. that a contractor will supply good and proper materials[355] and will provide

---

[352]*Hydraulic Engineering Co Ltd* v. *McHaffie, Goslet & Co* (1878) 4 QBD 670 CA and *Alexiadi* v. *Robinson* (1861) 2 F&F 679.
[353]*A. Davies & Co (Shopfitters) Ltd* v. *William Old Ltd* (1969) 67 LGR 395.
[354]But see, for example, s. 55.-(1), s. 55.-(2), s. 55.-(3) and paragraphs 11 and 12 of Schedule 1, Sale of Goods Act 1979.
[355]*Young & Marten Ltd* v. *McManus Childs Ltd* [1968] 2 All ER 1169.

completed work which is constructed in a good and workmanlike manner.

— A term will be implied where necessary to make the contract work, i.e. to give it 'business efficacy'. This is commonly referred to as 'the Moorcock' doctrine[356]. A more recent example might be a contract for driving lessons where the parties would not likely make any express stipulation as to the roadworthiness or other suitability of the car but it would, nevertheless, be necessary to imply such a term[357].

— A term will be implied where the contract as it stands is perfectly workable but where the term implied simply states what the parties obviously intended would be, but was not expressly said to be part of their bargain. A term to that obvious effect will then be implied. In deciding whether or not such a term should be implied the courts will establish the parties' intention objectively, applying what is commonly referred to as the 'officious bystander test'. Put simply, the question will be whether it can confidently be said that: 'if at the time the contract was being negotiated someone had said to the parties, "What will happen in such and such a case?" they would have replied: "Of course, so and so will happen; we did not trouble to say that; it is too clear".'[358]

— Unless the express terms of the contract provide otherwise, a term will be implied where the courts have already laid down that in particular types of agreement certain terms will automatically be implied so that, for example, under a building agreement embodying the standard JCT General Conditions there will be an implied term that the building owner, and on his behalf the architect, will do all that is necessary to enable the contractor to carry out the work and that the architect will provide the contractor with accurate drawings and information.

— By custom and usage, e.g. where it has invariably been the longstanding practice in a particular trade, profession or business context to conform, and to perform, in a certain manner then, unless the parties have expressly stated to the contrary, they will be presumed to have contracted with the intention of operating the agreement according to that custom.

— Terms will be implied where the parties have consistently, regularly and invariably dealt previously on certain terms and conditions then it may be taken that in future dealings, unless expressly provided to the contrary, they are conducting their business on similar terms to those used in their previous course of dealings (qv).

There are important and often ignored limits to when terms will be implied. A term will not, for example, be implied at common law merely because the court thinks it would have been reasonable to insert it into the contract, and even where one or other of the situations referred to above may otherwise arise,

---

[356]So named after the leading case of that name, *The Moorcock* (1889) 14 PD 64.
[357]*British School of Motoring Ltd* v. *Simms* [1971] 1 All ER 317, and see also, *Liverpool City Council* v. *Irwin and Another HL* [1976] 2 All ER 39.
[358]*Reigate* v. *Union Manufacturing Co* [1918] 1 KB 592; see also, *Shirlaw* v. *Southern Foundries Ltd CA* [1939] 2 KB 206.

terms will generally be implied only under certain conditions:

— An implied term must not be in conflict with or inconsistent with an express term.
— It must be based on the imputed or presumed intention of the parties.

Contractors' claims (qv) may be based on breach of some implied term, e.g. by the employer not to prevent completion and to do all that is necessary on his part to bring about completion of the contract.

See also: **Express terms.**

Imposition of restrictions/obligations As a general rule, unless the contract imposes specific obligations and restrictions on the contractor's hours of work and methods of working or the like, the contractor will be free to choose how and when he performs the works. Hence, the imposition of new restrictions or the significant alteration of existing contractual restrictions on the contractor during the course of the project will likely place the employer in breach of contract, unless and to the extent that the contract expressly sanctions such interference by the employer or on his behalf by the architect. Four categories of restriction are expressly sanctioned by JCT 98 (clause 13.1.2) and IFC 98 (clause 3.6.2):

— Obligations or restrictions in relation to the contractor's access to or use of any specific part(s) of the site.
— Obligations or restrictions that may limit the working space available to the contractor.
— Obligations or restrictions that limit the contractor's working hours.
— Obligations or restrictions in relation to the order of executing or completing the works.

MW 98 makes similar, but rather less specific provision (at clause 3.6) whereby the order of period in which the contractor is to carry out his work may be dictated.

Where the imposition or alteration of any such restriction(s) is required by the employer it must be done through the architect (or contract administrator), by means of a properly issued variation. Whereas under JCT 98 and IFC 98 the contractor has a right of reasonable objection against complying with such a variation instruction (JCT 98, clause 4.1.1.1 and IFC 98, clause 3.5.1 respectively), MW 98 gives the contractor no such express rights of objection. However, if he chooses not to object or is otherwise bound to comply, then both under the JCT 98 and IFC 98 contracts and under the MW 98 contract he will be entitled to have the consequences of the instruction valued and paid as a variation to the contract works. If and where appropriate, he will also be entitled to any resulting extension of time and consequential direct loss and expense.

Clauses such as those in the JCT and IFC contracts referred to above are most useful to enable an employer to accommodate the practical difficulties that often arise as work proceeds without being forced to breach the contract, and provided the contractor is given corresponding rights of reasonable objection and of entitlement to reimbursement in terms of both time and money for the consequences of such restrictions, it is difficult to see why the introduction of provisions such as these should be resisted.

Impossibility A contract which, at the time it is made, is made without realising that its performance will be physically impossible may be void (qv), and so will be of no effect on the grounds of the parties' mistake (qv). In such cases the parties are left to bear their own losses unless one of them can show that he was induced to enter into the contract by fraud (qv) or misrepresentation (qv). However, that situation must be distinguished from one where some intervening neutral event (qv) arising after the contract is made causes the contract to become physically or commercially impossible to fulfil, in which case it will be frustrated (qv) and thus becomes unenforceable.
See also: **Frustration; Voidable.**

Improper materials Materials which are not in accordance with the contract. The architect may instruct that such materials are to be removed from site, e.g. JCT 98, clause 8.4.

Inconsistency Generally, within all of the commonly adopted standard building contracts, and in any event as a matter of good drafting in all contracts of any significance, where the contract incorporates (qv) numerous different drawings, schedules or other documents, the contract will make express provision for what action should be taken where it appears that there is inconsistency in or between any part of the documents forming the contract (see, e.g. NEC clause 17). To give rise to inconsistency it is unlikely that the conflicting words or information need go so far as to be mutually exclusive. It will be sufficient if they can be considered as offering new or different meanings. Under most construction contracts, inconsistency in or between the various documents is usually a matter to be resolved by an instruction (qv) given by the project manager, architect, engineer or other contract administrator.
See also: **Discrepancies; Divergence.**

Incorporation A word with several meanings in law. It may refer to the process by which a corporation (qv) is constituted, i.e. to form an organisation with a separate personality in law. Also used when referring to the inclusion of specific contract terms or conditions in a contract.

The word is also used in many building contracts in the context of defining a time when property, title and/or risk in goods and materials used in the construction may be transferred from one party to another. By way of example, as a general rule when building materials are built into (or are 'incorporated' into) the structure, they become part of that structure with the result that, the maxim *quicquid plantatur solo, solo credit* ('whatever is affixed to the soil becomes part of the soil') may apply to defeat any retention of title clauses (qv) that might otherwise have been effective. It will be a question of fact in each case whether, and if so when, goods and materials can be considered to be incorporated to a sufficient extent to give rise to application of the *quicquid plantatur solo, solo credit* maxim but, as a general rule the test for those purposes will be a stringent one[359].

Similarly, where up until practical completion a sub-contractor might otherwise be liable for and would be required to insure against certain damage to materials for use in the works, as and when before practical completion those materials are 'fully, finally and properly incorporated into the Works' (qv), then that incorporation may have the effect of limiting or even wholly extinguishing that liability. (See, for example, the Construction Confederation's Domestic Sub-Contract DOM/1 (qv) for use with JCT main contracts (clause 8).)

The test of whether materials have been 'incorporated' sufficiently to satisfy the insurance provisions described above is likely to be less stringent than for the purposes of defeating a retention of title claim.

Incorporation of arbitration agreement It is entirely possible, merely by making reference to the incorporation of terms in another contract, to incorporate an agreement in that other contract requiring all disputes or differences to be referred to arbitration rather than litigation. However, for such an arbitration agreement to be taken as incorporated the form of words used for such incorporation must be clear and unambiguous, and must leave no doubt as to the parties' intention to agree upon arbitration as their chosen final dispute resolution process. Even where the effective incorporation of an arbitration agreement is indisputable, the parties must ensure that the words giving rise to incorporation of that basic agreement are not so vague as to leave in doubt precisely which disputes etc. do or do not fall within the scope of that arbitration agreement. Incorporation of an arbitration agreement by reference alone should, therefore, be avoided and where parties wish to rely upon arbitration as their sole means of dispute resolution then they should expressly say so and should fully set out terms that clearly establish the scope of their agreement.

See also: **Arbitration agreement; Incorporation of terms.**

Incorporation of documents Construction contracts are often complex with the parties' agreement and their respective rights and obligations being clear only from a review of numerous documents such as: bills of quantities, specifications, schedules, drawings, method statements, programmes and the like. Since extrinsic evidence of the parties' intentions is not usually admissible (see: **Admissibility of evidence**) to assist an arbitrator or judge in determining the parties' intentions in the case of a dispute, it is crucial that all such documents are properly incorporated into, and become part of, the contract documents (qv).

In the various JCT main contracts, such as JCT 98, IFC 98, MW 98, and in other associated standard forms of sub-contract in common use (such as

---

[359]For an interesting and most useful discussion on the scope and effect of incorporation in the context of the *quicquid plantatur solo, solo credit* maxim see John Parris, *Effective Retention of Title Clauses*, 1986, Collins.

DOM/1 etc.), this is generally achieved by a combination of various express terms within the standard conditions, coupled with specific annotation and signature by each party on the various documents that are to be treated as incorporated into the contract. See, for example, the combined effect of clauses 2.1, 1.3 and the first Recital of JCT 98 in relation to incorporation of documents such as bills of quantities, drawings and the like. A modified but equally effective approach is adopted in DOM/1 (qv) where the documents to be treated as incorporated are uniquely annotated and signed by the parties and are then identified as 'Numbered Documents' in the Appendix to the sub-contract.

Where a document is referred to as being incorporated for a particular purpose the effect of its incorporation generally will not be extended beyond that specific purpose.

Incorporation of documents should be distinguished from incorporation of terms and it should be noted that it is possible to incorporate the terms of a standard form of contract by reference to it in a properly worded but simple exchange of letters designed so that the letters then form part of the contract. For example, a request to a contractor to quote for a job on the basis of MW 98.

Incorporation of terms It is by no means essential that a contractual document needs to contain all of the relevant terms. Subject to certain particular requirements and safeguards, terms (see: **Term of the contract**) may be incorporated into what is otherwise a relatively short and simple contractual document merely by a reference made in that document to those other terms. This might, for example, be done where an exchange of letters refers to other particular standard terms or where a ticket, such as a lottery ticket, makes a reference on it to other terms and conditions which, although not specifically printed on the ticket, will generally have contractual force.

A contracting party wishing to rely on such incorporation must show that the standard terms were properly and effectively incorporated. To do this they must prove, on balance, that the parties intended the other document or terms in question to form part of the contract. This can be shown either by proving that the party specifically agreed to be bound by those terms by virtue of having signed the contractual document which makes reference to them, or that he entered into the contract having first been given proper notice of, and was fully aware that the terms concerned are to be treated as being incorporated into the contract. Where such a contractual document has been signed, then it is well established that as a general rule that will suffice to bind the party or parties who have signed it, irrespective of whether they read or understood them. Exceptions to this rather general and strict rule may, however, arise where, for example, statute prescribes for a 'cooling off' period during which a party to certain consumer contracts may withdraw from an agreement even though he has signed it. The general rule will also not apply where the person putting forward the document has misrepresented its contents[360] (see: **Misrepresentation**).

---

[360]*Curtis* v. *Chemical Cleaning Co Ltd* [1951] 1 All ER 837.

See also: **Unfair Contract Terms Act 1977.**

Although the incorporated terms must be notified at or before the making of the contract[361], that need not mean that they must have actually been read or that their importance and/or consequences must have been known or appreciated. Whether their existence and incorporation was sufficiently notified is to be tested objectively. Thus, a reference in a contractual document to the contract being made subject to general conditions 'available on request' may well be treated as sufficient to incorporate into the contract the current edition of those conditions whether or not the individual noted that particular reference or did, in fact, specifically request sight of the conditions to be incorporated[362]. This principle is of importance in the construction industry where, for example, an invitation to tender (qv) may refer to the contract conditions 'being available for inspection at the architect's office' — a common but nevertheless bad practice.

Incorporation of terms by reference was discussed by the Court of Appeal in *Modern Buildings (Wales) Ltd* v. *Limmer & Trinidad Co Ltd* (1975)[363]. The words 'in accordance with the appropriate form for nominated sub-contractors' were used in an exchange of correspondence between a main contractor and a nominated sub-contractor (qv). On the facts of that case, this was held sufficient to incorporate the terms of the then current FASS/NFBTE form of nominated sub-contract. It should, however, be noted that, where as commonly happens sub-contracts, or even sub-sub-contracts, are let on terms that seek to incorporate *mutatis mutandis* (with necessary changes) the terms (see: **Term of the contract**) of a head contract (qv) which includes provision for disputes or differences under the contract to be referred to arbitration, the courts will closely examine the words of the sub-contract or sub-sub-contract to establish whether there has been an unequivocal intention by the parties specifically to incorporate those arbitration provisions from the head contract[364].

See also: **Incorporation of arbitration agreement.**

In addition to the incorporation of terms by reference discussed above, terms may also on occasion be treated as incorporated by virtue either of the previous course of dealings (qv) between the parties concerned or where the parties' particular trade or profession suggests the habitual use of some particular standard terms. Although terms will not be taken as incorporated simply because parties have contracted on similar terms on a previous occasion, where they have previously and consistently contracted on such terms then that previous course of dealings may then be taken as good evidence that despite the absence of any express agreement to adopt them once again on a future occasion, it is nevertheless reasonable to believe that the parties intended to be bound by those same terms yet again in their latest course of dealing.

---

[361]*Olley* v. *Marlborough Court Ltd* [1949] 1 All ER 127.
[362]*Smith* v. *South Wales Switchgear Ltd* (1978) 8 BLR 1.
[363][1975] 2 All ER 549.
[364]*Aughton Ltd* v. *M. F. Kent Services Ltd* (1991) 57 BLR 1.

Incorporeal hereditament Property rights to which the law of real property and in particular certain specific property statutes, such as the Law of Property and Settled Land Acts, refer and apply. They include rights of way (qv) and other such rights over land.

See also: **Chattels; Goods.**

Indemnity clauses Since it is entirely likely that some negligence, act or default of a contractor during construction work may result in a claim made directly against the employer, most standard form building contracts and sub-contracts contain various indemnity clauses under which one party – usually the contractor – undertakes liability to make good defined losses, costs, etc. incurred by the other – usually the employer – on the occurrence of one or more specified events. Although drafted with varying degrees of complexity, examples of such provisions can be found in clause 20 of JCT 98 and clause 83 of the NEC (qv). Unless very clear words are used to show a contrary intention the courts will generally construe indemnity clauses strictly against the person seeking to rely on them so that:

> 'If a person obtains an indemnity against a consequence of certain acts, the indemnity is not to be construed so as to include the consequences of his own negligence unless those consequences are covered either expressly or by necessary implication'.[365]

Likewise indemnity clauses should be strictly construed where it is sought to hold a person liable for defaults of any persons other than themselves and over whom they have no control[366].

Although for the purposes of reckoning when an action for breach of contract may be statute barred the cause of action will generally arise when the alleged breach took place, for the purposes of the Limitation Act 1980, under an indemnity clause time does not begin to run until the party indemnified has suffered loss (i.e. has judgment entered against him)[367]. This may well have the effect of extending the period of liability to a considerable extent, with even more significant consequences in the context of sub-contracts.

See also: **Contra proferentem; Insurance; Limitation of actions.**

Indemnity contract Such contracts may be taken in the broad sense as one and the same as guarantee contracts. However, indemnity contracts are readily distinguishable[368]. A guarantee contract will exist and be made collateral to some other contract (such as, for example, where a party may undertake liability for due performance of another under a separate agreement). An indemnity contract, on the other hand, gives rise to an independent, stand alone, obligation to keep harmless against loss[369]. The person giving the indemnity is required to make good some specified loss suffered by the party to

---

[365]*Walters* v. *Whessoe Ltd and Shell Refining Co Ltd* (1966) 6 BLR 23 per Lord Devlin at 34.
[366]*City of Manchester* v. *Fram Gerrard Ltd* (1974) 6 BLR 70.
[367]*County & District Properties Ltd* v. *C. Jenner & Son Ltd* (1976) 3 BLR 41.
[368]For a useful analysis of the distinction, see *General Surety and Guarantee Co Ltd* v. *Francis Parkeer Ltd* (1977) 6 BLR 16, 21 per Donaldson J.
[369]*Yeoman Credit Ltd* v. *Latter* [1961] 2 All ER 294, 296.

whom the indemnity is given. Whilst invariably such indemnities are reduced to writing, a promise of indemnity need not be evidenced by a written and signed undertaking.

See also: **Indemnity clauses; Guarantee; Bonds.**

Independent contractor A person who works under a contract for services, as opposed to one who works under a contract of service where he would be an employee and in the legal context is commonly referred to as servant (see also: **Master**). The distinction may be important for a number of purposes such as, for example, deciding whether special duties owed by an employer to his employee exist, or determining whether a person is in insurable employment for purposes of National Insurance contributions, or accountability to HM Inspector of Taxes or the extent of any vicarious liability (qv) an employer may have. It is often difficult to distinguish between the two.

A general but now not altogether satisfactory test will be one of the degree of control that the employer retains over the way in which the work is to be carried out. An independent contractor is often distinguished by the fact that he is free to control the way in which work is done. An employee under a contract of service (i.e. a servant), on the other hand, will be one over whom the employer retains express or implied rights to control how the work is done and over whether or not he may discipline or even dismiss the employee. However, in a case of doubt (as to whether the contract is one for service or of service) no single element in the relationship can be regarded as conclusive. 'No exhaustive list has been compiled and perhaps no exhaustive list can be compiled of the considerations which are relevant ... nor can strict rules be laid down as to the relative weight which the various considerations should carry in particular cases.' The most that can be said is that control (i.e. instructions over when the work should be done; where it should be done; how it should be done) will no doubt always have to be considered ...'[370]. In practice, the most realistic test may be that there will be a contract for services 'if the work, although done for the business is not integrated into it, but is only accessory to it', e.g. the normal architect–client relationship[371].

The contractor under the normal building contract is an independent contractor. In general, a person is not liable for the negligence of his independent contractors or agents (see: **Agency**) to the same extent as he is liable for the negligence of his employees. It is commonly for this reason that the distinction between employees and independent contractors is important, although in many instances under building contracts the architect will be acting as the agent of the employer so as to make the employer vicariously responsible. In *Rees & Kirby Ltd* v. *Swansea City Council* (1983) the general position was aptly summarised at first instance: 'An architect is usually and for the most part a specialist exercising his special skills independently of his employer. If he is in breach of his professional duties he may be sued personally. There may,

---

[370]*Market Investigations Ltd* v. *Minister of Social Security* [1969] 2 QB 173 per Cooke J at 184–5.
[371]*Stevenson Jordan & Harrison* v. *Macdonald & Evans* (1952) 1 TLR 101 per Denning LJ.

however, be instances where the exercise of his professional duties is sufficiently linked to the conduct and attitude of the employers so as to make them liable for his default'[372].

Industrial building A term with particular relevance under planning legislation concerned with the use and development of buildings (qv). It refers generally to a building used for non-agricultural trade or business where the purpose is the carrying out of any process for or incidental to making, altering, repairing or adapting, breaking up or demolishing all or part of any article. The nature of the particular process involved and its impact on the amenity of the surrounding area will be factors in determining whether the industrial usage of the building can be classed as light, general or special.

Industrial dispute A dispute between workers and their employer which may otherwise be referred to as a trade dispute (qv) which either entirely or at least partly concerns issues over employment, non-employment termination and/or suspension of the duties or employment of employees or their terms and conditions of work or physical conditions under which they are expected to work. Other matters which can properly be described as coming within the scope of an industrial dispute would be issues over:
— Discipline.
— Allocation of duties.
— Trade union membership or non-membership and/or the rights and/or machinery for negotiations.

Industrial property A generic term applied to kinds of property rights of an intangible nature which are valuable in industry, e.g. patents (qv), trade marks and industrial 'know-how'.

Inevitable accident An accident 'not avoidable by any such precautions as a reasonable man, doing such an act then and there, could be expected to take' (Sir Frederick Pollock), e.g. a fire caused by lightning. Inevitable accident is sometimes said to be a defence to certain kinds of actions in tort (qv) but modern writers consider that the conception of inevitable accident has no longer any useful function and it is doubtful whether much advantage is gained by the continued use of the phrase. Where strict liability is imposed by statute or at common law, such as under the rule laid down in *Rylands* v. *Fletcher*[373], the inevitability or otherwise of the event will not in any event be a defence.

In the particular case of damage caused by fire, the Fires Prevention (Metropolis) Act 1774 – which applies to the whole country – provides that no action is maintainable against anyone on whose land a fire begins *'accidentally'*[374]. However, the Act gives no protection where the fire begins

---

[372](1983) 25 BLR 129 per Kilner Brown J at 147.
[373](1866) LR 1 Exch 265.
[374]*Collingwood* v. *Home & Colonial Stores Ltd* [1936] 3 All ER 200.

accidentally but the owner is negligent in letting it spread[375]. The burden of proving negligence is on the claimant. The defendant does not have to prove that the fire was accidental.

Information In the context of building contracts it refers to drawings (qv), schedules, instructions (qv) which are generally the responsibility of the architect to produce and which are necessary to enable the contractor to carry out and complete the works in accordance with the contract. The type and extent of the information which is intended to be provided during the course of the works and the date(s) by which that information is to be given may be conveniently scheduled. Under some JCT contracts (where it is not defined as a contract document), it may take the form of an Information Release Schedule (qv) and under ACA 3 as a Time Schedule (qv). Where such a schedule is used and is specifically given contractual effect then, unless altered by agreement, a failure to provide the particular information concerned and/or a failure to provide it on the specified date stated in the schedule may result in a claim for extension of time for completion and/or for additional loss and expense (see JCT 98, clauses 5.4.1, 25.4.6.1 and 26.2.1.1).

Under JCT 98, clause 29.1, provided that the contract bills contain 'information' in relation to work that the employer or his servant(s) or agent(s) intends will be carried out during the project and provided that 'information' is sufficient to enable the contractor to carry out and complete his works then, the contractor is duty bound to permit the employer or his servant(s) or agent(s) access to execute such other work. This can be a useful consideration for employers wishing to influence the choice and employment of certain specialist contractors on the works but who are reluctant to adopt the nomination procedure used under JCT contracts. If information is not provided in the bills or if, when provided, it is insufficient to enable the contractor to carry out and complete the contract works, then the consent of the contractor will be required before any direct work may be executed.

If, throughout the contract, information of any kind is expressly stated to be provided by either party then a failure to provide it properly or at the right time will be a breach of contract. Whether a claim can be made under the provisions of the particular contract will usually depend upon whether the party requiring the information applied for it at the right time, i.e. neither too early nor too late. Where the contractor's common law rights are preserved they are, however, unaffected by the timing of such an application.

Information Release Schedule A schedule referred to in JCT 98 and IFC 98, setting out specifically the information which the architect or other contract administrator intends providing to the contractor during the currency of the contract and stating when, during that time, the information referred to will be provided. Without such a schedule the obligation to provide information and the timing of its release are usually expressed in more general terms and, for

---

[375]*Goldman* v. *Hargrave* [1966] 2 All ER 989.

example, require simply that the architect or contract administer shall, from time to time so as to enable the contractor to carry out and complete the works, provide such further amplification or other information as reasonably necessary for that purpose (see clause 5.4.2 of JCT 98).

Under JCT 98 the employer may choose whether, and if so to what extent, he wishes to provide such a schedule (see footnote [e] to the Sixth Recital). The benefits of an Information Release Schedule as a useful means of providing certainty and of avoiding disputes over the interpretation of vague terms, such as those of the clause 5.4.2 referred to above, should be weighed against the fact that, where the employer elects to provide a schedule then, subject only to the architect being prevented by some act, default or culpable delay by the contractor or unless the contractor otherwise agrees to extend the scheduled times or dates for provision of the specified information, the architect must achieve the specified dates. Should he fail to do so, the contractor will be entitled to claim an extension of time and/or additional costs occasioned by the delay. The ability to make a claim also exists in the absence of a schedule; however, a contractor may face significant difficulties proving his claim.

Infringement of rights See: **Human Rights Act 1998.**

Injunction An order of the court whereby a person is prohibited from acting or required to act in a prescribed manner. To ignore such an order will be a contempt of court. Injunctions can be:
— Prohibitory, where the order forbids the act or omission complained of.
— Mandatory, where the order restrains further action or insists that some act shall be performed.
— Interim, where there is a temporary order which maintains the *status quo*.
— Of temporary effect where an order is made to have effect until a specified date.
— Final, where the order has perpetual effect.
See also: **Freezing injunction.**

Injury Harm done to persons or property. Injury need not be physical, it may be purely economic loss. It is generally actionable in contract (qv); however, the recovery of pure economic loss in tort (qv) is problematic.
See also: **Action; Damage; Damages; Insurance.**

Innocent misrepresentation An innocently made but nevertheless untrue statement of fact, past or present, made in the course of contractual negotiations and which is one of the causes inducing the other party to enter into the contract. A statement of opinion, unless it was not or on the known facts could not reasonably honestly be believed when it was expressed, will not otherwise amount to a statement of fact. Innocent misrepresentation is to be contrasted with a fraudulent misrepresentation (qv) and with a negligent misrepresentation (qv). The test is whether the statement would have affected the judgment of a reasonable man in deciding whether to enter into the contract. An innocent misrepresentation may entitle the innocent party to rescind the contract. Damages can be granted at the discretion of the court in lieu of rescission for innocent misrepresentation under s. 2 (2) of the Misrepresentation Act 1967. See also: **Misrepresentation; Rescission.**

Innominate term See: **Express term.**

Insolvency In general, a state where an individual or company is, in the short term, unable to meet its debts, having insufficient money coming in to meet the outgoing cash flow or where, in the long term, there is a shortfall in assets against liabilities. In such a situation:

An individual may:
— become or may be made bankrupt.
— seek to avoid bankruptcy by making a composition (qv) or arrangement (qv) with his creditors.

A company registered under the Companies Acts may:
— make a proposal for a voluntary arrangement (see: **Arrangement, deed or scheme of**) for a composition of its debts.
— make proposals for a scheme of arrangement (see: **Arrangement, deed or scheme of**).
— have a provisional liquidator appointed.
— have a winding up order made against it and be put into liquidation (qv).
— pass a resolution for voluntary winding up and thus be put into voluntary liquidation.
— have an administrator or administrative receiver appointed.

Insolvency may cause a contractor or employer to breach the contract where, for example, there will inevitably be a resulting failure to progress the works or a default in payment. However, insolvency, *per se,* may not amount to a breach of contract. In that case, whereas they may differ significantly as to the scope of the insolvency events giving express rights of termination of the contractor's employment under the contract, most standard form construction contracts will incorporate some such express terms. Under contracts made in the terms of JCT standard forms (e.g. clauses 27.3 and 28.3 of JCT 98), where either party makes a composition or proposal of the types referred to above, or where the employer becomes bankrupt or has an administrator, administrative receiver, liquidator or provisional liquidator appointed, then on proper notice the contractor's employment under the contract may be determined. Where it is the employer who becomes insolvent then, pending effective notice the contractor's obligations to continue with the works will also be suspended. Notice of termination is not required where the contractor is made bankrupt or liquidates since, in that case, termination will be automatic at that time. Contrasted with the rather long and detailed provisions in JCT contracts, similar provisions in the 4th Edition of FIDIC are drafted much more widely.

Since important consequences flow from any action terminating obligations under the contract, all notice and other provisions setting out the timing and procedure to be followed should be rigidly adhered to.

See also: **Bankruptcy.**

Inspection of documents A clause in the bills of quantities (qv) sent out with an invitation to tender may refer to drawings not included in the set sent to the contractor but available for inspection at the office of the architect. Depending upon the precise wording, such a clause may be sufficient to incorporate the documents in any subsequent contract. Similarly, bespoke and standard form contracts commonly provide that whether or not actually inspected, on execution of the contract the contractor or sub-contractor concerned will nevertheless thereby acknowledge that he has been given the opportunity to, and is deemed (qv) to have inspected the provisions of other incorporated contract terms or documents.

Also, in both arbitration and litigation, the second stage of disclosure (qv) whereby a party may have sight and take copies of an opponent's documents which have been set out in that party's list of documents and to which no objection to disclosure has been made. This may require a party's representative to attend the other party's offices to physically inspect the documents or the party may require photocopies of all or some of the disclosed documents to be provided. In the case of the provision of photocopies, there will usually have to be an undertaking (qv) to pay reasonable photocopying charges.

See also: **Incorporation of documents; Disclosure.**

Inspection of the works Where constant inspection is required, the employer should appoint a clerk of works as inspector (qv). Under the RIBA's Standard Form of Agreement (SFA/99) and Standard Conditions of Engagement (CE/99), responsibility for advising the employer whether the nature of the works warrants additional full or part time site inspector(s) rests with the architect (clause 2.5). The architect himself is not expected to be in constant attendance on site. Nor is he required to make constant inspections. He is required to inspect only as and when reasonably necessary[376] and to the extent required in the particular circumstances of the project, to ensure the work is done to the contractually agreed standards. Whereas previous editions of RIBA Standard Forms of Appointment for Architects set out the position in somewhat more explicit terms[377], SFA/99 and CE/99 express the architect's contractual duty to be that: 'The Architect shall make visits to the Works in accordance with clause 2.8'. Clause 2.8 goes on to state simply that; 'The Architect shall in providing the Services make such visits to the Works as the Architect at the date of the appointment reasonably expected to be necessary'. Where a standard form of appointment such as SFA/99 is not used, the matters that the architect must consider when deciding how extensive his inspection regime must be may be

---

[376]For guidance on what will be considered reasonable see *Jameson* v. *Simon* (1899) 36 Sc LR 883.
[377]See clause 3.10 of the 1982 RIBA Architects Appointment.

significantly increased and might even have to take into account the competence and experience of the contractor[378].

The position of the building control officer (formerly known as the 'building inspector'), and through him the local authority, with regard to inspections has exercised several judicial minds. It is now well settled[379] that any duty of a local authority to take care in securing compliance with building by-laws extends only to a duty to take reasonable care to avoid injury to persons or property other than the defective building itself. The architect will, nevertheless, still have the primary duty. The nature and extent of the local authority's duty of care (qv) towards building owners must also be considered in light of their statutory responsibilities for public health and there must be present or imminent danger to the health or safety of the occupiers for the local authority to be liable. It is also settled that local authorities, in the exercise of their building control functions, do not thereby owe a duty of care (qv) to a building developer to see that his property does not suffer damage[380].
See also: **Conditions of Engagement.**

Inspection of the site Before building work is contemplated and/or design work is properly begun, an architect will have a general duty to his employer to first personally inspect the site to satisfy himself that it is suitable for the proposed works. This general duty is expressed in contractual terms in the Services Supplements of the commonly used Standard Forms of Appointment for Architects, e.g. SFA/99, CE/99 and SW/99 (clause 1.5).

The precise extent to which the architect will be expected to inspect the site will vary depending on the circumstances of each project, but whatever the extent the architect must personally undertake the task and cannot simply rely on information obtained from third parties. As a part of his general duty he will also be expected, in each case, to take account of any impediments caused by obvious rights over the land on which the site is to be situated. He should consider, too, such matters as whether any particular or specialist soil or site investigations are necessary. If so, he should advise his employer accordingly to engage other specialist consultants to conduct those further investigations where the architect himself does not have the requisite expertise.
See also: **Inspection of the works.**

Inspector Someone who inspects, examines and checks. Many organisations have inspectors to ensure that work or duties are being carried out correctly. In the context of building contracts the architect, project manager, resident engineer or clerk of works have the role of inspector to varying degrees. JCT 98 clause 12 defines the role of the clerk of works (qv) as solely that of inspector on behalf of the employer under the direction of the architect. The clerk of works

---

[378]*Brown & Brown* v. *Peter Gilbert Scott and Mark Payne* (1992) 35 Con LR 120.
[379]Per the House of Lords in *Murphy* v. *Brentwood District Council* [1991] AC 398, overruling its own previous decision in *Anns* v. *London Borough of Merton* [1978] 5 BLR 1.
[380]Per the decision of the House of Lords in *Peabody Donation Fund* v. *Sir Lindsay Parkinson & Co Ltd* [1984] 3 All ER 529.

is not there, however, to undertake inspection for or on behalf of the architect and to reduce or avoid altogether the architect's own duty, such as it is, to inspect. He is there to provide further and additional inspections and the architect must still, in any event, carry out such inspections as he would otherwise be expected to carry out. Through the Institute of Clerks of Works a professional approach to the role of clerk of works is encouraged and guidance concerning the duties and record keeping to be expected of a good clerk of works is published by and can be obtained from that Institute.

The Architect's Standard Forms of Appointment (forms SFA/99 and CE/99 at Schedule 2 and clauses 2.5 and 2.8) lay down the limits and responsibilities to be expected of the architect so far as the matter of inspection is concerned. He is to advise the employer whether the nature of the works warrants additional, full or part time site inspectors and for his own part: 'The Architect shall in providing the Services make such visits to the Works as the Architect at the date of the appointment reasonably expected to be necessary'.

The architect's duty to inspect is often misunderstood by the employer. He is to carry out such reasonable supervision as would enable him to form an honest opinion that the works are being executed in accordance with the contract and if necessary to certify to that effect[381]. In considering the architect's role as inspector it is important that the employer should realise that: 'The architect is not permanently on the site but appears at intervals, it may be of a week or a fortnight ... It is the contractor who is responsible for progressing the work in accordance with the requirements of the contract and the architect's instructions'[382].

Other types of inspector who have relevance to the construction industry are inspectors appointed under the provisions of the Health and Safety at Work etc. Act 1974, and at Schedule 4 of SFA/99 other professions such as the following are also designated as inspectors:
— Project manager.
— Planning supervisor.
— Quantity surveyor.
— Structural engineer.
— Building services engineer.
— Site inspector.

For a fuller discussion of the duties involved in inspection and supervision and the position of the local authority with regard to the building regulations see: **Inspection of works; Supervision of works.**

Instructions A generic term for directions, orders and certain other categories of information which the contract expressly provides may or must be issued or given to the contractor. In the standard forms of contract the word is normally used to refer to orders given to the contractor by the architect, whose power in relation to the type and timing of such instructions will be restricted by the

---

[381] *Jameson* v. *Simon* (1899) 1 F(Ct of Sess) 1211.
[382] *East Ham Borough Council* v. *Bernard Sunley & Sons Ltd* (1965) 3 All ER 619 per Lord Upjohn at 637.

express provisions of the contract (see JCT 98, clause 4.1.1, and, in somewhat broader terms, NEC clause 29.1). JCT 98 clause 12 makes clear that where a clerk of works is employed for all material purposes any instruction given solely by him will be of no effect.

Instructions may or may not have a financial implication. For example, if the architect instructs the contractor to make some alteration or modification to the design, quantity or quality of the work specified in the contract documents (qv) or measured in the contract bills (qv), then the employer may have to pay extra or, where the instruction involves an omission, may pay less. He may, on the other hand, issue such an instruction involving no additional cost and, provided he does so in good time, there should be no financial consequences to the employer. Instructions may be said to fall broadly into five categories:

— The ordering of additions, omissions, alterations, modifications and/or substitutions to the design, quality, quantity of the work or the kinds and standards of materials for use in the work, e.g. JCT 98, clause 13.
— The provision of information, procedural or clarifying instructions necessary for the works to be carried out and completed, e.g. JCT 98, clauses 2.3, 6.1.3, 7, 8.6.
— Changes in the timing, sequence or method of working, e.g. JCT 98, clauses 13.1.2, 23.2, 34.2.
— Expenditure of sums which the employer has reserved the right to expend as the works progress, e.g. JCT 98, clauses 21.2.1, 22D.1, 13.3.1, 13.3.2, 35.
— Actions in relation to work found or at least thought not to be in accordance with the contract, e.g. JCT 98, clauses 8.4.1, 8.4.3, 8.5

Although not exhaustive, a list of the more commonly applicable empowering clauses under JCT 98 is given in **Table 2**.

JCT 98, clause 4 deals specifically with the powers of the architect in relation to instructions and the procedure for their issuing. Except only in the most limited and expressly stated circumstances, the contractor must comply forthwith (qv) with any instruction given by the architect, provided that:

— The contract expressly empowers the issue of the instruction.
— The instruction is in writing.
— The contractor does not make reasonable objection to the issue of an instruction as he is entitled to do under clauses 4.1.1.1, 13.2.2 and 13.2.1.

Although the contract specifies that all instructions must be issued in writing (clause 4.3.1), and although a convenient – but not mandatory – standard pro forma for the issuing of architect's instructions is published by the RIBA and wherever possible that should be used, the following clause (clause 4.3.2) contains detailed provisions regarding what is to happen if the architect purports to issue an instruction which is not in writing. This is a necessary provision and recognises the common situation where the architect visits the site and gives an oral instruction. If the architect confirms it, the instruction takes effect from the date of the confirmation. If the architect does not confirm, the contractor must confirm within seven days and it will take effect after the expiry of a further seven days if the architect does not by then dissent. If neither architect

Table 2 The more commonly used clauses empowering the issue of instructions under JCT 98 (Private With Quantities Edition).

Clause 2.2.2.2	Instructions correcting misdescriptions, errors, omissions in bills of quantities or rectifying their non-conformance with the Standard Method of Measurement.
Clause 2.3	Instructions removing discrepancies between or in contract documents and/or other instructions.
Clause 4	General powers and requirements in relation to giving of instructions.
Clause 6.1.3	Instructions to resolve discrepancies or divergence between statutory requirements and contract documents or other (variation) instructions.
Clause 7	Instruction absolving the contractor from the obligation to correct his incorrect setting out and to make an appropriate deduction from the contract sum.
Clause 8.3	Instruction requiring tests or inspections to be made.
Clause 8.4	Instructions for dealing with non-conforming work, materials or goods.
Clause 8.5	Instructions as are reasonably necessary as a result of the contractor's or any nominated sub-contractor's failure to carry out work in a proper and workmanlike manner.
Clause 13.2	Instructions requiring or confirming a variation.
Clause 13.3	Instructions for expenditure of provisional sums.
Clause 13A	Instructions under clause 13.2 requiring advance quotation.
Clause 17.2	Issuance – on expiry of defects liability period – of a Schedule of Defects requiring, or specifically not requiring, to be made good.
Clause 17.3	Instruction – during the defects liability period – requiring, or specifically not requiring, to be made good.
Clause 23.2	Instructions concerning postponement.
Clause 34.2	Instruction on actions to be taken following discovery of antiquities.
Clause 35	Instructions for and in connection with nomination of sub-contractors.
Clause 36	Instructions for and in connection with nominated suppliers.
Clause 42	Instructions in relation to performance specified works.

nor contractor confirms, the architect may confirm at any time prior to the issue of the final certificate (qv) and the instruction takes effect retrospectively from the date it was originally issued otherwise than in writing. If the contractor asks the architect to specify in writing the clauses empowering the issue of an instruction, the architect must do so immediately. The contractor may then:

— comply with the instruction and if he does so it will be deemed to be an instruction empowered by the provision specified by the architect and the contractor may not then refer the question to adjudication or arbitration (whether or not the architect is, in fact, correct); or

— seek immediate adjudication or arbitration.

If the contractor fails to carry out an instruction within seven days of receipt of a notice from the architect requiring compliance, the employer may employ and pay other people to carry out 'any work whatsoever' that is necessary to give effect to the instruction and, subject to any other notice provisions in the contract concerning deduction or withholding of money, may deduct the costs from monies due to the contractor. The employer is entitled to deduct all extra costs and not simply the money he pays to the third party he engages to do the work. Thus, architect's and quantity surveyor's fees in connection with the additional work resulting from the contractor's default become the contractor's liability. The employer would be wise to obtain competitive tenders for the work of others, if time allows, so that the contractor has little chance of succeeding in any claim that the employer could have had the work carried out at a cheaper rate.

Other contracts give the architect similarly wide powers in relation to the issuing of instructions.

Clause 1.1 of the ACA 3 states that the contractor must comply with and adhere strictly to the architect's instructions issued under the agreement. A full list of the matters on which the architect is empowered to issue instructions and the procedures for doing so are contained in clause 8 (see **Table 3**). Clause 8.1 gives the architect authority to issue instructions at any time up to the date of taking-over and to issue instructions about removal from site of defective work, dismissal from the works of incompetent people, opening up of work for inspection and testing, altering obligations or restrictions as regards working hours, space or site access or use at any time during the maintenance period (qv) or within ten working days of its expiry. All instructions must be given in writing (clause 23.1) except in the case of an emergency (clause 8.3) when the architect may give an oral instruction and confirm it in writing within five working days. Since the contractor must immediately comply with an oral instruction under clause 8.3, he will be in a difficult position if the architect forgets to, or will not, later confirm such an instruction. There seems to be no good reason why the architect cannot issue a written instruction on site or, if the oral instruction is given by telephone, by the same day's post. If the contractor neglects to carry out instructions, the employer's remedy is to employ someone else to carry out the work (clause 12.4) and/or to invoke his right of termination (clause 20.1 (d)).

GC/Works/1 (1998) refers to instructions in clauses 40 and 56. Clause 40 (2) lists the instructions which the PM is empowered to issue (see **Table 4**). Instructions must be issued in writing, with four stated exceptions, e.g. removal and/or re-execution of work, which may be given orally and confirmed in writing within seven days of the issue of the instruction. The contractor must comply forthwith (qv) with any and all instructions given by the PM. The PM may issue the contractor with a notice requiring compliance with an instruction within a specified period. If the contractor does not comply, the authority may engage others to carry out the work and recover additional expenses from the contractor. This provision is similar to JCT 98, clause 4.1.2. A failure to comply with an instruction may also give rise to a right to determine the contract.

MW 98, clause 3.5 enables the architect to issue written instructions. The extent of his authority is not precisely set out, but there must be an implication that the instructions will be in connection with the contract. Oral instructions must be confirmed in writing within two days (clause 3.5). The clause contains a similar provision to JCT 98, clause 4.1.2.

Table 3 Instruction empowered by clause 8.1 of the ACA Form of Building Agreement Third Edition 1998.

Clause 8.1 (a)	Removal from the site of any work, materials or goods which are not in accordance with the contract or with CDM Regulations.
Clause 8.1 (b)	Dismissal from the works of any person employed on them if, in the opinion of the architect, such person misconducts himself or is incompetent or negligent in the performance of his duties.
Clause 8.1 (c)	Opening up for inspection of any work covered up or the carrying out of any tests of any materials or goods or of any executed work.
Clause 8.1 (d)	Addition, alternation or omission of any obligations or restrictions in regard to any limitations of working space or working hours, access to the site or use of any parts of the site.
Caluse 8.1 (e)	Alteration or modification of the design, quality or quantity of the works as described in the contract documents, including the addition, omission or substitution of any work, the alteration of any kind or standards of any materials or goods to be used in the works and the removal from the site of any materials or goods brought on to it by the contractor for the works.
Clause 8.1 (f)	Any matter connected with the works.
Caluse 1.5	Clarification, removal and/or resolution of any ambiguity or discrepancy within or between drawings and documents comprising the contract documents.
Clause 1.6	The manner in which any infringement by the contract drawing, documents details or instructions of the requirement(s) of any Act, or any rule, order or instrument made thereunder, or of any regulation, rule, order or by-law of any local authority or any statutory undertaker with jurisdiction over the works may be resolved.
Clause 1.7	Absolving the contractor from what would otherwise be the contractor's obligations to give all notices and pay all fees required to be given or paid by virtue of stautory requirements.
Clause 2.6	(Where clause 2.6 applies) requirements for overcoming adverse ground conditions or artificial obstructions encountered by the contractor.
Clause 3.5	Requiring the contractor to provide samples of the quality of goods, materials and/or workmanship.
Clause 9.4	Naming of a sub-contractor from a list of persons named to carry out work which in the contract documents is made the subject of a provisional sum.
Clause 9.5	Instructing expenditure of a provisional sum by the nomination of a particular sub-contractor to be engaged by the contractor to carry out the work concerned.
Clause 10.2	Ordering the contractor to allow access to the works by others engaged to carry out work or supply goods not forming part of the building contract.
Clause 11.8	Instructing the contractor to accelerate or postpone take-over dates given on the time schedule for all or any specified sections of the works.
Clause 12.2	Requiring repairs or remedial works to be done immediately during the maintenance period.
Clause 14	Instruct on action to be taken on the discovery of fossils, antiquities, remains, structures and the like during the course of the works.

Table 4 Instructions empowered by clause 40 (2) of GC/Works/1 Edition 3.

Instructions may be given in relation to all or any of the following matters:

(a)   Variation or modification of all or any part of the specification, drawings or bills of quantities, or the design, quality or quantity of the works.

(b)   Resolving discrepancies in or between the specification, drawings and/or bills of quantities forming a part of the contract documents.

(c)   Removal from site and substitution of things brought on by the contractor for incorporation into the works.

(d)   Removal and/or re-execution of any work already done by the contractor.

(e)   The order of execution of all or part(s) of the works and/or the hours of work or extent of night time or overtime working to be adopted.

(g)   Suspension of execution of all or part(s) of the works.

(h)   Replacement of employee engaged in connection with the contract.

(i)   The opening up for inspection of any work covered up.

(j)   Altering and/or or making good defects under Condition 21 (Defects in maintenance periods).

(k)   Cost savings under Condition 38 (Acceleration and cost savings).

(l)   The execution of any emergency work as mentioned in Condition 54 (Emergency work).

(m)   The use or disposal of material obtained from excavations, demolition or dismantling on the site.

(n)   The actions to be taken following discovery of fossils, antiquities or objects of interest or value.

(o)   Measures to avoid nuisance or pollution.

(p)   Quality control accreditation of the contractor as mentioned in Condition 31 (Quality).

(q)   Any other matter which the PM considers necessary or expedient.

Generally, in order to qualify as a written instruction, there must be evidence in writing together with an unmistakable intention to order something. An instruction may be implied from what is written down, but it is safer from the contractor's point of view to ensure that the words clearly instruct. For example, a drawing sent to a contractor with a compliments slip is not an instruction to carry out the work shown thereon. It may be deemed to be an invitation to carry out the work at no cost to the employer. The same comment applies to copy letters sent under cover of a compliments slip. An instruction on a printed 'Architect's Instruction' form is valid if signed by the architect. An ordinary letter is a valid instruction. The minutes of a site meeting may be a valid instruction if the contents are expressed clearly and unequivocally and particularly if the architect is responsible for the production of the minutes. **Figure 8** is an example of an instruction.

Instrument A word with several meanings, but for the purpose of this book an instrument is a formal legal document, e.g. a statutory instrument (qv). The

Instrument

Issued by: E W Pugin
address: Gothic Buildings
Birmingham

Employer: The Duke of Omnium
address: Trollope House
Belstead

Contractor: Gerrybuilders Ltd
address: The Crescent
Belstead

Works: Mansion
situated at: Belstead Gardens, Belstead

Contract dated: 1 February 2001

**Architect's
Instruction**

Job reference: EWP/ABC

Instruction no: 5

Issue date: 3 September 2001

Sheet: 1   of 1

Under the terms of the above-mentioned Contract, I/we issue the following instructions:

	Office use: Approximate costs	
	£ omit	£ add

Pursuant to clause 13.1.1

OMIT Bill of Quantities item 35F (MS Brackets)
500 no.

ADD Stainless Steel Curvilinear Brackets
450 x 450 x 100 mm obtained from Fancy
Fittings Ltd Cat No FF1851/1c
500 no.

To be signed by or for
the issuer named
above

Signed *E W Pugin*.

Amount of Contract Sum	£
± Approximate value of previous Instructions	£
Sub-total	£
± Approximate value of this Instruction	£
Approximate adjusted total	£

Distribution
[x] Contractor [x] Quantity Surveyor [x] Clerk of Works [ ]
[x] Employer [ ] Structural Engineer [ ] Planning Supervisor [ ]
[ ] Nominated Sub-Contractors [ ] M&E Consultant [ ] [x] File

F809 for JCT 98/IFC 98/MW 98

© RIBA Publications 1999

Figure 8   Architect's instruction.

word is also used in a legal context to indicate an important factor in something, e.g. 'her evidence was an instrument in his arrest'[383].

Insurance A concept whereby a contract is made which provides one of the parties to the contract (the insured) with financial protection from the other party (the insurer) against the risk of one or more specified but uncertain event(s) occurring, which, on its occurrence, will generally have adverse consequences for the insured. The 'uncertainty' may be as to the happening of the event or as to the timing of an otherwise certain event (such as in the case of life insurance).

Contracts of insurance have evolved from beginnings which were almost exclusively concerned with marine, fire and life insurance to the present day where the scope of the risks that may be insured against are almost infinite (e.g. professional indemnity, all risks, specified perils, travel, life, illness, weather and even holiday insurance). Although the majority of insurance contracts are a matter of choice, some, such as, for example, motor vehicle and public and employer's liability insurance, are required by law. For the purposes of this book, insurance can be conveniently classified by risk as:

— Marine, aviation and transport insurance.
— Personal accident insurance.
— Life insurance.
— Property insurance.
— Liability insurance.
— Pecuniary loss insurance.
— War risks, terrorism and such insurance.

Insurance is a very specialised field and irrespective of the type required, the advice of a broker should always be sought. All the standard forms of contract contain insurance provisions but it should be noted that such standard provisions may not be adequate for the specific requirements of the individuals or projects concerned. Such standard terms should not simply be automatically taken as providing adequate cover for each and every project. The contract administrator on no account must simply pass details of the contractor's policy to the employer without comment. He has three options:

— give the advice himself; or
— obtain expert advice and pass this to the employer; or
— advise the employer to obtain expert advice[384].

Moreover, each employer's particular circumstances and the particular nature and requirements of each individual construction project should be considered closely in the context of the proposed standard insurance provisions to ensure no shortfall in the amount or nature of the cover that will be appropriate.

GC/Works/1 (1998), clause 8 provides alternative insurance provisions and requires insurance to be taken out. JCT 98, clause 21 refers to the arrangements for insuring against injury to persons and property and clause

---

[383]*Collins English Dictionary.*
[384]*Pozzolanic Lytag Ltd* v. *Bryan Hobson Associates* [1999] BLR 267.

22 makes provision for insurance of the works against what are termed 'All Risks' (qv) by either contractor or employer and where existing property is involved against what are referred to as 'Specified Perils' (qv). There are alternatives to suit the kind of developments being insured and who is to bear the cost of premiums. There is an optional provision (clause 22D) requiring insurance against the employer's loss of liquidated damages. IFC 98 has very similar insurance provisions to those in JCT 98 as does MW 98, albeit that the MW 98 provisions are less detailed. Clause 6 requires the contractor to insure against personal injury or death and injury or damage to property other than the works with insurance in joint names being required under that same clause 6 in respect of damage to the works by one or other of the perils specified in that clause 6. Where the works involve existing structures, then provided it can be obtained, that insurance and insurance of the existing structure along with its contents will be the responsibility of the employer.

Under MW 98, provisions similar to those in JCT 98 exist under clause 6 with alternative provision being made for either contractor or employer to take out 'All Risks' (qv), or such other definition of insurance cover as the employer instructs, under a joint names policy (qv). Where the work involves existing structures and their contents, provision also exists for the employer to take out and maintain a joint names policy to cover specified perils (qv). The NEC (qv) contract deals with the parties' liabilities to insure in a rather less orthodox, but nonetheless effective way by the combined provisions in clause 8 and the particular requirements stated in the contract data.

ACA 2 deals with insurance against injury to persons and property in clause 6.3. There are alternative provisions for insurance of the works in clause 6.4 with alternative 2 (insurance by employer) catering, also, for insurance of existing structures and their contents (if any) associated with the works. Clause 6.5 provides for there to be insurance cover, if so required, against injury to other property not caused by negligence etc. Clause 6.6 is a deletable clause to be used if the contractor is responsible for design of any work, goods and/or materials. It provides for professional indemnity cover and clauses 6.7 to 6.10 (inclusive) deal with premiums, breach and claims.

See also: **Uberrimae fidei.**

Insurrection A term meaning an uprising against state authority, rather less in ramification than outright revolution. It is expressly referred to in ACA 3, clause 11.5, alternative 2, as a ground for awarding an extension of time (qv). It is a ground for extending time in other forms of contract, usually under the head of *force majeure* (qv).

See also: **Civil commotion; Civil war; Commotion; Disorder; Riot.**

Interest on money Interest may describe the costs to a borrower of financing those borrowings (e.g. the cost of a bank overdraft) or the sum which accrues or would have accrued on amounts held on deposit with a bank (deposit interest). In the context of building contracts the term generally is used in the former context, that is to say where a claim is made by a contractor against an employer for the costs to the contractor of overdraft to fund an overdue payment, for so long as the payment remains unpaid. It is a long established principle that, in the absence of express provision in the contract giving an entitlement to payment of interest on overdue debts, interest is not payable as general damages (qv) resulting from that breach of contract[385]. This rule, although much criticised, has been reaffirmed by the House of Lords[386].

The rule applies only to claims for interest by way of general damages. A claim for interest may be made as a claim for interest as special damage (qv), that is to say: 'if a plaintiff pleads and can prove that he has suffered special damage as a result of the defendant's failure to perform his obligation under a contract, and such damage is not too remote, on the principle of *Hadley* v. *Baxendale* it is recoverable'[387].

The contract itself may provide for the payment of interest on amounts overdue, and this is commonly done in contracts for the sale of goods and in some construction contracts (e.g. JCT 98, clause 30.1.1.1; NEC clauses 51.4 and 51.5 and IFC 98, clause 4.2), by inserting an express term to that effect. Limited statutory rights to interest on late payment have also now been introduced under the Late Payment of Commercial Debts (Interest) Act 1998 (qv) which has been in force since November of that year. This is a radical departure from the common law rule described above. A contract for supply of goods or services coming within the scope of the act[388] will have implied terms to the effect that a 'qualifying debt' arising under the contract will attract statutory (simple) interest. Only if the parties' contract makes alternative express provisions for some other 'substantial remedy'[389] for late payment will it be possible to exclude the right to statutory interest. Otherwise a term purporting to do so will be void (qv).

Subject to the effect of any express contractual agreements made by the parties in relation to the treatment of interest and/or to the effect of any other statutory right to interest, the courts and arbitrators also have statutory powers to award interest on sums for which judgment is given or an award is made. Arbitrators have considerable discretion as to awards in respect of both pre and post award liability for interest. That discretion goes to matters such as the rate, the period and/or the basis – i.e. whether it should be compound or simple interest – that will be applied.

In arbitration, subject to any particular agreement made between the parties, the arbitrator's powers to award interest are governed generally by the Arbitration Act 1996, s. 49. In litigation, the court's discretionary powers to award interest are governed by a series of legislation and primarily by s. 3 of the Law Reform (Miscellaneous Provisions) Act 1934 as amended by the Supreme Court Act 1981, the County Court Act 1984 and the Civil Procedures

---

[385] *London, Chatham & Dover Railway Co* v. *South Eastern Railway Co* (1893) AC 429.
[386] *President of India* v. *La Pintada Compania Navegacion SA* [1984] 2 All ER 773.
[387] *Wadsworth* v. *Lydall* [1981] 2 All ER 401 CA; *Holbeach Plant Hire Ltd* v. *Anglian Water Authority* [1989] 14 Con LR 101.
[388] Those contracts to which the Act applies are defined at s. 2.
[389] For the meaning and definition of what will suffice as a 'substantial remedy' in this context see s. 8.-(4) and s. 9 of the Late Payment of Commercial Debts (Interest) Act 1998.

Act 1997. Additionally, the court has an equitable jurisdiction to award interest in relation to its equitable remedies, e.g. recission.

Subject to any contractual (or other statutory) provisions for interest, the courts may allow interest to be included in the sum for which judgment is given. Interest on judgment debts will run from the time prescribed by the rules of court[390]. The courts may also award interest alone where payment of an agreed debt has been made after commencement of proceedings, but before judgment is given.

Following the decision of the Court of Appeal in *F.G. Minter Ltd* v. *Welsh Health Technical Services Organisation* (1980)[391] it is beyond doubt that under the 'direct loss and/or expense' provisions of the JCT contracts (i.e. JCT 63, clause 24 (1), 11 (6), and 34 (3); JCT 80, clause 26.1 and 34.3; JCT 98, clause 26 and 34.3; IFC 84, clause 4.11 and IFC 98, clause 4.11) interest or financing charges must be included by the architect or quantity surveyor in an ascertainment of loss and/or expense. The same principle applies to similar wording in other standard form contracts. Under ACA 3, clause 7.1 the phrase used is 'damage, loss and/or expense', and clause 7.5 (dealing with the contractor's failure to submit estimates) refers specifically to interest or financing charges and debars the contractor from recovering them between the date of the contractor's failure to submit the estimates required by clause 7 and the date of the final certificate. In general, interest is to be calculated from the date when the loss etc. was incurred but the actual wording of the clause must always be considered. The wording in JCT 98, clause 26, for example, requires that the ascertainment must relate to loss and/or expense incurred or likely to be incurred – and thus covers future losses – whereas under JCT 63 the wording referred only to losses in the past.

GC/Works/1 (1998), clause 47 deals specifically and at some length with the subject of interest and/or finance charges. It provides that neither will be paid or allowed unless the employer or his agents cause payment to be made late or the QS changes an earlier decision on payment. The clause also deals at length with the method of calculating any interest or finance charges that might be due and gives the QS discretion to take into account any previous over-payments when calculating such interest or finance charges as would otherwise be due.

So far as the rate of interest is concerned, if the contract documents do not provide a rate, it is suggested that the correct thing to do is to take the rate at which contractors in general could borrow, disregarding the special circumstances of the particular claimant[392]. The whole common law position regarding interest and finance charges (qv) is complex and for a further useful guide on the subject see Powell-Smith and Sims' *Building Contract Claims,* 3rd edn, 1998, pp. 176–179, Blackwell Science.

See also: **Claims; Direct loss and/or expense.**

---

[390]SI 1998/2940.
[391](1980) 13 BLR 1.
[392]*Tate & Lyle Food and Distribution Co Ltd* v. *Greater London Council* [1982] 1 WLR 149.

Interference A wrongful act or omission which generally impedes or inhibits arrangements for the progress or administration of the contract. Subject to any express terms to the contrary, it will be an implied term (qv) of building and engineering contracts that the employer will not interfere with progress[393] or with the certification[394] or administration processes required under the contract. If the employer is guilty of interference, the contractor may be able to repudiate the contract or claim damages at common law, depending upon the circumstances. The employer will also be liable for the interference of third parties for whom he is responsible unless provision is made in the contract. The employer will not be liable if the wrongful interference is caused by a third person for whom he is not responsible in law, e.g. an adjoining owner[395]. JCT 98, clause 28.2.1.2, expressly states that if the employer interferes with or obstructs the issue of any certificate (not only financial certificates), the contractor has grounds to determine his own employment under the contract. Where an employer attempts to influence the contents of, or otherwise interferes with or prevents the issuing of a certificate for payment, then in addition to any other right the contractor may have under the contract, he will generally be able to insist on payment since the employer cannot rely on the absence of that certificate to avoid payment[396]. If he so wished, the contractor could alternatively claim the sum due as damages for the breach of contract caused by the employer's interference. Proving that the employer has interfered may not be easy but an employer will be deemed to have interfered with the certification process where, for example, he attempts to influence the timing of the issue of a practical completion certificate or the granting of an extension of time or attempts to influence the timing or contents of a payment certificate or issues directions to his architect or quantity surveyor as to adopting a method of valuation not provided for under the contract.

Interference may be by a positive act, such as where, contrary to the express or implied terms of the contract, the employer seeks actively to impose restrictions or limitations on progress or administration. Alternatively, it can be the result of some omission or inactivity where, for example, the express or implied terms of the contract require some action to be taken by the employer to allow unimpeded progress or administration and the employer fails to take the appropriate action[397]. Interference must be distinguished from express rights under the contract which entitle the employer to influence the progress or carrying out of the works. For example, JCT 98, clause 29 allows the employer in certain circumstances to engage others to do work on site. Any problems which the contractor may encounter as a result can be dealt with by extension of time (clause 25.4.8.1) and/or by financial recompense (clause 26.2.4. 1). Dictating the order in which the contractor is to carry out the work

---

[393]*London Borough of Merton* v. *Stanley Hugh Leach* (1985) 32 BLR 51 per Vinelot J at 79.
[394]For example, *Croudace Construction Ltd* v. *London Borough of Lambeth* (1986) 6 Con LR 70.
[395]*Porter* v. *Tottenham Urban District Council* [1915] 1 KB 776.
[396]*Roberts* v. *Bury Commissioners* (1870) LR 5 CP 310 per Blackburn J at 326.
[397]*Ellis-Don Ltd* v. *The Parking Authority of Toronto* (1978) 28 BLR 98.

would also normally amount to interference, but clause 13.1.2.4 allows the architect to vary such requirements albeit with financial implications.

Similar terms concerning interference appear in many of the standard form contracts commonly in use. ACA 3, clause 20.2 (b) allows the contractor to terminate his own employment if the employer obstructs the issue of any certificate. Clause 10 permits the employer to engage others to carry out work on site and there is provision (clause 10.4) for the contractor to be paid damage, loss and/or expense if such other persons cause disruption (qv) to the regular progress of the works. The architect may issue instructions amending limitation of working space, working hours, access to the site or use of any part of the site (clause 8.1 (d)).

Under GC/Works/1 (1998) express provision for the contractor to determine his employment due to the interference of the authority in the issue of a certificate is given by clause 58 (3) (b). Clause 65 allows the authority to execute other works on the site and if any damage is done to the contractor's works thereby, the authority must take financial responsibility. MW 98, clause 7.2.2 gives the contractor the contractual right to determine his employment if the employer interferes with or obstructs the carrying out of the works.

The provisions in the various standard forms are intended to avoid the danger of the contractor seeking to repudiate the contract in the event of what may be considered to be some fairly normal occurrences, e.g. the employer bringing specialists on to the site to carry out certain works and thereby causing the contractor some delay or loss. Contractual provisions for the contractor to determine his own employment in the event of certain specified kinds of interference are in addition to the contractor's normal rights at common law. Many employers do not appreciate the legal position and it is something which architects should be most careful to clarify at the beginning of the contract. See also: **Hindrance or prevention.**

Interim certificates A term found in most standard forms referring to the periodic certification of money due to the contractor. All 'construction contracts' (qv) within the meaning of the Housing Grants, Construction and Regeneration Act 1996 (qv) must make provision for payment by instalments. Provisions for interim certification of money within all JCT contracts comply with that requirement. JCT 98, clause 30.1 has long and detailed provisions requiring the architect to issue interim certificates stating the amount due to the contractor from the employer and, subject to any agreement to the contrary providing for stage payments to be made, the amount certified will comprise:

Amounts that are subject to retention:
— Value of properly executed work, including work for which a priced statement (qv) has been accepted.
— Value of materials and goods properly and timeously delivered to and stored on site.
— Value of any listed items of materials or goods, including those that may be of a prefabricated nature which may be stored off site.

— Value of nominated sub-contract works and materials and contractor's profit thereon.
— Fluctuations due under clause 40.

Amounts that are not subject to retention:

— Payments or costs incurred by the contractor related to statutory charges, setting out, opening up and testing, royalties, defects liability, insurance under clause 21.2.3, 22B.2 and 22C.3.
— Loss and/or expense.
— Amounts ascertained in respect of any restoration, replacement or repair, etc. associated with claim made under all risks policies of insurances taken out by the employer.
— Final payments to nominated sub-contractors.
— Fluctuations due under clause 38 or 39.
— Certain specified nominated sub-contract costs.

Less fluctuations due to the employer (not subject to retention) and previous amounts certified.

Subject to there being no obligation on the architect to issue interim certificates within less than one month of each other, interim certificates must be issued under JCT 98 at the intervals which the parties have specifically agreed upon and which they have stipulated in the Appendix (qv) to the contract. If no period is specified, then by default the period will be set at monthly intervals. The issue of regular interim certificates ends at practical completion (qv). After that, they may be issued:

— As and when further amounts are ascertained (qv) as payable.
— After the end of the defects liability period (qv) or when the certificate of making good defects is issued, whichever is the later.

The dates for the issuing of those further interim certificates are likewise subject to the proviso that the architect is not in any case required to issue any such interim certificate within one calendar month (qv) of a previous issue.

The provisions of IFC 98 are somewhat shorter but very similar to those of JCT 98. Again they comply with the Housing Grants, Construction and Regeneration Act 1996 as do the relatively simple provisions for interim certification of payments under the NEC (qv). In essence, by clause 51 of NEC the project manager (qv) is required to certify interim payments due to the contractor within one week of him having asssessed the amount to be certified. The assessment intervals must be specified in the contract data when the contract is made and the periods begin to run from when the project manager first considers an assessment should be made following the start date. The period(s) should not, however, exceed five weeks.

ACA 3, clause 16 sets out the procedure for the issue of interim certificates under that contract. On the last working day (see: **Day**) of each calendar month, the contractor must present to the architect an interim application stating the total amount due in accordance with clause 16.1 together with supporting documentation. Within ten working days of receipt, the architect must issue an interim certificate stating the amount due to the contractor. Clause 16 and clause 18 (if used) state what must be included and what may be deducted when calculating the amount and the contractor is entitled to payment on the tenth working day after the date of each interim certificate. That will be the final date for payment for the purposes of the Housing Grants, Construction and Regeneration Act 1996.

Under GC/Works/1 (1998) provision is made, under clause 48, for what are termed 'Advances on Account' (qv). A detailed certification process relating to those advances is described under clause 50. The contractor is entitled to be paid advances on the contract based either on a stage payment chart (clause 48, alternative A) or on the value of the work done, materials supplied to site and other matters specified (under clause 48, alternative B), calculated and certified at monthly intervals. A further alternative, using a milestone payment chart, is also available for use under clause 48, alternative B. The PM must certify payments but no certificate is expressed to be conclusive that the work or things to which it relates are in accordance with the contract.

MW 98 provides for 'progress payments' under clause 4.2. The architect must certify the payments at intervals of not less than four weeks. The clause specifies what must be included and what must be deducted from the certificate.

The amounts included in interim certificates under all the above mentioned forms are subject to revision in the next certificate. This means that if the amount certified is too much, the next certificate can reduce it and vice versa. The process is usually simple because the value of work done is cumulative. That does not mean that the architect should not take great care in certifying interim certificates[398] since there is always the danger that the contractor may go into liquidation (qv) or may otherwise leave the site before completion (qv). In *Townsend* v. *Stone Toms & Partners*[399] it was held to be a clear breach of contractual duty for the architect to certify work which he knows has not been done properly.

Following the enactment of the Housing Grants, Construction and Regeneration Act 1996, the provisions in standard form construction contracts concerning interim payment have undergone significant changes. In addition to detailed provisions for valuation and certification a number of other important provisions affecting the parties' rights in relation to interim certification and payment have been introduced. They include:

— Rights concerning the making of interim applications for payment by the contractor.
— Rights and obligations in relation to the giving of notices further and in addition to the certification process in relation to amounts stated in certificates to be due to the contractor.
— Rights and obligations in relation to the giving of notice prior to making any set-off or withholding of any sums from payments otherwise due to the contractor.
— Rights of suspension by the contractor in the event of default on payment by the employer.

---

[398] *Sutcliffe* v. *Thackrah* [1974] 1 All ER 319.
[399] (1985) 27 BLR 26.

See also: **Certificates; Final certificate; Housing Grants, Construction and Regeneration Act 1996; Interim Payment.**

Interim Payment A phrase referring to payment made by instalments, periodically or in stages during the progress of a building contract. Where the works are carried out under a qualifying construction contract (qv) as defined by the Housing Grants, Construction and Regeneration Act 1996[400], in the absence of express terms to that effect there will be implied terms (qv) that interim payments (referred to in the Act as instalments, stage or periodic payments) shall be made to the contractor during the course of the works. Although differing widely in approach, all currently used standard form construction contracts must, therefore, include as a minimum, terms for:

— Periodic payment where the contract work is specified to last 45 days or more.
— An adequate mechanism for determining the amount of each periodic payment.
— An adequate mechanism for determining the intervals at which, or circumstances in which, periodic payments will become due.
— A final date for payment in relation to each periodic payment that becomes due.
— Notice of payments due to be made and of any intention to withhold any or all of the amount due.
— Rights for the contractor to suspend further performance of the works unless and until the employer meets his payment obligations.

All such contracts must also make provision for the contractor to receive written notice[401] of the amount the employer proposes to pay along with details of how that amount is calculated. If the employer then intends to withhold all or any part of a payment otherwise due to the contractor, he must give written notice of his intention to do so[402], setting out the amount being withheld, the ground(s) for such withholding and the calculation(s) that go to make up the sum(s) being withheld. If no or no suitable express terms to that effect are written into the contract, then suitable terms, found in the Scheme for Construction Contracts (see: **Housing Grants, Construction and Regeneration Act 1996**), will be implied.

There are strict time limits imposed for the giving of those statutory notices. According to the Act, subject to any different agreement as to timing, notice of the amount due to be paid must be given not later than five days after the due date for payment. Where it is proposed to withhold any amount after the final date for payment, proper advance notice to that effect (i.e. a 'withholding notice') must also be given. The amount of advance warning to be given is again open to be agreed by the parties, but in default of agreement the statutory period will be not later than seven days before the final date for payment. A failure to meet the time constraints laid down in the contract or, if

---

[400]Sections 104.-(1) to 107.-(1) of the Act.
[401]Section 110.-(1) of the Act.
[402]A withholding notice under s. 111 of the Act.

applicable, the Act, will have serious implications for the employer. Irrespective of the validity of the employer's counterclaim or set-off in principle, a failure to meet those time constraints or other procedural irregularity in relation to such notices will result, at least in the short term, in the contractor being able to raise a potentially indefencible claim to payment in full, without deduction; added to which the contractor will have a right to suspend further performance of the works unless and until he is paid properly, in accordance with the contract.

By Amendment No. 2 issued January 2000, clause 30.1.3 of JCT 98 now provides for specific dates to be stated for the issue of interim certificates, thus replacing the former provision whereby such certificates were to be issued by reference to periods stated in the Appendix. The Appendix entry relating to clause 30.1.3 has likewise been amended to give effect to this change. Where the parties neglect to specify a date, the Appendix entry provides a fall-back position so that, in that case certificates must be issued 'at intervals not exceeding one month up to the date of Practical Completion'. Interim payment provisions meeting the various requirements of the Act are likewise embodied in the IFC 98 by means of an Amendment No. 2 to clause 4.2 (a) and the Appendix entry (and/or the Appendix (sectional completion) entry) on clause 4.2 (a).

Periodic payments are also provided for under all other commonly used standard form contracts including, for example, MW 98, clause 4.2.1, the NEC, clauses 50 and 51, and GC/Works/1 (1998) clauses 48 to 50 (A).

See also: **Interim certificates.**

Interim remedy (or application) A new term introduced by the Civil Procedure Rules (CPR) (qv) to describe any transitory, temporary or intermediate order, application, hearing, etc. which is designed to have effect only until the occurrence of a specified event (usually a trial) or expiry of a defined period of time or is designed to regulate the procedural aspects of the litigation (e.g. an order for disclosure (qv)).

Interlocutory A word formerly used to describe the various applications, hearings, etc. which are stages in litigation or arbitration. An *interlocutory judgment* is one which is not final or which disposes only of part of the matter at issue.

Intermediate Form of Contract (IFC 98) First produced by the Joint Contracts Tribunal in 1984 in response to a general demand for a form that was less complex than the JCT 80 Standard Form but more comprehensive than MW 80. A new 1998 Edition based on the previous 1984 Edition incorporating Amendments 1 to 12 but with other various amendments and corrections was published by the JCT in November 1998.

The IFC 98 now incorporates provisions suitable for use where partial possession will arise (at clause 2.11) and other provisions have also been added to incorporate agreement for the use of electronic data interchange (EDI) (qv). The Articles of Agreement have now also been restructured and footnote lettering has been revised.

Those various revisions, additions and consequential alterations and omissions in the IFC bring it into line with recent developments in the common law and legislation. However, the form remains suitable for and is intended to be used where contract and proposed building works are:
— To be administered by a professional consultant engaged by the employer.
— Of simple content involving the normally recognised basic trades and skills of the industry.
— Without any building service installations of a complex nature, or other specialist work of a complex nature.
— Adequately specified, or specified and billed, as appropriate prior to the invitation of tenders.

The standard form is intended for use in England and Wales but, if to be used in Northern Ireland it must be modified by incorporation of an Adaptation Schedule published by the Royal Society of Ulster Architects. If for use in Scotland appropriate amendments will again have to be made[403].

The form is normally most suitable for use if the contract period is not more than 12 months. Although not to be taken as a deciding factor, the value of work for which this contract may generally be considered suitable will be in the region of £90 000 to £380 000 (at 1998 prices). It may be suitable for rather larger or longer contracts but it should be borne in mind that the provisions are less detailed than the JCT 98 standard form and, if used for unsuitable works, there may be cases where the equitable treatment of the parties could be prejudiced. Contracts to the value of £90 000 (1998 prices) would normally be carried out under MW 98. Guidance on the use of the form is provided in Practice Notes 20 and IN/1. Conversely, the form is not significantly less complex than JCT 98 and this latter form does provide more flexibility.

The form may be used with drawings and either specification (qv), or schedules of work, or bills of quantities (qv). It is a lump sum contract (qv) with provision for interim payments (qv). The clauses are as follows:
— Recitals.
— Articles of Agreement.
— Attestation provisions.
— Conditions, which make express provision concerning:
  1. Intentions of the parties.
  2. Possession and completion.
  3. Control of the works.
  4. Payment.
  5. Statutory obligations, etc.
  6. Injury, damage and insurance.
  7. Determination.
  8. Interpretation, etc.
  9. Settlement of disputes – adjudication, arbitration, legal proceedings.
— Appendix.

---

[403]See, for example, footnote [hh] to the IFC 98.

— Annex 1 to Appendix – Terms of bonds for:
  • Advance payment bond.
  • Bond in respect of payment for off-site materials and/or goods.
— Supplemental Conditions:
  A. Value Added Tax.
  B. Statutory tax deduction scheme.
  C. Contributions, levy and fluctuations.
  D. Use of price adjustment formulae.
— Annex 2 to the Conditions.
— Supplemental Provisions for EDI.

Two further forms are produced by the JCT for use with ICF 98:

— JCT Form of Tender and Agreement (NAM/T) for use between contractor and a sub-contractor named by the employer.
— Sub-contract Conditions (NAM/SC).

The RIBA has produced a RIBA/CASEC Form of Agreement to be used by the employer when inviting tenders or approximate estimates for the sub-contract works. For a useful and comprehensive guide to the terms of the IFC 98 see, Chappell & Powell-Smith, *The JCT Intermediate Form of Contract,* 2nd edn, 1999, Blackwell Science.

Interpretation clauses To assist in the interpretation of the intentions of the parties, many standard forms include a clause defining particular words and phrases used in the contract. Examples are JCT 98, clause 1; IFC 98, clause 8; GC/Works/1, clause 1; ACA 3, clause 23.
See also: **Interpretation of contracts; Identified terms.**

Interpretation of contracts Technically, the process of interpreting the meaning and legal effect of the words used in a written contract where their meaning is unclear or ambiguous is referred to as 'construing' the contract. It is the expressed intention of the parties which is important and this is to be found by ascertaining the meaning of the words actually used. The courts have no power to modify the contract in any way. Extrinsic ('parol' (qv)) evidence is not normally admissible, although there are well-defined exceptions to this rule. The first source of reference to discover the meaning of a word is a dictionary, but both courts and arbitrators must give effect to any special, technical, trade or customary meaning which the parties intended the word to bear. All but the simplest contracts will generally contain a definitions clause which will assist in establishing precisely the intended meaning to be given to certain important words and phrases used throughout the contract, e.g. JCT 98, clause 1.3.

The main basic rule of interpretation is that the contract must be read as a whole – a particular clause must be seen in context and cannot be read in isolation: 'The contract must be construed as a whole, effect being given, so far as practicable, to each of its provisions'[404]. This point is often overlooked by those without formal legal training, who will seize on a particular word or phrase out of context.

---

[404]*Brodie* v. *Cardiff Corporation* [1919] AC 337.

In the building industry, the definition of the 'contract documents' (qv) is important. All well-drafted contracts will give a clear definition of them. Thus the JCT contracts give a comprehensive definition and in interpreting the contract it is to these documents that one looks. The wording of the contract may also attempt to introduce interpretative rules of its own. This is so with JCT 98, clause 2.2.1, which has the effect of making the printed conditions prevail over typed or handwritten documents where there is a conflict between them. This reverses the normal and logical rule. The validity of this provision has been upheld time and time again[405] and the printed conditions will prevail over any typed provisions in the bills or specifications. At best, the bills etc. may be used not in the interpretation of the contract but in order to follow exactly what is going on and presumably as part of the surrounding circumstances. ACA 3 overcomes this difficulty by providing (clause 1.3) for 'priority of documents' (qv). The court will disregard completely meaningless words and phrases. But the judicial task is to interpret the intentions of the parties and not to write a contract for them. Apparent inconsistencies between contract clauses will be reconciled if it is possible to do so, otherwise the court will give effect to the clause which, in its view, expresses the true intention of the parties.
See also: **Construction.**

Intervening cause A happening or event which breaks the chain of causation (qv).

Invalidate To put an end to the validity of something. The word is used in MW 98 (clause 3.6). No instruction of the architect requiring additions, omissions or alterations to the works will invalidate the contract. It is merely a statement of the position at common law since variations (qv) are provided for in the contract and complying with a provision can never alone invalidate the contract. ACA 3 makes no reference to the point whilst JCT 98 and IFC 98 use the term 'vitiate' (qv) which amounts to the same thing.
See also: **Vitiate.**

Invitation to tender A preliminary procedure to the formation of a building contract. The architect or project manager (the client's representative in the BPF System (qv)) is normally responsible for inviting tenders from interested contractors. Unless it states to the contrary (such as, for example, where the employer specifically states that he will accept the lowest tender or where the invitation is a final part of some ongoing negotiation), an invitation to tender generally does not amount to an offer (qv) in contractual terms. It is merely an invitation to the contractor to make an offer and thus it is generally unnecessary to add a specific note stating that the employer will not be bound to accept the lowest or any tender that may be submitted. As a general rule, the costs incurred by an unsuccessful contractor in preparing his tender, although possibly quite significant, will be at the contractor's risk and will not be recoverable from the employer. However, circumstances can arise where, on the particular

---

[405]See, for example, *M.J. Gleeson (Contractors) Ltd* v. *London Borough of Hillingdon* (1970) 215 EG 165.

wording of the invitation, the employer may be liable for those costs where he has undertaken to but has not given the tender proper consideration[406]. Where tenders are to be invited in relation to the procurement of Public Works Contracts care must be taken to ensure compliance with all of the requirements and procedures of EEC Directives concerning the co-ordination of procedures for the award of public works contracts[407] and with their UK counterparts[408]. See also: **Code of Procedure for Single-stage Selective Tendering 1977; Code of Procedure for Two-stage Selective Tendering 1983; Tender.**

Invitation to treat An invitation by one party to another to make an offer (qv) which, if accepted, becomes the basis of a binding contract (qv). The most common example is the display of goods in a shop window. Even if price tags are attached to the goods, it is not an offer by the shop but an invitation to treat, i.e. an invitation to the passer-by to go into the shop and offer to buy the goods at the price shown (or indeed at any price). The shop may refuse to accept the offer and no contract results in that case.
See also: **Contract.**

Invitee A person who is invited on to an occupier's premises with the occupier's consent, express or implied, and to whom a common duty of care is owed under the Occupier's Liability Act 1957, which defines the occupier's duty towards his 'visitors'. Everyday examples of an invitee are the milkman, postman and newspaper boy, as well as guests and tradesmen.
See also: **Occupiers' liability.**

---

[406] *Blackpool & Fylde Aero Club* v. *Blackpool Borough Council* [1990] 1 WLR 1195.
[407] Such as, for example, Council Directive 71/305/EEC.
[408] For example, The Public Services Contracts Regulations 1993, SI 1993/3228.

JCLI Form for Landscape Works A form of contract suitable for landscape works produced by the Joint Council for Landscape Industries. Modelled generally on the JCT Minor Works Form of Contract the form also includes a number of other useful provisions some, although not all, of which are of particular relevance to landscaping operations, e.g.:
— Partial possession.
— Plant failures.
— Objections to nomination.
— Disturbance to regular progress.
— Retention.
— Fluctuations.
— Malicious damage and theft.

Since first published, the form has undergone numerous thorough revisions with the current edition, published in 1999, now drafted to take account of the various obligations arising from the introduction of the Construction Design and Management (CDM) Regulations (qv) and the Housing Grants, Construction and Regeneration Act 1996 (qv). Under the latter, landscaping contracts are specifically stated to be 'construction contracts' (qv) within the meaning of that Act.

Despite the additional provisions described above, the JCLI contract remains closely comparable with the Agreement for Minor Building Works (qv), albeit that the term 'Landscape Architect', which is a designation not protected by the Architects Registration Acts, is used in place of the term 'Architect'. See also: **Architect's Registration Board.**

On larger landscape contracts, a JCLI Supplement for use with the JCT's IFC 98 Agreement may be used and like the JCLI form itself, the Supplement covers a number of landscape specific matters, including:
— Partial possession.
— Maintenance of trees, shrubs and plants, etc.
— Responsibility for making good loss or damage to landscape works occurring before practical completion.
— Failure of trees and shrubs, etc.

JCT Arbitration Rules 1988 Procedural rules for the conduct of arbitrations. First issued in July 1988 the rules were incorporated in all the JCT forms of contract but they are now outdated and are no longer relevant to any arbitration conducted under any of the now current JCT forms. Following enactment of the Arbitration Act 1996 and the Housing Grants, Construction and Regeneration Act 1996 (qv), new rules – the Construction Industry Model Arbitration Rules (CIMAR) (qv) – are now applicable and have been written into the JCT contracts either by means of separate amendment, for use with the 1980 editions of JCT contracts (i.e. by JCT 80, amendment 18, IFC 84, amendment 12 and MW 80, amendment 11), or by incorporation in the case of 1998 Editions of each of the JCT forms.

JCT contracts The first form of building contract agreed between architects and builders was published in 1903. By 1931, after several editions of the form, the body known as the Joint Contracts Tribunal (qv) was set up to keep the form under constant review. Revised editions were published in 1939, 1963, 1980 and 1998 with the 1980 and 1998 Editions in particular being subject to ongoing amendment intended specifically to reflect judicial decisions affecting the interpretation and effect given to the contract terms and generally to improve the forms. A distinct advantage claimed for the JCT contracts is that they are negotiated documents, agreed by representatives of all sides of the construction industry[409]. Thus, a contract in JCT form is not an employer's 'standard form of contract' for purposes of s. 3 the Unfair Contract Terms Act 1977 and ambiguities will not be construed *contra proferentem* (qv) by the courts. Employers who use outdated editions of standard forms that no longer have JCT sanction may, however, find themselves caught by the provision of the Act.

A number of variants of the Form are published and the range of contracts is being increased constantly. At the time of writing, they are:
— Standard Form of Building Contract JCT 98:
  - Local Authorities With Quantities;
  - Local Authorities Without Quantities;
  - Local Authorities With Approximate Quantities;
  - Private With Quantities;
  - Private Without Quantities;
  - Private With Approximate Quantities.
— JCT Intermediate Form of Building Contract for works of simple content 1998 Edition (IFC 98).
— Agreement for Minor Building Works 1998 Edition (MW 98).
— Fixed Fee Form of Prime Cost Contract 1998 Edition (PCC 98).
— Standard Form of Management Contract 1998 Edition (MC 98).
— Standard Form of Building Contract With Contractor's Design 1998 Edition (WCD 98).
— Agreement for Renovation Grant Works (where an architect is employed).
— Agreement for Renovation Grant Works (where no architect is employed).

A useful electronic version of the most commonly used standard forms is now available on CD-ROM, published under the title *JCT Forms on Disk* by RIBA Publications, Construction House, 54–64 Leonard Street, London EC2A 4LT. This electronic format provides editable templates of JCT 98, IFC 98, MW 98, WCD 98 main contract forms and of the nominated sub-contract forms NSC and NAM designed for use with JCT 98 and IFC 98.

---

[409]See **Joint Contracts Tribunal** for a list of the constituent bodies that comprise the Joint Contracts Tribunal.

A large number of tender documents, agreements, supplements and sub-contracts have also been produced in print for use with the main forms of contract. The more significant of them include:

— Fluctuations Supplement for use with Local Authorities versions.
— Fluctuations Supplement for Private versions.
— Sectional Completion Supplement (SCS 98) for use with the With Quantities, Without Quantities and Approximate Quantities versions of JCT 98.
— Contractor's Designed Portion Supplement (CDPS) for use with the With and Without Quantities versions of JCT 98.
— Government Department Supplement.
— NSC/T 98: Part 1 Architect's/Contract Administrator's Invitation to a Sub-Contractor to Tender.
— NSC/T 98: Part 2 Tender by a Sub-Contractor.
— NSC/T 98: Part 3 Particular Conditions to be agreed by Contractor and a Sub-Contractor Nominated under JCT 98.
— NSC/A 98 Standard Form of Articles of Nominated Sub-Contract Agreement between a Contractor and Nominated Sub-Contractor.
— NSC/N 98 Standard Form for Nomination Instruction for a Sub-Contractor.
— NSC/W Employer/Nominated Sub-Contractor Agreement.
— NSC/C 98 Nominated Sub-Contract Conditions. For use with Articles NSC/A.
— NSC 98: Specimen (sample) copies of NSC/T 98: Parts 1, 2 and 3, NSC/A 98, NSC/N 98 and NSC/W 98.
— NAM/T 98: Tender and Agreement Sections 1, 2 and 3 comprising: Invitation to proposed Named Sub-Contractor, Tender by Sub-Contractor and Agreement between Contractor and Sub-Contractor respectively.
— Works Contracts 98 in three sections, comprising:
Wks/1: Section 1: Invitation to Tender;
Wks/1: Section 2: Works Contractor's Tender;
Wks/1: Section 3: Articles of Agreement.
— Works Contract/2: Conditions of Contract.
— Works Contract/3: Employer/Works Contractor Agreement.

Comprehensive practice notes and standard contract administration forms are also available and those, too, are updated regularly. Those and all of the other JCT contracts and supporting documents are published by RIBA Publications.

Where the forms are intended for use in Scotland, special supplements making the forms suitable for use there are published by the Scottish Building Contracts Committee and where intended for use in Northern Ireland an Adaptation Schedule for each of the forms is published by, and is available from, the Royal Society of Ulster Architects, 2 Mount Charles, Belfast BT7 1NZ.

See also: **ACA Form of Building Agreement; Engineering and Construction Contract (NEC); Government contracts.**

Joint Contracts Tribunal (JCT) Now an incorporated limited company going under the title of The Joint Contracts Tribunal Limited, the JCT was first formed, as a committee, in 1931. The constituent bodies now are:
— Royal Institute of British Architects.
— Royal Institution of Chartered Surveyors.
— Local Government Association.
— Construction Confederation.
— Association of Consulting Engineers.
— National Specialist Contractors Council Ltd.
— British Property Federation.
— Scottish Building Contract Committee.

The JCT's standard forms are published by RIBA Publications but copyright now vests in the Joint Contracts Tribunal Limited. The terms of reference under which the Tribunal functions are stated briefly as being, to review and update the Standard Forms of Building Contract, to produce, approve and update as necessary other ancillary forms and agreements, to issue practice notes and to liaise with other bodies. Previously, an important part of its constitution was the power of any of the constituent bodies to veto amendments, etc. This ensured that all JCT publications were published with the agreement of all members. More recently the Tribunal has undergone and continues to undergo a number of significant changes tending to have a limiting effect on that power.

Joint Fire Code A comprehensive Code of Practice, first published in May 1992 and since then regularly updated. The Code is aimed specifically at providing guidance for the effective protection of new construction, demolition, alteration, repair and renovation works from risk of accidental or malicious fire during the construction process. Although published jointly by the Building Employers Confederation and the Loss Prevention Council, the Code has been developed following consultation with, and its implementation is supported by, a very broad range of interested parties that includes: end user clients, contractors, developers, insurers (through the Association of British Insurers) and leading fire authorities (through the Chief and Assistant Chief Fire Officers Association and London Fire Brigade).

It is a 'stand alone' Code and is not of statutory effect. Nevertheless, it should be read in the context of and together with relevant health and safety legislation and in particular should be seen as complementing the provisions of the Construction (Design and Management) Regulations (CDM) 1994 (qv) and the Health and Safety at Work Act 1974.

Non-compliance with the Code by those responsible for procurement, design and/or construction may give rise to the refusal, withdrawal or repudiation of insurance cover, and under construction contracts making compliance with the Code an express term of the contract (such as the JCT 98 and IFC 98 standard forms), non-compliance by either party will also amount to a breach of contract.

The Code is concerned both with permanent and temporary works and materials (including temporary accommodation) and covers:
— Design phase.
— Construction phase.
— Emergency procedures.
— Permanent and temporary fire protection measures.
— Security against arson and malicious fire damage.
— Temporary accommodation.
— Storage of materials generally and of flammable liquids and gas etc. in particular.
— Safeguards against fire from electricity or gas services, from 'hot works', waste and on site smoking.

Under the JCT Standard Forms JCT 98 and IFC 98, provision is made in the Appendix for the parties to state whether the Joint Fire Code will or will not apply to the works and if applicable then the parties must state, too, whether for the purposes of the Code the project should be categorised as a 'Large Project'. If so, then additional requirements for the appointment of a fire marshall and specific ongoing liaison with emergency services will also apply.

Where the Code is stated to apply, the parties undertake by other provisions in the contract (JCT 98, clause 23FC and IFC 98, clause 6.3FC) to observe the requirements of the Code and to indemnify each other against the consequences of any breach or non-observance of it. In addition to that blanket indemnity, the contractor further specifically undertakes (see, for example, clause 22FC.3.1 and 23FC.3.2 of the JCT 98 contract) to carry out any 'Remedial Measures' insisted on by the insurers where a breach of the Code has arisen. A failure to do so within the stipulated time will entitle the employer to engage others to do so or do so himself and to recoup the resulting costs from the contractor by way of set-off (qv), or as a debt.

If after the base date (qv) the Code is amended so that the Joint Names Policy applicable to the works is affected, any resulting additional costs to the contractor will be an addition to the contract sum (qv).

Joint liability Parties who are jointly liable will share a single liability with the result that each party may thus be held entirely liable. In some cases liability for a tort (qv) may arise jointly between two or more defendants, e.g. in employment law, where an employer is vicariously responsible for the torts of his employees, or under the rules of agency (qv). Under the Civil Liability (Contribution) Act 1978 the courts may apportion liability. Section 1 (1) of the Act provides that 'any person liable in respect of any damage suffered by another person may recover contribution from any other person liable in respect of the same damage (whether jointly with him or otherwise)'. In other words, in the case of joint liability where only one wrong doer is sued, he may bring in a co-defendant who is jointly liable. Alternatively, he issues subsequent proceedings under the Act to recover a contribution. The amount of the contribution is to be such as the court finds 'just and equitable having regard to the extent of that person's liability for the damage in question'.

In practice, an injustice sometimes occurs because one of the parties sharing liability may be left in a position where he has to pay the whole of the resultant damages due to insolvency or unavailability of the others. As a consequence, it is becoming more common for professional terms of engagement (e.g. SFA/99) to contain what has come to be referred to as a 'net contribution' clause. This provides that a party will only be liable to pay the same proportion of any damages as the proportion of his liability.

Joint names policy Defined for the purposes of JCT contracts (e.g. JCT 98, clause 22) as 'a policy of insurance which includes the Contractor and the Employer as the insured'. A policy in joint names will have effect to avoid rights of subrogation (qv) that the insurers might otherwise have against the contractor or the employer as the case may be. A policy under which both parties are named as insured and each has mutual rights and benefits of cover under that single policy will also avoid common provisions and restrictions in non-marine policies whereby insurance cover will be refused or, if already given, may be avoided if the interest insured or to be insured under that policy is already the subject of other insurance cover under another separate policy.

All commonly used standard form contracts will generally make express provision for relevant policies of insurance to be in joint names (see: JCT 98, clause 22, IFC 98, clause 6, NEC contract, clause 84.2, GC/Works/1 (1998), clause 8 (6)). In all versions of the JCT main contract the provisions go further in that the employer and contractor must also obtain for each nominated and domestic sub-contractor the benefit of the All Risks Joint Names Insurance Policy. In the case of domestic sub-contractors they will not, however, be entitled to the benefit of any Joint Names Policy taken out in respect of existing structures under clause 22C.1.

The contractor under the JCT standard form main contract may meet his obligation to insure by means of his usual annual works insurance policy rather than taking out a separate policy. But, he may do so only if that annual works policy conforms both to the definition of a Joint Names Policy and to those other provisions for sub-contractors referred to above.

Joint tortfeasor A joint wrongdoer. Certain torts may be committed jointly and the tortfeasors are jointly liable, e.g. directors with a limited liability company. See also: **Joint liability.**

Joint venture contracting A form of contracting where a general building contractor forms a joint company with a major sub-contractor (usually one specialising in mechanical and electrical services installation) for the purpose of undertaking a building contract jointly. Each of the parties is normally supported by a guarantee (qv) given by a parent or holding company. It avoids a conflict of interest between the two but can limit competition.

Judgment The decision of a court in legal proceedings which determines the rights of the parties. It is also the reasoning of the judge in arriving at his decision. This may be reported and cited as an authority. The judgment is based on the judge's decision as to what are the important and relevant facts of the case *and* statements of the applicable rules of law. The parties are estopped (qv) from re-opening any disputed matter covered by the judgment, except that they may have the right of appeal (qv).
See also: **Appeal; Judicial precedent; Law reports.**

Judgment debt The sum of money which a judgment debtor has been ordered to pay as a result of court proceedings. A judgment debt bears interest (qv) at a statutory rate which varies from the date of judgment. Unless the judgment debtor has obtained from the court a stay of execution pending an appeal or trial of a counterclaim (qv) the judgment creditor may proceed to enforce the judgment in various ways.
See also: **Garnishee.**

Judicial notice Notorious facts which are recognised as being within common knowledge and which the judge (or in the case of jury trial, the jury) can be expected to already know from their own experience or which can readily be established from their own enquiries. Such facts need not otherwise be proved, e.g. it is common knowledge that the streets of London are full of traffic[410].

Judicial precedent The doctrine of judicial precedent is an important feature of the common law system. In general terms, a judge in a lower or the same court is bound to follow the decision of a previous judge in similar circumstances.
   Not all of the judgment is of binding force in subsequent cases but only the legal principle which is necessary for the actual decision. This is known as *ratio decidendi* (the reason for decision) which is, in effect, the legal principle upon which the decision rests. Judges often make general statements about the principles involved which are not germane to the facts before them. Such remarks are called *obiter dicta* and are not binding on another court, although they may be of persuasive authority in a subsequent case.
   There may be several *rationes decidendi* in a judgment, in which case all are binding unless they are inconsistent with each other. Judges have limited power to distinguish cases they do not wish to follow and sometimes exercise considerable ingenuity in doing so. By distinguishing a case a judge finds, for example, that the facts of the earlier case are not sufficiently similar to those before him for the *ratio decidendi* to be applied.
   The general rule is that every court in the judicial hierarchy binds all lower courts by its decisions; some courts bind themselves as well. A decision of the House of Lords is binding on all other courts. The Court of Appeal binds itself and the courts below it. In general, the decisions of the House of Lords are

---

[410]*Dennis* v. *White & Co* [1917] AC 479.

binding upon the House itself but, in rare cases, the House of Lords is free to depart from its own decisions if there is sufficient reason[411].

A higher court has power to *overrule* an earlier decision of a lower court and thus declare that it does not in fact represent the law.

The rules about judicial precedent are very complex, and too rigid adherence to precedent may lead to injustice in a particular case and sometimes restrict the proper development of the law. However, judicial precedent provides some degree of certainty and a basis for the orderly development of legal rules.

See also: **Courts; Law reports; Obiter dictum; Ratio decidendi; Stare decisis.**

Jurisdiction (1) The power or authority of a court or tribunal to take cognisance of and to decide matters put before it. In the UK, the jurisdiction of the High Court derives from the Crown, in whose name and by whose authority the judges exercise jurisdiction. In the County Court jurisdiction is based upon the County Courts Act 1984. Unlike the judiciary, arbitrators have no inherent power to determine their own jurisdiction[412].

Arbitrators' jurisdiction and the powers bestowed on them derive from and are limited by the terms of the parties' arbitration agreement. The parties are free to agree otherwise, but in the absence of a contrary agreement an arbitrator may rule on whether there is in existence a valid arbitration agreement, whether the arbitration is properly constituted and what matters have been submitted to arbitration in accordance with the arbitration agreement[413]. Where an arbitrator elects to rule on his own substantive jurisdiction his decision will be open to challenge in the courts. Alternatively, where the question of jurisdiction arises the arbitrator may, and if the parties so agree he must, stay the proceedings and make application to the court for determination of a preliminary point of jurisdiction.

In either event, a party intending to raise objections and to contend that from the outset the arbitrator lacks substantive jurisdiction on any matter must raise such objections before taking any steps to contest the issues that he says are outside the jurisdiction of the arbitration[414].

So far as other tribunals are concerned, broadly similar restrictions generally apply. In particular, in relation to adjudication, an adjudicator must ensure that he does not act outside the limits of his statutory (or express contractual) jurisdiction. If he does so, his decision will be a nullity provided the party seeking to contest the adjudicator's (or expert's) jurisdiction to hear and decide the issue concerned has objected at the outset and has thereafter maintained that objection throughout the entire proceedings. See also: **Housing Grants, Construction and Regeneration Act 1996.**

(2) The territorial limits within which the judgments or orders of a court etc. can be enforced.

---

[411]Practice Statement (Judicial Precedent) [1966] 1 WLR 1234 (HL).

[412]E.g. *Hyundai Engineering and Construction Co Ltd* v. *Active Building and Civil Engineering Construction Pte Ltd* (1988) 45 BLR 62, 70.

[413]Arbitration Act 1996 s. 30.-(1).

[414]Arbitration Act 1996 s. 31.-(1).

King's enemies See: **Queen's enemies.**

Knowledge Strictly, that which is known. In a legal context you may have actual, imputed or constructive knowledge.

A person may have actual knowledge of a particular fact if he has seen or experienced the fact itself or evidence of it. Knowledge is imputed where he is deemed to know that particular fact, e.g. where an agent has been informed of the fact, the law imputes that knowledge to his principal. Constructive knowledge arises where a party should have known the fact, but did not carry on sufficient investigations to prove or disprove the existence of the fact. It is also an important concept in assessing the proper measure of damages which are said to flow from any alleged breach of contract[415].

---

[415]*Hadley* v. *Baxendale* (1854) 9 Ex 341.

Labour In the context of building contracts, the term is generally given its ordinary meaning and applies to workpeople or operatives, skilled or unskilled. JCT 98 makes the contractor's inability to secure essential labour properly to carry out the works a ground for extension of time provided the inability was outside the control of the contractor.

Laches Negligence or unreasonable delay in asserting or enforcing a right. In rare cases it may be pleaded as a defence (qv), but only where there is no statutory time-bar. In the case of performance bonds (qv) conduct of the employer which prejudices the surety's position may discharge the obligation[416] and this is another type of laches. It has been said that the validity of the defence 'must be tried upon principles substantially equitable. Two circumstances always important in such cases are the length of the delay, and the nature of the acts done during the interval, which might affect either party and cause a balance of justice or injustice in taking the one course or the other, so far as relates to the remedy'[417].

For example, if an adjoining owner (A) waited until building work was almost complete before seeking an injunction to prevent the contractor (B) from gaining access over part of A's land, B may be able to plead laches successfully on the grounds that A had delayed unreasonably and was acting with malice.

Landfill Tax A tax introduced from 1 October 1996 for which operators of licensed landfill sites are liable to account to HM Customs and Excise and which can have significant cost implications on the removal and disposal of excavated and other materials arising during the course of construction operations. The substantive law governing the introduction of the tax is found in the Finance Act 1996[418] which received Royal Assent in April 1996 and has thereafter been updated, revised and amended through secondary legislation in a variety of Regulations and Orders (such as Landfill Tax Regulations, Landfill Tax (Contaminated Land) Orders and Landfill Tax (Qualifying Materials) Orders) made by Statutory Instrument.

Any operator of a licensed landfill site accepting or intending after 1 October 1996 to accept taxable disposal of waste must, as from 31 August 1996, be registered with HM Customs. Failure to notify liability to register, evasion of the tax, deliberate or persistent non-compliance with the taxation process or failure to otherwise comply with the regulations or to preserve the appropriate records required by HM Customs will render the operator liable for a range of civil and criminal penalties although the latter are generally reserved for cases of serious fraud.

---

[416]*Kingston-upon-Hull Corporation* v. *Harding* [1892] 2 QB 494.
[417]*Lindsay Petroleum Co* v. *Hurd* (1874) 22 WR 492 per Lord Selbourne at 510.
[418]Finance Act 1996, s. 39 to 71 and Schedule 5.

Currently two rates of tax apply depending on the nature of the material disposed of. A lower rate (which at November 2000 is set at £2/tonne) is payable for those inert or inactive material listed under the Landfill Tax (Qualifying Material) Order 1996 whilst a standard rate (which at November 2000 is set at £11/tonne) is payable for all other active materials.

A range of useful information notes on matters relating to Landfill Tax is issued by HM Customs and Excise, in addition to which helpful and authoritative advice on every aspect of the tax is available from a Landfill Tax Help Desk set up by HM Customs and Excise at Dobson House, Regent Centre, Gosforth, Newcastle upon Tyne NE3 3PF, tel. 0191 128484.

Lands Tribunal A tribunal created by the Lands Tribunal Act 1949 to deal with the following matters:
— Questions relating to compensation for the compulsory acquisition of land. If the acquiring authority's offer is unacceptable to the expropriated owner, either party may refer the case to the Lands Tribunal, the decision of which is final as to the merits of the case.
— The discharge or modification of restrictive covenants (qv) affecting land. In some cases such covenants are outmoded in modern conditions, but this power can only be exercised on very limited grounds.
— Appeals from decisions of local valuation courts relating to rating assessments.

Procedure and practice before the tribunal is governed by special procedural rules. Its membership consists of a president and several nominated members, who usually sit singly. They are either lawyers or chartered surveyors. The tribunal gives a written and reasoned decision, and appeal on point of law (qv) lies only direct to the Court of Appeal.

See also: **Courts; Sealed offer.**

Latent Damage Act 1986 In *Pirelli General Cable Works Ltd* v. *Oscar Faber & Partners*[419] the House of Lords laid down that 'the date on which the cause of action accrues' in the case of negligent design and construction of a building is the date when the physical damage occurs, even though that damage was not reasonably discoverable until a later date. Following the *Pirelli* ruling, on 29 November 1984, the Law Reform Committee published a report on latent damage (Cmnd 9390) which concluded that, following the ruling the law of limitation gave rise to uncertainty and might cause injustice to both claimants and defendants. The committee made two main recommendations:
— In negligence cases involving latent defects, the limitation period of six years should be extended to allow the claimant three years from the date of discovery or reasonable discoverability of significant damage.
— There should be a long-stop which should bar a claimant from starting proceedings more than 15 years from the date of a defendant's breach of duty in negligence cases involving latent damage.

---

[419](1983) 21 BLR 99.

The Latent Damage Act 1986 attempts to give effect to the committee's recommendations. It came into force on 18 September 1986 and does not apply to any action commenced before that date. In the normal tort case, the limitation period would have been six years from the occurrence of the damage, and the Act does not enable actions to be brought in respect of damage which occurred before its coming into force. The Act does not affect limitation for contractural claims. Importantly, in light of the inability to recover damages for pure economic loss in tort[420], the Act appears to have little impact on the building industry[421].

The Act's three main provisions are:

— In the case of latent damage not involving personal injuries, it introduces a special three year time limit which runs from the date of knowledge (qv) if this is later than the usual six years from the accrual of the cause of action. The three years can be extended where the claimant is under a disability.

— The Act inserts a new section (s. 14A) into the Limitation Act 1980. Curiously, 'latent damage' is not defined. Section 14A applies to 'any action for damages for negligence' for latent damage not involving personal injuries. 'Negligence' is also not defined in the Act, but it is thought that it covers not only actions for the tort of negligence but also the negligent breach by local authorities of their duties under the building regulations.

— There is an overall 15 year long-stop for the date of the breach of duty: Limitation Act 1980 Act, s. 14B. This protects defendants from stale claims. The 15 year long-stop runs from the date of the breach of duty, which is not necessarily the date of the completion of the building. 'Breach of duty' is defined as the date (or last date) on which there occurred 'any act or omission which is alleged to constitute negligence and to which the damage is attributable'.

— In the case of fraud, concealment or mistake, time does not begin to run until the claimant has discovered the fraud etc. or could with reasonable diligence have discovered it. In all other cases, however, once the 15 years have expired, no action can be brought. This applies whether or not the relevant facts were known and even if the damage has not yet occurred.

Latent defect A defect which is not discoverable during the course of ordinary and reasonable examination but which manifests itself after a period of time. If the defects could have been discovered by a reasonable examination by the architect before certifying practical completion or making good of defects, they will not be latent defects. In *Victoria University of Manchester* v. *Hugh Wilson*[422] Judge John Newey QC defined latent defects as those which could not have been discovered on examination by a reasonably careful person skilled in building. In building work the most common application is defects

---

[420]*Murphy* v. *Brentwood District Council* [1991] 1 AC 398.
[421]Quaere the effect of the House of Lords' decision in *Merrett Syndicates Ltd* [1995] 2 AC 145.
[422](1984) 2 Con LR 43.

becoming apparent after the making good of defects certificate has been issued. In such circumstances, where a defect becomes apparent only after the issue of the making good of defects certificate but before the final certificate is issued and the contractor refuses to rectify such latent defects, the architect will often withhold his satisfaction with the work and hence withhold the issue of the final certificate (qv), until the defects have been corrected. This was a prudent course of action since, before publication of amendments 15 and 9 to JCT 80 and IFC 84 respectively (being amendments to clause 30.9.1.1 of JCT 80 and clause 4.7.1 of IFC 84), the terms of those contracts were held to give conclusive effect to the final certificate as evidence that the works were constructed to the satisfaction of the architect and in all respects in accordance with the contract. This conclusivity extended not only to latent defects becoming apparent after the making good of defects certificate and remaining unrectified prior to issue of the final certificate but also to those latent defects appearing long after the certificate had been issued. Consequently, the final certificate had the effect of absolving the contractor from his liability under the contract.

Provisions amending the effect of the final certificate are now incorporated (at clauses 30.9 and 4.7.1 respectively) into the 1998 editions of the JCT and IFC contracts but, even then, the position remains that the final certificate will still create a bar to certain future claims.
See: **Final certificate.**

If the contractor refuses to rectify latent defects, the employer's only remedy is to arbitrate (or if appropriate sue) for damages within the limitation period.
See **Limitation of actions; Latent Damage Act 1986.**

In contracts for the sale of goods, there are implied terms that the goods will be in conformity with the description and with the sample, if any. If the goods supplied appear to, but do not in fact, conform, the defects will be latent and the supplier will be liable. This principle applies even if the goods conform to the sample and the sample itself contains hidden defects. It is the 'apparent sample', i.e. one without hidden defects which is to be taken as the true sample[423].
See also: **Patent defect.**

Latham Report Abbreviation of the name given to the final report of Sir Michael Latham and inspired, in the first instance, jointly by the construction industry and Government. The report was drawn up by Sir Michael Latham in 1994 under the title *Constructing The Team*. It undertakes a review of procurement and contractual arrangements in the industry aimed at improving industry performance and teamwork and in particular reviews and makes recommendations concerning[424]:

— The processes by which clients' requirements are established.
— Methods of procurement.
— Responsibility for the production, management and development of design.

---

[423] *Adcock's Trustees* v. *Bridge RDC* (1911) 75 JP 241.
[424] Appendix 1 — Terms of reference for the review.

— Organisation and management of the construction processes.
— Contractual issues and methods of dispute resolution.

The report undertakes detailed consideration of each of those topics and offers recommendations for improvements of the current approach taken by the industry towards:

— The design process: consultants and specialist sub-contractors.
— Contract choice for clients.
— Selection/tendering procedures.
— Issues determining performance.
— Team work at site level.
— Dispute resolution.
— Insolvency and security for payment.
— Liability post completion.

The report concludes by setting out the actions which it sees are essential to the effective implementation of its recommendations and suggests a time scale for each.

Although of no statutory authority or of any other binding effect the report and the recommendations made by it have generally received Government approval and in particular the recommendations relating to improvements in payment procedures and for the introduction of more speedy and cost effective dispute resolution procedures have largely been adopted and now have statutory force in Scotland, England and Wales through Part II of the Housing Grants, Construction and Regeneration Act 1996 (qv) and the associated Scheme for Construction Contracts (SI 1998/649). Those same recommendations are given similar force in Northern Ireland under the provisions of the Construction Contracts (Northern Ireland) Order 1997 (qv) (SI 1997 No. 274 (NI.1)).

Law of the contract Parties are generally free[425] – subject to any statutory limitations, such as those created by virtue of the Unfair Contract Terms Act 1977 – to choose and to stipulate in their agreement which legal system shall govern the contract between them (see, for example, clause 12 and part one of the contract data under the NEC).

In the absence of an express agreement between the parties as to the proper law of the contract and, where appropriate, the procedural law by which any 'dispute' will be settled, there are complex rules for deciding where a dispute should be resolved. For contracts signed on or after 1 April 1991, the Contracts (Applicable Law) Act 1990 adopts an agreed approach to these throughout the EU:

— The court may give effect to the parties' implied choice of law where this choice can be 'demonstrated with reasonable certainty'[426]. It is unclear whether reference may be made to the subsequent conduct of the parties in demonstrating such choice.
— The parties may choose the law of the contract, even post-contract[427].

---

[425] Article 3 (1) of the Rome Convention (19 June 1980).
[426] Article 3 (1) of the Rome Convention (19 June 1980).
[427] Article 3 (2).

— Where the contract has connection with only one country, the parties can still submit to foreign law. This ability cannot be used to avoid so-called 'mandatory rules', such as consumer legislation[428].

— Where no such choice can be found, the contract will be governed by the law of the country to which 'it is most closely connected' unless and to the extent that the contract is severable, whereupon each part will be governed likewise[429]. The 1990 Act usually applies a test relating to the location of the performing party's habitual residence or a company's central administration[430]. In other words, where a contractor agrees to build a warehouse in Scotland, if the contractor is an English company, English law is likely to apply.

Not uncommonly, contracting organisations trading from offices registered in England will undertake projects situated in Scotland or it may be that the employer under the contract trades and resides there. In that case, since Scots Law differs significantly from the law of England and Wales, the parties must ensure that they agree upon and expressly stipulate which law is to govern their contract and must adopt the form of contract suitable to their choice (see, e.g. JCT 98, clauses 1.10 and 41 and footnotes [q] and [yy] respectively, IFC 98, clause 1.15 and footnote [l], WCD 98, clause 1.7 and footnote [f] and MW 98, clause 1.7 and footnote [j]).

Law reports Reports of decided cases are essential for the operation of the doctrine of judicial precedent (qv). From the time of Edward I (1272–1306) until today we have had law reports in some form, although their quality and reliability varies.

Law reporting rests on private initiative. There are no 'official' law reports, although since 1865 the Incorporated Council of Law Reporting has published a continuous series of reports known simply as 'The Law Reports', divided for convenience into volumes to cover the divisions of the High Court (see: **Courts**).

Cases in the Court of Appeal are reported in the volume containing reports of cases in the Division in which the case was first heard. Decisions of the House of Lords are reported in a separate volume.

The Council is a private body but has semi-official status and if a case is reported in the Law Reports that report will be cited to the court in preference to any other. The transcripts of the judgments are revised by the judge concerned. There are many other series of reports, e.g. the All England Law Reports, but until recently many decisions of importance to the building industry went unreported and specialist building contract lawyers had to rely on privately circulated transcripts. Since 1976, however, *Building Law Reports* (BLR) have been published and contain reports of cases of interest to the building industry – including a number of decisions from the Commonwealth. Full coverage of all the major decisions of the Technology and Construction Court (qv) (known previously as the Official Referees' division of the High

---

[428]Article 3 (3).
[429]Article 4 (1).
[430]Article 4 (2).

Court (qv)) and appeals therefrom which are of relevance to the building industry are fully reported in *Technology Construction Law Reports* (TECLR) published four times a year by Butterworth & Co (Publishers) Ltd. These used to be called the *Construction Law Reports* but the name change reflects the court's new title. This series contains the only complete coverage of these important judgments. Important judgments of the Technology and Construction and other civil divisions of the High Court, the Court of Appeal as well as Scottish cases are all now also available on the Internet.

References to law reports are given by standardised abbreviations which indicate the volume and the series of reports wherein the case is reported. So, *D. & F. Estates Ltd* v. *Church Commissioners for England* (1988) 15 Con LR 35, means that the case will be found in the fifteenth volume of *Construction Law Reports* at page 35. Round brackets are used to enclose the date where the reports are in sequentially numbered volumes, such as in the example above where the case appears in the fifteenth volume of the series. Where the reports are recorded annually with their volumes being numbered sequentially in each year beginning with volume 1 each time, the report is cited by reference to the date − given in square brackets − followed by the volume number of the year in which the case is reported, e.g. *Amalgamated Building Contractors* v. *Waltham Holy Cross UDC* [1952] 2 All ER 452, being volume 2 of the 1952 All England Law Reports at page 452. These are legal conventions which should be, but are not always followed consistently. A list of commonly used abbreviations for various reports is given with the Table of Cases at the back of this book.

Legal tender A creditor is entitled to demand payment of a debt in legal tender, i.e. money. Legal tender consists of Bank of England notes for payment of any amount in England and Wales, gold coin of the realm to any amount, cupro-nickel or silver coins of more than 10p for any amount up to £10, cupro-nickel or silver coins of 10p or less up to £5, and bronze coins to an amount not exceeding 20p.

Scottish and Northern Ireland bank notes are not legal tender in England and Wales. Only Bank of England notes of less than £5 are legal tender in Scotland[431].

In practice, a court would require little evidence to be satisfied that a creditor had waived his legal right to payment in legal tender, e.g. past dealings where payment by cheque (qv) had been made and accepted[432].

Letters of intent A document sent before entering into a contract. It often expresses a firm intention to enter into a contract, sometimes requiring work to be put in hand. Usually, a letter of intent merely expresses an intention to enter into a contract in the future. Such letters are usually sent by the employer to a prospective contractor, nominated sub-contractor or nominated supplier. If sent by the architect, the letter must clearly state that it is sent on behalf of his client; otherwise the architect may find himself financially accountable if the

---

[431]Section 1 (2) of the Currency and Bank Notes Act 1954.
[432]See also the reference to acceptability of cheques under CPR Part 36.

contract does not proceed. The client must see the letter and agree its contents, preferably in writing. Legal scrutiny of each letter of intent is advisable because each case has its own peculiarities.

The whole process is fraught with difficulties, e.g.:

— The main contractor, when appointed, may object to a nominated sub-contractor who has been given a letter of intent unless he is named in the contract tender documents.

— The employer may have to pay costs even if the contract does not proceed.

— The courts sometimes consider that a full binding contract has been created.

Letters of intent should be avoided if at all possible. The object of a letter of intent is to ensure that there is a limited or no contractual liability, but whether or not the sender has attracted liability depends upon the facts and surrounding circumstances of each case.

The difficulties arising from the use of letters of intent are avoided, so far as nominated sub-contractors are concerned, if the JCT Standard Form of Employer/Nominated Sub-Contractor Agreement (NSC/W (see: **Collateral contract**)) is used in conjunction with the JCT 98 main contract nomination procedure, since the matter of preliminary design, fabrication and allied work, and payment for it, is dealt with by clause 2.2. (In Scotland, NSC/W/Scot applies.) The case law concerning letters of intent is conflicting. In general, the courts look at the substance of each transaction rather than its form. In *Turriff Construction Ltd* v. *Regalia Knitting Mills Ltd*[433], the contractors undertook pre-contract design work provided they were given 'an early Letter of Intent . . . to cover (them) for the work they will now be undertaking'. The employer sent the letter requested and it concluded that 'the whole to be subject to an agreement on an acceptable contract'. It was held that the employer was liable for the work carried out, the court ruling that the proviso applied only to the full main contract and not to the preliminary work carried out by the contractor which was done pending the conclusion of a formal contract.

In contrast, in *British Steel Corporation* v. *Cleveland Bridge & Engineering Co Ltd*[434], a letter of intent was held to negative contractual liability but to give rise to liability in restitution or quasi-contract (qv). The judgment of Robert Goff should be studied carefully, but in general it seems that the sender of such a letter is likely to be under a measure of liability, save in exceptional circumstance. **Figures 9** and **10** illustrate letters of intent.

Levels and setting out The architect is responsible for showing accurately all necessary levels on the drawings and all dimensions to set out the building on the site. The contractor is responsible for transferring the levels and setting out the building on site. For a fuller discussion of the implications.

See: **Setting out.**

---

[433](1971) 9 BLR 20.
[434](1981) 24 BLR 94.

Dear Sir,

My client *(insert name)* has instructed me to inform you that your tender of the *(insert date)* in the sum of *(insert amount in figures and words)* for the above project is acceptable and that I intend to prepare the main contract documents for signature subject to my client *(insert the provisos appropriate to the particular situation)*.

It is not my client's intention that this letter, taken alone or in conjunction with your tender, should form a binding contract. However, my client is prepared to instruct you to *(insert the limited nature of the work required)*.

If, for any reason whatsoever, the project does not proceed, my client's commitment will be strictly limited to payment for *(insert the limited nature of the work required)* at the cost reasonably and properly incurred. No other work included in your tender must be carried out without a further written order. No further obligation is placed upon my client and no obligation whatever, under any circumstances, is placed upon me.

Yours faithfully.

Figure 9   Letter from architect to contractor — letter of intent.

Dear Sir,

My client *(insert name)* has instructed me to inform you that your tender of the *(insert date)* in the sum of *(insert amount in figures and words)* for *(insert the nature of the works)* is acceptable and that I intend to instruct the main contractor to *enter into a sub-contract* with you after the main contract has been signed.

It is not my client's intention that this letter, taken alone or in conjunction with your tender, should form a binding contract. However, my client is prepared to instruct you to undertake the following limited work, namely *(insert the limited nature of the work required)*. If, for any reason whatsoever, the project does not proceed, my client's commitment will be strictly limited to payment at cost reasonably and properly incurred for the limited works described above.

No other work[†] included in your tender must be carried out without a further written order. No further obligation is placed upon my client and no obligation whatever, under any circumstances, is placed upon me.

Yours faithfully.

*Substitute 'place an order' in the case of a supplier.
[†]Substitute 'work or materials' in the case of a supplier.

Figure 10   Letter from architect to sub-contractor or supplier — letter of intent.

Liability A person is said to be liable when he is under a legal obligation to act or to suffer an action of another. Liability may be criminal (where a person may suffer fines or imprisonment) or civil (where a person may suffer various sanctions, e.g. payment of damages). Civil liability may arise by the operation of statute (qv) or because parties have entered into a contract or in tort (qv) by virtue of common law (qv). Thus, in a building contract, the parties incur liabilities which they have decided upon themselves. The principal ones are that the contractor must carry out the work in accordance with the contract documents (qv) and the employer must pay the contractor for doing the work. Common law will also imply certain liabilities into contracts, such as that the contractor must use the kind of skill and care which the average contractor would use in the same circumstances. An architect will be liable for the consequences of his negligence (qv). If he is proved negligent he will be required to pay damages (qv). Liability may be *strict* – in other words, a person may be liable even though he is not negligent and has no intention to commit a tort as is the case under the rule laid down in the case of *Rylands* v. *Fletcher*[435]. Liabilities under certain Acts of Parliament fall into this category and sometimes also into the category of *absolute liability* where failure to carry out a duty imposed will render the person responsible liable quite irrespective of the amount of care taken or intention. See also: **Absolute; Care, duty of; Care, standard of; Strict liability.**

Libel Defamation (qv) in permanent form, e.g. in writing. Libel is actionable without proof of actual damage in contrast to slander which, in general, requires the claimant to prove loss.

Licence Permission or authority to do something, e.g. to enter on land. The law on the subject is complex. Under the ordinary building contract, the contractor has a licence to occupy the site for the purposes of the contract, i.e. a contractual licence[436]. In general, the employer is not entitled to revoke the contractor's licence before completion, although all well-drafted building contracts deal with the situation should the contract be determined.

Although an architect almost invariably[437] retains exclusive copyright (qv) in the designs and drawings that he prepares, on proper payment of his fees a licence to his client – and perhaps to other consultants and contractors providing services to the project – will be implied to the extent that they may 'use and copy drawings, documents and bespoke software produced by the architect in performing his services for purposes related to the project'. SFA/99, clause 6.2 and CE/99, clause 6.2 each provide express terms to that effect. Such a licence will generally extend to use for the purposes of:

— Operation, maintenance, repair and reinstatement of the whole or any part of the project.
— Alteration of all or any part of the project.

---

[435](1866) LR 1 Exch 265.
[436]*London Borough of Hounslow* v. *Twickenham Garden Developments Ltd* (1970) 7 BLR 81.
[437]But, for an exception to the general rule, see *Cala Homes* v. *Alfred McAlpine* (1995) CILL 1083.

— Extending the project (but excluding the reproduction of the designs for any part of any extension of that or any other project).
— Promotion, leasing and/or sale of the project.

It may be expressly extended further by agreement and it is not uncommon for such extensions to encompass other interested parties via collateral warranties that the architect or other designer may be asked to enter into (see, e.g. the standard BPF financial warranty agreements).

See also: **Forfeiture clause.**

Licensee A person who enjoys the benefit of a licence (qv), e.g. a contractor who enters land under a building contract. ACA 3, clause 10 refers specifically to 'Employer's Licensees'. JCT 98, clause 29 contains similar provision referring to, among other things, work that is carried out by 'persons employed or otherwise engaged by the employer' during the currency of the contract. Such clauses normally permit the employer's licensees to enter upon the site notwithstanding that the contractor has been given possession. The contractor then generally has a contractual claim for any resulting loss and/or expense that he suffers and/or will have a right to an extension of time (qv) on account of any resulting delay. Clause 29.1 of the JCT contract in effect gives the employer such express contractual rights of access for those directly employed by him, provided the contractor had prior notice and sufficient information (within the specification or the like), concerning their likely intervention on site. Information given will be sufficient if it is such as would enable the contractor to organise himself around those others and will allow him to carry out and complete his works notwithstanding their intervention. Where insufficient prior information is given to the contractor, under JCT 98 he has a right of reasonable refusal and may disallow access to the employer or his direct employees (clause 29.2).

Lien A lien is the right to retain possession of property otherwise lawfully belonging to another so long as an existing claim remains unsatisfied, or to retain specific goods in relation to charges incurred in respect of that property pending payment. Liens can be categorised as:

Possessory liens: The person claiming the lien must be in lawful possession so that if the true owner has a right to obtain possession, there is no lien[438].

General liens: These are rare. The courts do not favour such liens and are loth to accept new instances. An accepted example is where a banker has a general lien over all a customer's security[439].

Particular liens: Where work has been done on or has created particular goods, e.g. an architect's drawings or solicitors' papers. It should be noted that the 'goods' must have been 'improved' not merely maintained. For example, training a horse will create a lien, but mere stabling will not[440].

---

[438]See, for example, *Scarfe* v. *Morgan* (1838) M & W 270.
[439]*Brando* v. *Barnett* (1846) 12 Claimant & F 787.
[440]*Bevan* v. *Waters* (1828) Mood & M 235.

Equitable liens: An equitable lien does not require possession and is in the nature of a charge. Such a lien requires registration as a land charge under s. 205 of the Law of Property Act 1925. It is impossible to define where the law will impose an equitable lien.

Life cycle cost analysis (LCCA)

A technique deriving from the research of the Quantity Surveyors' Research and Development Committee and published by the Royal Institution of Chartered Surveyors in July 1983. It seeks to examine the total costs of a building throughout its useful life in order to evaluate and compare alternatives to achieve optimum long-term cost benefits. Two specific applications are:
— To embrace construction and running costs at the design stage.
— To evaluate the running costs of existing premises.

The idea has been in circulation in the field of building maintenance for some time. LCCA refines and codifies it by providing a sophisticated methodology to arrive at a system of cost comparisons. A fundamental part of the system employs a form of cost on a comparable basis. The process is complicated and four distinct categories of information should be assembled:
— Cost in use or running costs, including fuel, maintenance and management.
— Physical information regarding the construction and fittings of the building.
— Quality of finishes and fittings.
— Performance of the building.

LCCA can be used to plan the cost management for the entire life of a building or for any shorter period desired. Tax implications can also be assessed. Properly applied, the system should transform the building owner's approach to new building and in particular the relationship between initial building design costs and costs in use (see **Figure 11**). The capital cost of a building has been found to be, on average, about one third of the total cost of the building throughout its life. That does not mean that a cheaper capital cost will give a reduced total of running costs. Generally, the reverse is true. Careful, comprehensive cost planning at design stage is the essence of the system.

Limitation clause

See: **Exemption clause.**

Limitation of action

This term covers the rules prescribing the periods of time within which actions to enforce legal rights must be started, either by the issue of a claim form (formerly called a writ) (qv) or by serving notice of arbitration (qv) or adjudication (qv).

In England and Wales a limitation period is prescribed by statute largely (although not exclusively[441]) under the provisions of the Limitation Act 1980.

---

[441]Certain special limitation periods apply under other enactments in relation to, for example, shipping, carriage by sea, air, road, rail and to employment contracts.

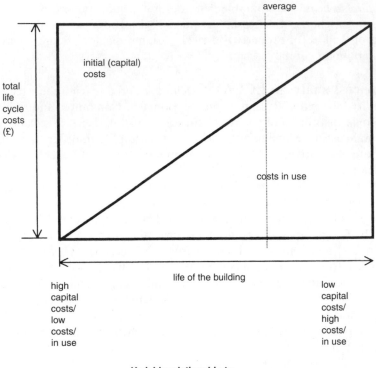

Figure 11 Relationship between initial building design costs and costs in use.

Parties are generally free to agree and to expressly provide in their contract that actions under the contract (whether brought in litigation or arbitration) must be commenced within a shorter period than prescribed by statute (see, for example, the facility provided in the Appendix to the Conditions SFA/99 relating to clause 7.2 and 7.4). However, if no valid[442] agreement to that effect is made then the prescribed statutory periods will apply as follows:

— The time limit for actions founded on a simple contract is six years from the date of the breach of contract: s. 5.
— The time limit for actions founded on a specialty contract (qv) is 12 years: s. 8.
— The time limit for actions founded on tort (qv) such as negligence, is six years (s. 2), except in the case of actions for damages for personal injuries when it is three years: s. 11.
— In the case of latent damage (other than in personal injury cases) the period is either six years from the date on which the damage occurred or three years from the date on which the claimant knew about the material and relevant facts: s. 14A.

---

[442]Subject to, for example, the effects of the Unfair Contract Terms Act 1977. See also, *Moores* v. *Yakely Associates* (1998) CILL 1446.

— There is a prohibition on the bringing of an action for damages for negligence (except in personal injuries cases) of 15 years after the expiry of the date of the negligent act or omission. This long stop applies whether or not the material or relevant facts were known, and even if the damage occurs: s. 14B.

It is not always easy to establish the date 'on which the cause of action accrued' in the case of claims in tort, particularly where defective building work is covered up. The leading case is the decision of the House of Lords in *Pirelli General Cable Works Ltd* v. *Oscar Faber & Partners* (1983)[443] which established that in actions alleging negligence in regard to the erection of a building, time ordinarily begins to run not from the date of the alleged negligence, nor from when it ought to have been discovered, but from the date when the damage occurred. The Latent Damage Act 1986 (qv) has altered the position as regards latent defects (qv).

It must be noted that the Limitation Act 1980 does not extinguish the right to sue. It merely sets the time limits within which the claimant must begin his action, and so if a defendant pays up after the limitation period has expired, the payment is valid. This is in contrast to the situation in Scotland where the right to sue is completely extinguished (per the Prescription and Limitation (Scotland) Acts 1973 and 1984 which prescribe a limitation period of five years in respect of actions for breach of contract, delict (tort) or breach of statutory duty). This period runs from the time when the pursuer (claimant) first knew, or ought reasonably to have discovered, the loss or damage. In England, in cases of fraudulent concealment (i.e. deliberate concealment of defects), time does not begin to run until the fraud is discovered or could have been discovered with reasonable diligence[444].

Limited company A company may be limited by guarantee or by shares[445] thereby limiting liability (qv) of the shareholders to the nominal value of the guarantee or share holding respectively. Characteristics of a limited company are:
— It can only be formed under the rules laid down by the Companies Act 1985.
— A limited company comes into existence when it has been registered with the Registrar of Companies. Transactions carried out before registration may be taken to be the transactions of a partnership with unlimited liability.
— The powers of a limited company are constrained by the 'objects' clause of the Memorandum of Association.
— Accounts must be filed with the Registrar of Companies and they may be inspected by the public.
— There are certain statutory constraints on the running of the company, e.g. at least one Annual General Meeting must be held for all shareholders each year.

---

[443](1983) 21 BLR 99.
[444]1980 Act, s. 32: *William Hill Organisation Ltd* v. *Bernard Sunley & Sons Ltd* (1982) 22 BLR 1.
[445]Section 1 (2) Companies Act 1985.

— A company normally comes to an end by being liquidated (qv) in accordance with the Companies Act 1985. It is a formal and possibly lengthy process.

— A shareholder cannot bind the company by his actions.

— Dividends must be apportioned strictly in accordance with the share holding.

— Changes in share holding do not bring the company to an end.

— The company is run by a board of directors. They may or may not be shareholders. Normally, they will carry no personal liability for the actions of the company.

— A private limited company must put the word 'Limited' or the abbreviation 'Ltd' after the company name. Public limited companies must put the words 'Public Limited Company' or the initials 'PLC' after the company name.

— The minimum number of members in each case is two.

See also: **Corporation; Ultra vires.**

Limited Liability Partnership Act 2000 An Act which came into force on 6 April 2001 and designed to introduce a new status of partnership – the Limited Liability Partnership (or LLP) – whereby, when properly formed by incorporation under the Act the partnership will have corporate status and will be a separate legal entity, distinct from its members (owners). As such, it will have unlimited capacity to undertake the full range of business activities which a traditional partnership may undertake and will more closely resemble a company than it will a partnership. Except to the extent otherwise provided for in the Limited Liability Partnership Act or in any other relevant enactment, the existing law relating to partnerships will not apply to an LLP.

In addition to governing the naming and legal status of the LLP, the means of its registration and incorporation, its past, present and future membership – including the relationship between members – and the need for it to have registered offices situated in England, Wales and/or Scotland, the Act also makes provisions as to:

— Generally ensuring the preservation of an income tax and capital gains taxation regime applicable to members of the LLP as if they were partners carrying on business in a traditional partnership, notwithstanding that the LLP is, in fact, a separate body corporate.

— Ensuring that present corporate insolvency and winding up procedures such as company voluntary arrangements, winding up, receivership and administration etc. are adapted and extended to encompass LLPs.

— Making certain non-conformance with regulations made by or under the Act can give rise to criminal liability.

See also: **Partnership.**

Liquidated damages A sum of money stated in a contract as the damages payable in the event of a specified breach. The sum must be a genuine pre-estimate of the loss likely to be caused by the breach or a lesser sum. The genuine purpose must be to compensate the employer (qv) rather than punish the contractor (qv). There is no need to prove actual damage after the event and it does not matter that the actual loss is greater or less than the stated sum and the specified sum is recoverable even if in the event there is no loss[446].

All the common forms of building contract include a liquidated damages clause to calculate the amount payable if the contractor fails to complete by the completion date (qv) or any extended date. A sum is included to represent the damages on a weekly or daily basis as appropriate. If no figure was stated, the employer would need to prove his actual loss and recover it by way of 'unliquidated damages' through arbitration or court action. Where there is a liquidated damages provision, this constitutes an exhaustive agreement as to the damages recoverable for the breach of the late completion. Where the Appendix entry in a JCT 80 contract is completed '£NIL' this will preclude the employer from gaining any damages, even at large[447]. On the present wording the position would be unlikely to be any different under the JCT 98 Edition.

The advantages of liquidated damages are:

— They do not require proof after the event.
— They can be simply deducted by the employer under the contractual mechanism.
— They are agreed in advance and stated in the contract so that the contractor knows the extent of his potential liability.

Liquidated damages clauses are likely to be construed *contra proferentem* (qv)[448], although this is probably not the case if the contract is in a negotiated form, e.g. if a current edition of the JCT contract is used[449]. But, handwritten or typewritten insertions which are inconsistent with the printed provisions will, it seems, be so construed[450]. It is essential that a careful calculation be made at pre-tender stage taking the relevant factors on the particular job into account. **Figure 12** shows a possible format for such a calculation. In the public sector, where it is difficult to estimate the loss, it is usual to make use of a formula calculation, and it is thought that this is an acceptable approach. There is some confusion among members of the construction industry regarding what constitutes a penalty. A penalty is not enforceable. It is either a predetermined sum which is not a realistic pre-estimate of damage or a sum which is payable on the occurrence of any one of a number of different kinds of events. It is of no consequence whether the sum is described as a penalty or not. It is the real nature of the sum which matters. Even if a sum is held to be a penalty, the employer may still pursue an action for his actual (unliquidated) damages at common law.

---

[446] *BFI Group of Companies Ltd* v. *DCB Integration Systems Ltd* (1987) CILL 348.
[447] *Temloc Ltd* v. *Errill Properties Ltd* (1987) 39 BLR 30.
[448] *Peak Construction (Liverpool) Ltd* v. *McKinney Foundations Ltd* (1970) 1 BLR 114.
[449] *Tersons Ltd* v. *Stevenage Development Corporation* (1963) 5 BLR 54.
[450] *Bramall & Ogden Ltd* v. *Sheffield City Council* (1983) 1 Con LR 30.

Figure 12 Calculation of liquidation and ascertained damages: typical format.

Contract . . . .
Client . . . .
Architect . . . .

	Costs/Week
1. *Supervisory staff (current rates)*	
Architect: Estim. hrs/wk.............. × time charge of £................. per hour.	£
Quantity surveyor: Estim. hrs/wk................ × time charge of £............... per hour	£
Consultants: (as above for each one)	£
Clerk of works: Weekly salary (yearly) / 52	£
TOTAL (1)	£

2. *Additional costs* (current rates)*	
Rent and/or rates and/or charges for present premises	£
Rent and/or rates and/or charges for alternative premises	£
Charges for equipment	£
Movement of equipment	£
Additional and/or continuing and/or substitute staff	£
Movement of staff (include travel expenses)	£
Any site charges which are the responsibility of the client	£
Extra payments to directly employed trades	£
Insurance	£
Additional administrative costs	£
TOTAL (2)	£

	Costs/Week
3. *Interest*	
Interest payable on estimated capital expended up to the contract completion date, but from which no benefit is derived. Estimated expenditure taken as 80% of contract sum and fees	
Contract sum:	£
Architect's fees (90%)[†]	£
Quantity surveyor's fees (90%)[†]	£
Consultant's fees (90%)[†]	£
Salary of clerk of works	£
(£/wk × contract period)	
Total	£
Interest charges at current rate of .... %	
Interest/wk (80% capital expended × interest) / 52	
TOTAL (3)	£

4. *Inflation*	
Current rate of inflation............... %/year	
TOTAL (1) × .... % × contract period (years)	£
TOTAL (2) × .... % × contract period (years)	£
TOTAL (4)	£

(*Continued*)

5. *Total liquidated and assertained damages/week*

TOTAL (1)	£
TOTAL (2)	£
TOTAL (3)	£
TOTAL (4)	£
FINAL TOTAL	£

*It is essential that all costs noted are additional, i.e. they would not be incurred if the contract was completed on the contract completion date. The headings given are examples only. Every job is different.

†Professional fees are taken as 90% of total because some professional work remains to be done after practical completion.

The FINAL TOTAL should be examined to see if it appears reasonable in all the circumstances. It should be appreciated that the calculation can only be approximate. If in doubt about the figure, reduce it. It is sound procedure for the architect to calculate the totals with the client, have the calculations typed out and send it to him for signature.

If sectional completion is to be used, the amounts of liquidated damages should be apportioned bearing in mind:

— The value of each section.
— The implications in cost to the client of each section. For example, the clerk of works may be required to stay on site until the last section is completed or his attendance may be reduced, at some stage, to half time.

In *Dunlop Pneumatic Tyre Co Ltd* v. *New Garage & Motor Co Ltd*[451], Lord Dunedin noted the principles by which the court decides whether a clause provides for liquidated damages or a penalty:

'(i) Though the parties to a contract who use the words penalty or liquidated damages may *prima facie* be supposed to mean what they say, yet the expression used is not conclusive. The court must find out whether the payment stipulated is in truth a penalty or liquidated damages...

(ii) The essence of a penalty is a payment of money stipulated as *in terrorem* of the offending party; the essence of liquidated damages is a genuine covenanted pre-estimate of damage.

(iii) The question whether a sum stipulated is penalty or liquidated damages is a question of construction to be decided upon the terms and inherent circumstances of each particular contract, judged as at the time of the making of the contract, not as at the time of the breach.

(iv) To assist this task of construction various tests have been suggested, which, if applicable to the case under consideration, may prove helpful or even conclusive. Such are: (a) It will be held to be a penalty if the sum stipulated for is extravagant and unconscionable in amount in comparison with the greatest loss which could conceivably be proved to have followed from the breach ... (b) It will be held to be a penalty if the breach consists only in not paying a sum of money, and the sum stipulated is a sum greater than the sum which ought to have been paid ... (c) There is a presumption (but no more) that it is a penalty when a single lump sum is made payable by way of compensation, on the occurrence of one or more or all of several events, some of which may occasion serious and others but trifling damages. On the other hand (d) it is no obstacle to the sum stipulated being a genuine pre-estimate of damage that the consequences of the breach are such as to make precise pre-estimation almost

[451](1915) All ER 739.

an impossibility. On the contrary, that is just the situation when it is probable that the pre-estimated damage was the true bargain between the parties.'[452]

However, hypothetical situations cannot be used to defeat a liquidated damages clause. The court will take a pragmatic approach[453].

Liquidated damages clauses are usually linked with an extension of time (qv) clause and the position has been clearly stated by the House of Lords[454], insofar as:

— The general rule is that the contractor is bound to complete the work by the date for completion stated in the contract, as extended. If he fails to do so, the employer is entitled to recover liquidated damages.

— The employer is not entitled to liquidated damages if he by his acts or omissions has prevented the contractor from completing by the due date, and if this occurs time may become 'at large' (see: **Time at large**).

— These general rules may be amended by the express terms of the contract and are so amended by the common standard forms. These provide for extensions of time to be granted in appropriate cases.

— Failure by the architect properly to extend time for acts etc. of the employer not covered by the events listed in the extension of time clause will result in time being at large (see: **Time at large**) and liquidated damages being irrecoverable. The contractor's obligation is then to complete within a reasonable time and the employer is left to sue for unliquidated damages at common law.

See also: **Damages; Extensions of time; Penalty.**

Liquidated prolongation costs A fixed daily rate (LPC) to reimburse the contractor for extra costs incurred by delays caused by variations, etc. An LPC clause is sometimes found in construction contracts and is sometimes called a 'Brown clause' (qv). Tenderers are asked to specify the LPC required in their tender so as to cater for delays or extended use of items in the bill preliminaries. Although superficially attractive as a cost-effective means of settling prolongation claims – and in one sense a reverse form of liquidated damages – LPC is inflexible; for example, it takes no account of disruption which does not result in prolongation of the contract period.
See also: **Additional variation percentage.**

Liquidation Also known as 'winding-up'. The legal process for terminating the existence of a company which is registered under the Companies Act 1985. The primary aim of the process is that of realising all debts due to the company and realising all of its assets so that they may be converted to cash for the purposes of distributing that cash fund to the creditors of the company in their order of priority.

---

[452] *Dunlop Pneumatic Tyre Co Ltd* v. *New Garage & Motor Co Ltd* (1915) All ER 739 per Lord Dunedin at 742.
[453] *Philips Hong Kong Ltd* v. *Attorney General of Hong Kong* (1990) 50 BLR 122.
[454] *Percy Bilton Ltd* v. *Greater London Council* (1982) 20 BLR 1.

There are three types of winding-up:
— Winding-up by order of the court.
— Creditors' winding-up.
— Members' voluntary winding-up.

The first two apply to insolvent companies and are creditors' procedures brought about either as a compulsory or voluntary 'winding-up'.

A compulsory winding-up by order of the court may be commenced by the company, a creditor or the receiver (qv) presenting a petition to the court to wind-up the company. If the court makes a compulsory winding-up order, the official receiver becomes a provisional liquidator and he may apply to the court for the appointment of a special manager. A meeting of the creditors called by the provisional liquidator decides whether or not to apply to the court for the appointment of both a liquidator and a committee of inspection.

A 'creditors voluntary winding-up' has the advantage that the creditors can settle matters without recourse to the court, although they may apply to the court if they deem it necessary. This voluntary winding-up procedure starts with the shareholders of the company passing an extraordinary resolution that it cannot, by reasons of its liabilities, carry on its business and that it is expedient that it should cease trading and be wound up. A meeting of creditors must be called on the same or the following day to appoint a liquidator and a committee of inspection.

A liquidator may only carry on the business if it is beneficial to the winding-up; for example, if the overall capital available is likely to be increased. The object of liquidation and the law governing it is to ensure equal distribution of the company's assets among the creditors, subject to the following order of preference:
— Fixed charges.
— Costs of liquidation.
— Preferential creditors (e.g. rates, taxes, national insurance etc. for a fixed period and wages for the previous four months to a statutory maximum per employee).
— Floating charges.
— Unsecured creditors (they may well be the creditors who force the winding-up).

A members' voluntary winding-up will arise where the directors of the company take the informed decision that, if wound up the company will nevertheless (within 12 months from the beginning of the liquidation) be able to meet all of its debts in full plus interest. In that case the shareholders may pass a resolution for liquidation of the company without reference to or the involvement of the creditors. Before any such resolution may be passed the directors must swear a statutory declaration as to the company's ability to meet its debts (with interest) in the prescribed period of 12 months and if it transpires that they did not have reasonable grounds for presuming that to be the case then they may be liable to a fine or imprisonment[455].

---

[455] *Re William Thorpe & Sons Limited* (1989) 5 BCC 156.

Most building standard form contracts (see, for example, JCT 98, clauses 27.3.1 and 28.3.1) make express provision entitling either employer or contractor as the case may be to determine their employment under the contract in the event that the other party:
— has a provisional liquidator appointed, or
— has a winding-up order made, or
— passes a resolution for voluntary winding-up (except for the purposes of amalgamation or reconstruction).

If a company transfers the whole of its interest to a new company, it is known as 'reconstruction'. Useful practice notes on the matter of insolvency and for use in connection with both the Scottish and English versions of the JCT contracts are published by the Scottish Building Contract Committee (SBCC) and JCT respectively.

See also: **Insolvency.**

Liquidator A person (who must be a licensed insolvency practitioner[456]) and who is appointed by a company or by the court to carry out the liquidation (qv) of the company's assets for the benefit of creditors.

Litigation The process of resolving a legal dispute before a court. The term is used in contrast to 'arbitration' (qv) which is the settlement of disputes before a private judge of the parties' choosing and to adjudication which is a private process commonly adopted to achieve a speedy, cost-effective, binding but nevertheless interim decision over the parties' disputes or differences. The great jurist Sir Frederick Pollock defined litigation as a game in which the court is an umpire.

Local Act A statute (qv) which is of purely local operation. Many local Acts are of relevance in the field of building control (qv).

Local authority Local authorities are statutory corporations charged with a range of functions over a limited geographical area. They are subject to the doctrine of *ultra vires* (qv). There are special local authority editions of some standard form building contracts, e.g. JCT 98, which deal in a particular way with advance payment insurance of the works and retention, etc.

Lockout Lockout may occur either where the employees' place of work is closed or may refer to circumstances where a number of employees are refused further employment as a result of a dispute where the purpose of refusing employment is to put pressure on those employees to accept terms and conditions relating to their employment. In either case it will usually be the result of an industrial dispute. JCT 98 (clause 25.4.4) and IFC 98 (clause 2.4.4) expressly refer to a lockout as one ground entitling the contractor to an extension of time (qv).

---

[456]Section 390 Insolvency Act (1986).

Locus sigilli The place of the seal. This latin expression is often abbreviated to LS and commonly appeared beside the attestation (qv) clause of a document requiring to be sealed where executed as a deed (qv).

By the Land Registration (Execution of Deeds) Rules 1990 (SI 1990/1010) the execution of deeds concerning registered land by individuals must no longer be effected by use of a seal and by s. 1.-(1) of the Law of Property (Miscellaneous Provisions) Act 1989, 'any rule of law which requires a seal for the valid execution of an instrument as a deed by an individual' has been abolished. Provisions allowing British companies to execute deeds by alternative means have also been introduced[457] so that, in the case of companies the traditional method of executing a deed under seal may now be replaced by alternative means provided certain specific requirements are met. Attestation provisions in current editions of the JCT standard form contracts are now drafted to meet those alternative means and there is now no *locus sigilli* such as used to appear in previous editions of the JCT contracts such as the JCT 80 and IFC 84 Editions.

Prior to introduction of the alternative provisions for execution of deeds referred to above, it has been held by the Court of Appeal[458] that, where a defendant had signed his signature across the circle which bore the printed letters LS and there was an attestation clause (qv) signed by a witness, a bank's printed mortgage form could be considered as having been properly executed as a deed despite the absence of any wafer or other seal. Thus, it was not strictly necessary for a physical seal to be attached to or impressed on the document.

Today, where the option for attestation by the traditional method can be and is intended to be used, it is thought that a document similarly lacking in a proper seal will not now be taken as legally capable of being a deed. In any event, it is clearly preferable that any document, intended to be under seal, should have a seal attached, stamped or impressed to remove any possibility of later dispute on the matter.

Loss and/or expense A phrase used loosely to refer to the damage suffered by the contractor and for which he might be expected to bring a claim (qv). GC/Works/1 (1998) refers only to 'expense', which it defines (clause 46 (6)) as meaning 'money expended by the contractor' but excluding 'any sum expended, or loss incurred, by him by way of interest or finance charges however described'. It is clear that the contractor may not claim under clause 46 of this contract for loss of profit (qv), i.e. he may claim for any expense over and above that which he might have expected but cannot claim for money he expected to receive but because of the event cited did not. Most other contracts allow loss and expense which equate to what is recoverable as damages at common law. More common heads of claim for loss and expense under JCT contracts will include losses associated with:
— On site establishment charges.
— Head office overheads.

---

[457]By the Companies Act 1989.
[458]*First National Securities Ltd* v. *Jones* [1978] 2 All ER 221.

— Inefficient or increased use of labour and plant.
— Costs of unforseen winter working.
— Costs of inefficient use of or additional plant and equipment.
— Finance charges.
— Profit.

For a useful review of the law and practice relating to claims for loss and expense under the most commonly used standard forms of building contract see: *Powell-Smith & Sims' Building Contracts Claims*, 3rd edn, 1998, published by Blackwell Science.

See also: **Direct loss and/or expense.**

Loss of productivity Loss of productivity is a permissible part of a claim under the money claims clauses of most standard form building contracts. Thus, JCT 98, clause 26 entitles the contractor to recover 'direct loss and/or expense' if 'regular progress of the Works' has been or is likely to be affected by the matters listed in the clause.

Some authorities have argued that this must involve *delay* (qv) in progress and that the contractor's entitlement is limited to the effects of delayed completion. A careful reading of the clause (and similar provisions in other contracts) does not support this view.

In principle the contractor is entitled to recover for loss of productivity, i.e. the effect of the event upon the cost of the work, by labour, plant and other resources having been used less efficiently during the original contract period, even if no extension of time (qv) is involved. Regular progress of the works can be 'materially affected' without there being any delay at all to completion (qv) and the additional cost (if proven) falls within the rule in *Hadley* v. *Baxendale*[459] – as being foreseeable. It is the natural consequence of the specified act and must be something which the parties had, or should have had, in mind.

In broad principle, loss of productivity is easy to establish, but it is difficult to prove and quantify in detail, and at the very least the contractor must be able to isolate the various items of cost which have been affected by the particular disruptive events on which he relies.

See also: **Claims; Foreseeability.**

Loss of profit Loss of profit is a recoverable part of a claim under the money claims clauses of the various standard forms of contract in common use, as well as being recoverable as a head of damages for breach of contract at common law, assuming of course that the contractor would have earned it had it not been for the event giving rise to the claim[460].

It is, however, only the 'normal' profit which is recoverable because such a loss is within the contemplation of the parties. Exceptionally high profit which the contractor might otherwise have earned will not be similarly recoverable

---

[459](1854) 9 Ex 341.
[460]*Hadley* v. *Baxendale* (1854) 9 LJ Ex 341; *Wraight Ltd* v. *PH & T (Holdings) Ltd* (1968) 13 BLR 26.

unless the other party of the contract knew, at the time of the contract, of facts which would bring the abnormal profit within his contemplation[461].

There is no *automatic* right to recover lost profit: 'The better view is that such a claim is allowable only where the contractor is able to demonstrate that he has been prevented from earning profit elsewhere in the normal course of his business as a direct result of the disruption or prolongation'[462].

See also: **Claims; Damages.**

Low performance damages A pre-estimated amount stated by way of liquidated damages (qv) which the contractor will be liable to pay to the employer in the event that a defect in the works is such that some prescribed performance level stated in the contract is not achieved. Under the NEC terms and conditions incorporation of such a provision is optional and may be used by adopting Option S of the optional clauses to the contract. Where the clause is used, the contract data (qv) must be completed so that the amount of damages to be applied in the event of low performance is inserted. However, it should be remembered that, when calculating and inserting the requisite amount the clause is, in effect, no different to any other provision for liquidated damages and to be enforceable it will be subject to the same tests as for any other liquidated damages provisions.

See also: **Liquidated damages.**

Lump sum contract When one party carries out work for a stated and fixed amount of money payable by the other. Almost without exception, all of the main forms of building contract are considered to be lump sum contracts even though they contain provisions for the adjustment of the contract sum (qv) for such things as fluctuations and variations. The important point is that the original contract sum is stated for a given amount of work. Some contracts are expressly not lump sum contracts, e.g. JCT 98 Private Edition, With Approximate Quantities. If the contract expressly provides for remeasurement, it is not a lump sum contract.

See also: **Firm price contract; Fixed price contract.**

---

[461] *Victoria Laundry (Windsor) Ltd* v. *Newman Industries Ltd* [1949] 1 All ER 997.
[462] Chappell (1998) *Powell-Smith and Sims Building Contract Claims*, 3rd edn, p. 159, Blackwell Science, referring to *Peak Construction (Liverpool) Ltd* v. *McKinney Foundations Ltd* (1970) 1 BLR 114.

Main contract A term sometimes given to a contract made between employer and contractor to distinguish it from any sub-contracts under which the contractor engages others to perform certain elements of the entire work to which the main contract refers. Thus the contractor (qv) is also referred to as the 'main contractor' or 'principal contractor', although the latter term should be reserved exclusively for a contractor carrying out that function under the Construction (Design and Management) Regulations (CDM) 1994 (qv). Sometimes also referred to as a 'head-contract' (qv). Many standard form main contracts, such as the JCT 98 and IFC 98, have associated standard form sub-contracts drafted for use with them to ensure, so far as possible, that main and sub-contracts sit 'back to back' with each other (for example: DOM/1 (qv) – domestic sub-contract form – for use with the JCT 98, DOM/2 (qv) – domestic sub-contract – for use with the WCD 98 and the NSC/C sub-contract documentation for use with the JCT 98 main contract, where it is proposed that the architect will nominate one or more sub-contractor(s)).
See also: **ACA Form of Building Agreement; BPF System; JCT contracts.**

Maintenance A term which, in connection with building contracts, can be used in a variety of contexts. It may be used to describe the work to be done under a contract where the contractor is engaged to carry out regular repair or replacement works as necessary over a period of time, possibly the life of the building, or is contracted to keep existing property to a defined standard of repair.

The word is used by GC/Works/1 (1998) clauses 21 and 49 and in the ACA 3, clause 12.2 to denote what might more properly be referred to under other contracts as the 'defects liability period' (JCT 98) or 'correction of defects period' (NEC). Use of the term in the manner adopted in the GC/Works/1 (1998) is regrettable because to use the word in this way does violence to its ordinary meaning. If the contractor were indeed required to 'maintain' a building for a period of six months after completion, it would involve his keeping it in pristine condition despite occupation and the passage of time. In fact, GC/Works/1 (1998) defines the contractor's responsibility more narrowly.
See also: **Defects liability period; Maintenance clause; Maintenance period.**

Maintenance clause A clause included in the ACA 3 and GC/Works/1 (1998) forms of contract (at clauses 12.2 and 21 respectively) designating a period of time after the works are completed during which the contractor is to make good defects.
See also: **Defects liability period.**

Maintenance period A misleading phrase in some contracts referring to the period of time after the works are completed during which the contractor is to

make good defects. For example, ACA 3, clause 12.2 refers to the maintenance period as does GC/Works/1 (1998), clauses 21 and 49.
See also: **Defects liability period; Maintenance clause.**

Management contract A loose term covering a wide variety of contractual situations. In appropriate circumstances it may be seen as a useful means of providing a 'fast track' approach to the project in circumstances where responsibility for design and contract administration is to be left with directly engaged professionals (see, for example, the Second Recital of MC 98). The term generally refers to a type of contract where the main contractor is selected at a very early stage and is appointed to manage the construction process and input his own expertise during the pre-contract stages. The contractor receives a fee for his services which is agreed between the parties before the contractor is appointed.

The Joint Contracts Tribunal (qv) produced a Management Contract package, first issued in 1987, in response to demands for standard documentation to suit this particular procurement method. Since then, the package has been subject to various amendments designed generally to improve the drafting and effectiveness of the documentation and to take account of changes in the law affecting its provisions. A fully revised edition (the Standard Form of Management Contract 1998 Edition) has now been published. The total package consists of:

— Standard Form of Management Contract 1998 Edition (MC 98).
— Amendment 1 to MC 98 – Construction Industry Scheme.
— Standard Form of Works Contract 1998, for use between works package contractors (qv) and management contractor, comprising:
    Works Contract/1, being the Invitation, Tender and Articles of Agreement;
    Works Contract/2, being the Conditions;
    Works Contract/3, being the Standard Form of Employer/Works Contractor Agreement: Amendment to Works Contract documentation applicable to Wks/1:1; Wks/1:2; Wks/1:3; Works Contract/2 – Construction Industry Scheme;
    JCT Sub-Contract/Works Contract Formula rules.
— Phased Completion Supplements for Management Contract and Works Contract.
— Practice Notes MC/1 and MC/2 for 1987 Edition (currently under review).

The contract has two phases: the pre-construction and the construction period. The management contractor is intended to be involved in both phases subject to a break clause permitting the employer to stop the project at the end of the pre-construction period. Both management contract and works contract follow the structure of IFC 98 and MW 98. There are nine sections to the head-contract (qv) and the drafting of the works contract owes much to JCT 98 and NSC/C. Unlike the position under some contractor inspired management contracts, under MC 98 there is no provision for the management contractor to carry out any work on site himself. His obligations during the construction stage are essentially to:

— Set out the project, including all work done by works contractors.
— Manage the project, including all work done by works contractors.
— Organise the project, including all work done by works contractors.
— Supervise and secure the carrying out and completion of the project, including all work done by works contractors.

Competitive tendering is usual for the various works contract elements. Points to note are:

— The contractor is responsible to the employer for the construction process.
— The system is most useful for large and complex contracts when a considerable degree of co-ordination of specialists is required and where early completion is vital.
— Accurate programming and cost planning is essential for success.
— The selection of a suitable contractor to undertake the management work is not an easy process.

See also: **BPF System; Cost reimbursement contract; Design and build contract; Directions; Project management.**

Mareva injunction See: **Freezing injunction.**

Master (1) The traditional legal term for an employer of labour, i.e. the relationship of employer and employee. The major distinction between the relationship of master and servant and that of employer and independent contractor (qv) appears to be that in the former case the employer has the power to direct and control how, when and what work is to be done. An employer is vicariously responsible for acts done by his employee in the course of his employment. (See also: **Vicarious liability.**)

(2) Masters of the Supreme Court are officers of the High Court of England and Wales who perform certain judicial work and issue directions on matters of practice and procedure. With the introduction of the Civil Procedure Rules the term 'Master' is now used in a rather more limited context. For example, whereas hearings for assessment (qv) of costs in legal proceedings (formerly called taxation) previously went before a 'Taxing Master' that title no longer applies and the process is now administered by a 'costs judge' (qv).

Master cost plan Under the BPF System (qv) this is a schedule prepared by the client's representative (qv) showing the total expenditure required to complete the project. 'At all times it should provide the best possible estimate of the final cost of the project, of the future cash flow, and of the future cost of the building.' The BPF manual contains, in Appendix A2.3, a checklist of the information which the master cost plan should include, arranged under the following headings:

— Description of project.
— Basis of cost plan.

> — Forecast tender price.
> — Other costs.
> — Target cost.
> — Development cost.

Master programme (1) A term to be found in JCT 98, clause 5.3 where it refers to the contractor's overall programme for the execution of the works. The reference in the contract merely formalises what has long been the practice in most contracts through an appropriate clause in the bills of quantities (qv) or specification (qv). The clause does not state the form the programme should take. The type of programme required should be specified in the bills of quantities or specification as appropriate, e.g. bar chart, network analysis, etc. Except for the smallest jobs, it is advisable to request a network analysis to be prepared, because it clearly highlights delay and disruption. Note that, if no programme is specified it is probable that the contractor is under no obligation to supply one. The footnote to the clause points out that the provision may be deleted but this is an unwise practice because, although the master programme is not a contract document (qv) it is an invaluable aid to monitoring contract progress for both architect and contractor. A good programme should show not only the start and finish dates and relationships of the key operations, but also the way in which sub-contractors of all kinds will be integrated into the work.

It is good practice for the contractor to note (realistic) dates for receipt of information – drawings, schedules, nominations, etc. – from the architect. The architect should not 'approve' the contractor's master programme, although it may have little significance even if he does so[463]. He must treat it as mere information from the contractor of what he intends to do and when he intends to do it.

Where under clause 5.3 of JCT 98 a master programme is prepared and has been provided by the contractor and any amendment or revision to that programme arises to take into account an extension of time given under clause 25, or resulting from a clause 13A quotation, that amended master programme must also be provided to the architect within 14 days of its revision or amendment.

(2) Under the BPF System (qv) for building design and construction the same term is used to describe the schedule prepared by the client's representative (qv) of the main activities required to complete the project. This master programme is produced at an early stage in the development of the project, and is updated by the client's representative as the project progresses to tender stage.

(3) Under the NEC (qv), the programme that the contractor proposes to work to should be identified in the contract data (qv). If not already prepared and identified there, the contract data should stipulate a period of time within which the programme should be prepared and given to the project manager for his acceptance (or reasoned non-acceptance) which should be given within two weeks thereof. It will then become the 'accepted programme'.

---

[463] *Hampshire County Council* v. *Stanley Hugh Leach Ltd* (1991) 8-CLD-07-12.

Clause 31 of the NEC (qv) sets out the minimum requirements that should appear on any such initial programme or on any revised version of it and in the contract data facility is provided for the employer to stipulate the precise intervals at which the programme must be revised and re-submitted throughout the course of the project.

See also: **Regularly and diligently; Time schedule; Accepted programme.**

Materially affected A phrase used in particular in the JCT 98 and IFC 98 forms of contract. Clause 26 of JCT 98 refers to regular progress of the works being 'materially affected' and makes such affect a condition precedent to the contractor being able to claim loss and/or expense. Clause 27.1.3 refers to the works being 'materially affected' by the contractor's refusal or neglect to comply with the architect's instructions requiring him to remove defective work or materials. It is a ground for the employer to determine the contractor's employment. The addition of the word 'materially' in each case makes clear that it is not sufficient if it can be said that the works or progress (as the case may be) are *affected*. They must be affected in some important or significant way or to a substantial extent. The word is not precisely defined in the contract. It is clear that trivial disruptions are excluded and whether progress can be said to be materially affected will depend upon the exact circumstances of each case.

GC/Works/1 (1998) clause 46 (1) – which is to the same effect as the JCT terms noted above – refers, instead, to the works being 'materially' disrupted or prolonged.

Materials Although most building contracts draw a distinction between 'goods' and 'materials' there is no distinction in law. Both are 'goods' for the purposes of the Sale of Goods Act 1979. In building practice, the things used to construct the building – bricks, sand and cement, timber, screws, etc. which are the raw elements of the building before any work has been done – are called 'materials' in contrast to such things as door furniture and sanitary fittings which are normally described as 'goods' (qv).

The Supply of Goods and Services Act 1982 (qv), which applies to, among other things, building contracts, implies certain conditions and warranties with regard to materials and goods supplied under such contract. These implied terms (qv) parallel those implied by the Sale of Goods Act 1979 in relation to matters such as, for example, the satisfactory quality of the goods supplied.

Under some forms of contract (e.g. JCT 98, clause 25.4.10.2) the contractor's inability for reasons beyond his control and which could not reasonably have been foreseen at the date of tender to secure essential goods and materials may provide grounds for an entitlement to an extension of time. At common law, such inability would not, other than in exceptional circumstances, excuse late completion.

See also: **Sale of goods.**

Measure and value contract A general name given to a contract where there is no fixed (lump sum) contract price but where the work is measured and valued by the quantity surveyor generally as the works proceed, by reference to a contractual schedule of rates and/or prices (qv) – in order to arrive at the price to be paid to the contractor.

See also: **Measurement contract; Schedule of Rates.**

Measure of damages See: **Damages.**

Measurement Generally, the method of ascertaining length, breadth or height, volume or area of objects, buildings, land, etc. in terms of a particular system of measurement, e.g. metric.

In building contracts, measurement of the work is carried out by the quantity surveyor either before work begins, from the drawings prepared by the architect, or during the progress and after completion of the work. The quantity surveyor generally works to a set of rules embodied in a Standard Method of Measurement (qv). Under JCT 98, except to the extent otherwise expressly stated, the quantities in the contract bills will be deemed to be measured according to that Standard Method of Measurement (qv). In that case, the quantities in any such bills will be taken to accurately reflect the quality and quantity of the works priced for and to be carried out.

See also: **Bills of quantities; Measurement contract.**

Measurement contract Normally used where the precise quantity (and sometimes type) of work cannot be accurately determined at the time of tender. A basis is provided for tendering purposes and the completed work is measured and payment made in accordance with the tendered rates. There are two main types of measurement contract used:

— Where approximate quantities are used. This type is suitable where the type of work is known but the quantity is unknown (see also: **Bills of quantities**).
— Where a schedule of rates and/or prices (qv) is used. This type is suitable where even the type of work is not known for certain.

Merchantable quality The former standard, imposed by s. 14 of the Sale of Goods Act 1979 by means of an implied term (qv) in contracts for sale of goods, that the goods are of 'merchantable quality'. 'Merchantable quality' is defined as meaning that the goods 'are as fit for the purpose . . . for which goods of that kind are commonly bought as it is reasonable to expect having regard to any description applied to them, the price (if relevant) and all the other circumstances . . .' The term 'merchantable' is a relative one, but the goods must remain 'merchantable' for a reasonable time. If the buyer examines the goods he will not be protected against defects that examination ought to have revealed, i.e. patent defects (qv). A similar provision is made by the Supply of Goods and Services Act 1982 (qv). In business transactions – which includes sales of building materials – the term can be excluded so far as it is reasonable to do so.

See also: **Satisfactory quality; Unfair Contract Terms Act 1977.**

Minutes of meeting The official record of a meeting. It is essential that all meetings, even of the most informal kind, which have any relevance to a contract, should be recorded in some way. Under NEC clause 16.3, where an 'early warning meeting' (qv) is convened, any proposals discussed and considered and any decisions made must be minuted by the project manager (qv) with a copy of those minutes then being sent to the contractor for his records and action as appropriate.

Generally, all short and/or informal meetings or telephone calls may be recorded simply by means of a brief personal note of all the important points made and that note then being kept in parties' own personal files. Meetings on a more formal basis, such as pre-start, design team or site meetings, should be minuted. All such meetings should have an agenda to ensure that necessary points are discussed and, if possible, a time limit so as to concentrate minds. The minutes of such meetings should be the responsibility of one person, often the architect or project manager (qv). They must record only the important items which, in practice, may mean recording only decisions made. A format for a typical site meeting is shown in **Figure 13.**

It is essential in the formal context to circulate minutes to all participants as soon as possible after the meeting. Indeed, under some forms of contract there is not only an express obligation to do so but an obligation also rests with those receiving a copy to notify any dissent from the minutes as a true record of

Job Title: Ref. No:
Location:
Site Meeting No:
Present:

1.0 The minutes of Site Meeting No. .... held on the ....
    are agreed as a true record
2.0 Matters arising from the minutes of the last meeting
3.0 Contractor's progress report
4.0 Clerk of Works' report
5.0 Consultants' report.
    5.1 Structural Engineer
    5.2 Heating and Ventilating Engineer
    5.3 Mechanical Engineer
    5.4 Electrical Engineer
    5.5 Landscape Architect
6.0 Final report by Quantity Surveyor
7.0 Any other business
8.0 Date and time of next meeting

Circulation of Minutes to:

Employer	_ Cps	Mechanical Engineer	_ Cps
Contractor	_ Cps	Electrical Engineer	_ Cps
Quantity Surveyor	_ Cps	Landscape Archietect	_ Cps
Structural Engineer	_ Cps	Clerk of Works	_ Cps
Heating and Vent. Engineer	_ Cps	File	_ Cps

Figure 13 Typical format of minutes of a site meeting.

events within a stipulated time. That is a procedure to be encouraged. In any event, any disagreements as to the accuracy of the minutes should, at the very latest, be recorded at the next meeting, if there is to be a series of meetings. Otherwise, some note must be put at the beginning of each meeting recording that the minutes are agreed as a true record. Where a contract calls for a certificate to be issued, a notice given or an application made, it is not thought that a note in any minutes will suffice, and certainly a note in the minutes about information supply cannot amount to a 'specific application in writing' for that information by the contractor. However, if the application is part of a contractor-generated progress report, the position would be different.

Misconduct Conduct falling below the standards required in the circumstances. It is particularly serious in the case of professional persons who have a duty to conduct themselves with complete integrity. Thus an arbitrator (qv) who hears one party in the absence of the other without good reason will have misconducted himself in the administration of the proceedings.

Under the Arbitration Act 1996, s. 24.-(1), provided substantial injustice has been or will be caused as a consequence, a party to arbitral proceedings may apply to the court to remove an arbitrator on grounds of what may broadly be termed misconduct where, for example, circumstances exist that give rise to justifiable doubts about the arbitrator's impartiality or where he has refused or failed properly to conduct the proceedings or to reasonably expedite the conduct of the proceedings[464]. Misconduct need not necessarily involve moral turpitude and it may best be defined for these purposes as amounting to any such mishandling of the arbitration as is likely to give rise to some substantial miscarriage of justice[465].

An architect could be guilty of misconduct by favouring one contractor during the tendering process.

Misrepresentation A misrepresentation is an untrue statement of fact made during the course of pre-contractual negotiations and which is one of the factors which induces the other party to contract. If the misrepresentation becomes a term of the contract, then liability depends on whether it is a condition (qv) or a warranty (qv), although in either case the innocent party will have a remedy for breach of contract (qv). Misrepresentations which do not become part of the contract — as is normally the case — may also give rise to liability at common law and under the Misrepresentation Act 1967, as amended. A misrepresentation may be:
— Fraudulent (qv) when it is made without honest belief in its truth.
— Innocent (qv) where it is made without fault.
— Negligent (qv).

---

[464]cf s. 23 (1) of the Arbitration Act 1950.
[465]*Williams* v. *Wallis & Cox* [1914] 2 KB 478.

In all cases the innocent party may seek to rescind the contract (see: **Rescission**) and/or claim damages (qv). By s. 2 (2) of the Misrepresentations Act 1967, damages can only be granted as an alternative to rescission in the case of *innocent misrepresentations*. The award of damages in that case is discretionary[466].

Mistake Where the contracting parties are at cross-purposes about some material fact this may make the purported contract void (qv). Lawyers call this an 'operative mistake' and it must be distinguished from 'mistake' in the popular sense. Operative mistake is classified as:
— Common mistake; where both parties make the same mistake.
— Mutual mistake; where the parties are at cross-purposes about some essential fact.
— Unilateral mistake; where only one party is mistaken.

An operative mistake may either nullify or preclude consent, but the cases establish that this is extremely limited in scope, although in some cases the courts have intervened to prevent hardship by giving equitable relief[467]. Operative mistake has not proved important in the field of building contracts, its main application being that the employer could not accept the contractor's tender if he knew that its terms were not intended by the contractor, such as, for example, where an employer 'accepted' a tender in the knowledge that the contractor had omitted its first page, which contained a fluctuations clause[468].

Where, under a JCT contract, there has been some common mistake made by the parties such that the contract does not in fact reflect their true mutual intention then the arbitration clauses in JCT standard form contracts (including JCT 98, IFC 98 and MW 98) each provide that: 'without prejudice to the generality of his powers the arbitrator shall have power to rectify the contract so that it accurately reflects the true agreement made by the parties'. See also: **Contract; Equity.**

Mitigation of loss Someone seeking to recover damages for breach of contract (or any other reason) should do everything reasonably possible to reduce the amount of his loss for which he will claim compensation from the defaulting party. For example, where an employer has various defects rectified by others and seeks to recover the cost paid to those others, he may fail to do so if the court holds that the employer has not mitigated his loss by allowing the builder to carry out remedial work himself. In such circumstances, the employer may discharge his obligation to mitigate the loss if he can show that it would not have been reasonable for him to give the builder such an opportunity. However, in general an unreasonable refusal constitutes a failure to mitigate loss and therefore reduces or eliminates the damages recoverable[469].

---

[466]See, for example, *Howard Marine & Dredging Co Ltd* v. *A. Ogden & Sons (Excavations) Ltd* (1977) 9 BLR 34.
[467]*Solle* v. *Butcher* [1950] 1 KB 671.
[468]*McMaster University* v. *Wilchar Construction Ltd* (1971) 22 DLR (3d) 9.
[469]*City Axis* v. *Mr Daniel P. Jackson* (1998) CILL 1382.

That approach to the rule must be balanced with the fact that the employer must not simply sit back and wait for the costs associated with the builder's default to increase. For example, if a builder constructs a roof badly and there are defects which allow water to enter the building, the architect will ask him to put the matter right. If the builder refuses, the employer will, no doubt, take whatever steps are open to him either within the contract or at common law, to recover damages (qv). However, in doing so he is still bound to take reasonable measures to avoid unnecessarily adding to the amount of damage suffered, in which case it may be appropriate to engage and pay others to do the work without delay so that further extensive internal damage to the building by the water may be avoided. The employer would be unable to recover through the courts the loss which he could have avoided by taking that prompt action. In some cases it would be reasonable to postpone remedial work until damages were recovered[470]. Mitigation of loss does not, however, cover the situation where the employer or his architect might, by minute and careful inspection, have discovered defects at an earlier date than they did. It seems clear that there should be no reason why 'a negligent builder should be able to limit his liability by reason of the fact that at some earlier stage the architect failed to notice some defective work'[471].

See also: **Inspection of the works; Supervision of works.**

Mobilisation A term which, in the context of construction projects, is generally taken to refer to the action taken by the contractor immediately prior to commencing the works on site and will include such things as organising labour, arranging for the transportation of plant and equipment to site, the delivery of initial materials and the delivery to site and establishment of his temporary welfare, storage and other accommodation facilities in readiness to begin construction. The term may also occasionally be used to refer to the readying of armed forces for war such as, for example, where reference may be made in the contract to the consequences of an outbreak of hostilities (e.g. JCT 98, clause 21.2.1.6).

Moiety A legal term meaning a half or one of two equal parts. It is found in some forms of contract, particularly in relation to retention money and its release.

Monopoly Where the supply of certain goods or services is controlled by one or a group of manufacturers and traders. There are statutory restrictions on monopoly situations but perhaps of more direct relevance to the construction industry is the effect of EC legislation which has the effect of enforcing free trade across the community and prevents EC members exercising a monopoly over the use of nationally manufactured goods and materials.

---

[470] *Dodd Properties (Kent) Ltd* v. *Canterbury City Council* (1979) 13 BLR 45.
[471] *East Ham Borough Council* v. *Bernard Sunley Ltd* [1965] 3 All ER 619 per Lord Upjohn at p. 637.

Month A *lunar month* is a period of 28 days in contrast to a *calendar month* which is a period of 30 or 31 days (28 days in February, or in a leap year, 29 days). Unless a contrary intention is expressly stated, in all statutes, contracts and deeds, etc. made or coming into effect after the commencement of the Law of Property Act 1925, 'a month' means a calendar month unless the contrary is indicated: Interpretation Act 1978, s. 3; Law of Property Act 1925, s. 6l. Where time is expressed in a given period of months from a stipulated date, the general rule will be that the calculation of that period will begin from the day following the stipulated date and ends on midnight of the resulting date of expiry of the period.

Mutual dealings Under Rule 4.90 of the Insolvency Rules[472] there are certain rules regarding set-off to insolvent limited companies. They apply to a situation 'where there have been mutual credits, mutual debts or other mutual dealings'. In such a situation 'an account shall be taken of what is due from one party to the other in respect of such mutual dealings...' and that only the balance due on the taking of such an account shall be claimed. The rules embrace debts and credits arising out of any number of contracts between the same parties. Sums owing on one contract may be set-off against sums due on another, including claims for unliquidated damages[473].

This is of great importance in the construction industry where such dealings take place between main contractors and their sub-contractors or suppliers[474] and as between employers and their contractors. In the normal course of things, each party would be liable to pay money owing to the other and any set-off between unconnected contracts[475] would be by agreement only[476]. This would mean that, if liquidation occurred, the solvent party would be liable to pay his debts to the party in liquidation and, in turn, could only expect to receive whatever dividend was finally declared. For example, A owes £100 to B; B owes £80 to A. A becomes insolvent. If B pays the £80 he owes, he may then have to wait until A's affairs are settled when he may receive 1p in the pound, i.e. £1. When mutual dealings are taken into account, however, B would owe nothing and expect to receive, eventually, 1p in the pound of £20 (the balance), i.e. 20p. B's loss in the first instance would be £99, in the second instance £19.80. The procedure is a protection for the solvent party. The operation of a building contract between employer and contractor may be held sufficient to establish that there had been 'mutual dealings' between them[477].

---

[472]Insolvency Rules 1996 (SI 1986/952).
[473]*Peat* v. *Jones* (1882) 8 QBD 147, CA.
[474]See, for example, *Rolls Razor Ltd* v. *Cox* [1967] 1 All ER 397.
[475]In the absence of agreement, there is no set-off between unconnected contracts between the same parties: *Anglian Building Products Ltd* v. *W. & C. French (Construction) Ltd* (1972) 16 BLR 1 – There was no set-off and therefore no stay granted.
[476]See *Modern Engineering (Bristol) Ltd* v. *Gilbert-Ash (Northern) Ltd* [1974] 1 BLR 73.
[477]See, for example, *Willment Brothers Ltd* v. *North-West Thames Regional Health Authority* (1984) 26 BLR 51.

Named sub-contractors and suppliers A term used in the BPF System (qv) to refer to specialists whose advice has been sought during the design stages. They are named in the invitation to tender to the main contractor with an indication of whether the client requires that they be invited to tender for their part of the work. Provision for named sub-contractors is made in clause 9 of ACA 3. Likewise, IFC 98 refers to 'Named Sub-contractors' in clauses 3.3.1 to 3.3.7 inclusive where a distinction is also made between the 'first' and 'second' named sub-contractor in situations where there is cause to determine the employment of the first named sub-contractor and where that sub-contractor must then be replaced by another, on the instruction of the architect. The provisions for named sub-contractors under IFC 98 are closer to those of JCT 98 than to the ACA 3, but named sub-contractors must be distinguished from the nominated sub-contractors (qv) referred to in JCT 98, for which clause 35 of that form makes detailed provision.

In relation to named sub-contractors and suppliers, JCT 98, clause 19.3 provides a mechanism whereby the architect may detail, in the main contract documents, work which the contractor is to price but which is to be executed by a sub-contractor chosen by the main contractor from a list provided by the employer. The bills of quantities (qv) must provide to that effect and the contractor's right to select from the list is at his sole discretion. Notwithstanding such provisions as those of JCT 98 allowing the employer (or the architect on his behalf) to name sub-contractors in that way, any sub-contractor ultimately appointed will, nevertheless, be a domestic sub-contractor (JCT 98, clause 19.3.3).

National House Building Council The NHBC is a non-profit making insurance company recognised under statute. Its Chairman is appointed by the Secretary of State for the Environment[478]. Its principal aim is to improve the private house-building industry in the UK.

To achieve this end, the NHBC:
— Undertakes research into housing and construction in order to improve its standard building specification (called 'the Council's Requirements') with which all builders registered with it must comply.
— Carries out a spot check system of inspection of all dwellings registered with it. On completion NHBC issue a 10 year notice (BM 4) stating that the dwelling appears to have been designed and constructed substantially in accordance with requirements.
— Operates an insurance scheme which guarantees the performance of the builder or developer to complete dwellings to satisfactory standards and to remedy all defects which occur within, broadly, the first two years

---

[478]Now the Secretary of State for Transport, Environment and the Regions.

and thereafter to insure the property for a further eight years against major damage caused by structural defects or subsidence, settlement or heave affecting the structure. This is called Buildmark (qv) and the scheme was revised and improved with effect from 1 April 1999.

Houses built by registered builders are exempt from the provisions of s. 1 of the Defective Premises Act 1972 (qv). The contractual arrangements between the builder and the house purchaser are set out in an offer of cover which incorporates the relevant cover and there is provision for arbitration (qv) in respect of disputes arising under the scheme. NHBC will honour an arbitrator's award if the builder fails to do so. The NHBC has more than 250 field staff responsible for inspecting dwellings and investigating claims. The Ten Year Insurance Scheme has been in operation since 1967 and currently it pays 3000 claimants, on average, more than £6 million a year.

By careful monitoring of the cause of insurance claims, the NHBC is then able to amend its standard building specification in order to prevent such claims from arising in the future. In 1975, for instance, it was established that over 50% of claims related to defective infill which caused sinking floor slabs. The NHBC introduced a requirement in 1975 which specified that if more than 600 mm of infill were used, the builder must put in a suspended floor construction. As a result of this change, claims for foundation failures are now less than half of what they would otherwise have been.

The NHBC publishes an extensive list of both technical publications and information booklets for purchasers, builders and the professions generally. They are all available, most of them without charge, from the Information Office, NHBC, 58 Portland Place, London W1N 4BU. Comprehensive information about NHBC, what it does and the services it provides, the benefits of registration and how to register are also available on the internet address www.nhbc.co.uk

NEC See: **Engineering and Construction Contract.**

Negligence A category or branch of the law of tort (qv). Negligence is not the same as carelessness or mistake: it is conduct and not a state of mind. It is the omission to do something that a reasonable man would do, or the doing of something that a prudent and reasonable man would not do.

A claimant (qv) suing in negligence must show:
— The defendant (qv) was under a duty of care (qv) to him.
— The defendant was in breach of that duty.
— As a result the claimant suffered damage (qv) which must normally be damage to persons or property.

Purely financial loss or other forms of economic loss in the absence of reliance by the claimant on the defendant as in *Hedley Byrne & Co Ltd* v. *Heller & Partners Ltd*[479] cannot be recovered in negligence[480].

---

[479][1963] 2 All ER 575.
[480]*Murphy* v. *Brentwood District Council* [1991] AC 398. But it is unclear what effect, if any, the decision in *Henderson* v. *Merrett Syndicates Ltd* [1995] 2 AC 145 will have on the question of recovering pure economic loss in negligence.

In theory, the situations in which negligence may arise are endless. In *Donoghue* v. *Stevenson,* Lord Macmillan said: 'The categories of negligence are never closed'[481] but recent developments indicate that the courts are taking a more restrictive view. It nevertheless appears that liabilities in both contract and tort can co-exist[482]. Questions of public policy must be taken into account and so in *Ryeford Homes Ltd* v. *Sevenoaks District Council*[483] it was held that it would be contrary to public policy for a planning authority to be liable to an applicant in negligence because the authority's overriding duty is not to individuals but to the public at large. A police authority when investigating a crime is similarly protected[484].

In relation to building contracts, the most usual negligence actions are in relation to:

— Professional or other negligence actions where, for example, the architect may be negligent in designing a building or in his supervision; the quantity surveyor or architect may be negligent in preparing estimates of cost[485].

— Negligence action where, for example, the contractor may be negligent in carrying out the work. The contractor is not normally liable in negligence to third parties for defective construction unless there is injury to persons or damage to property other than the actual structure itself[486]. There is a large and fast growing body of case law dealing with the negligence of architects, contractors and local authorities.

See also: **Care, duty of; Contributory negligence.**

Negligent misstatement/misrepresentation
Since 1963 it has been the law that a negligent misstatement which is acted upon may give rise to liability in tort (qv)[487]. This is so even if only economic or pure financial loss results, as opposed to physical damage to persons or property. It appears that there must be some 'special relationship' between the maker of the statement and the recipient as well as reliance on the statement. In *Thomas Saunders Partnership* v. *Harvey*[488] a nominated sub-contractor's director was held personally liable for giving a false post-contractual assurance of compliance with a specification. The director was in breach of the personal duty of care (qv) which he owed to both the architects (to whom the assurance was given) and the employer under *Hedley Byrne* principles. He had specialist knowledge and skill and assumed responsibility for what he was saying[489]. Liability under the *Hedley Byrne* principle is not confined to factual statements; it extends to all forms of negligent advice, legal and financial, even if these are matters of opinion, e.g.

---

[481][1932] AC 562.
[482]*Henderson* v. *Merrett Syndicates Limited* [1995] 2 AC 145.
[483](1989) 16 Con LR 75.
[484]*Hill* v. *Chief Constable of West Yorkshire* [1988] 2 All ER 238.
[485]*Nye Saunders & Partners* v. *Alan E. Bristow* (1987) 17 Con LR 73.
[486]*Murphy* v. *Brentwood District Council* [1991] AC 398.
[487]*Hedley Byrne & Co Ltd* v. *Heller & Partners Ltd* [1963] 2 All ER 575.
[488](1989) CILL 518.
[489]*Esso Petroleum Ltd* v. *Mardon* [1976] 2 Lloyds Rep 305.

advice as to probable building costs. Indeed, the principle extends beyond the provisions of information and advice to include the performance of other services[490]. For there to be a liability for negligent misstatement it is not necessary for there to be a voluntary assumption of responsibility by the person giving the advice[491].

There can also be liability for negligent misrepresentations where they amount to misrepresentations (qv) under the Misrepresentation Act 1967. Under the Act, once a claimant shows a representation to be false it is for the person making the representation to disprove his negligence, in contrast to the position at common law where the claimant bears the burden of proving negligence (qv)[492].

See also: **Misrepresentation.**

Negotiated contract A contract which is not put out to tender, but where the price is agreed by negotiation between the parties. Although EU directives generally impose a strict regime on the tendering process for the procurement of public sector contracts, the legislation still provides that, in certain limited circumstances, negotiation of certain contracts is still sanctioned provided various strict criteria are met.

Nemo dat quod non habet One cannot give what he has not got. A fundamental principle of law which is of great importance so far as the ownership of goods and materials is concerned.

Although most standard contracts provide for ownership in goods and materials to pass when their value is included in interim certificates (qv) this is effective only insofar as the contractor owns the goods and materials. If they are sold to him subject to a retention of title (qv) clause of which he has actual or constructive knowledge, for example, ownership will not generally pass from supplier to contractor until the terms of the relevant clause are satisfied.

JCT 98, clause 16.1 is a typical vesting clause (qv) which provides for the property in goods and materials stored on site and intended for the works to pass to the employer when the contractor has received payment. Clause 16.2 makes similar provision for title to pass to the employer in respect of listed items held off site. Such clauses are not binding on those who are not parties to the contract[493] and will not defeat the maxim *nemo dat quod non habet*. The position is expressly acknowledged in clause 70 of the NEC (qv) which is a less typical example of a vesting clause. In that case, the clause expressly limits the employer's rights to title and he will only have such rights where and to the extent that the contractor lawfully has title to give. Beyond that, the clause is drafted in rather wider terms in that the employer not only has the express right to such title in materials but may also take title in the contractor's plant and equipment. Moreover, unlike JCT contracts under the standard NEC form, transfer of title does not depend on

[490] *Henderson* v. *Merrett Syndicates Limited* (1995) 2 AC 145.
[491] *Smith* v. *Eric S Bush* [1989] 17 Con LR 1.
[492] Section 2 (1) Misrepresentation Act 1967.
[493] The Contracts (Rights of Third Parties) Act 1999 does not disturb this position.

the contractor first receiving payment for the materials, plant and equipment concerned. When and for so long as the materials etc. are brought to and remain within the 'working area' transfer of title occurs at that time. Moreover, the employer will similarly take title in materials, plant and equipment held outside the working area provided the contract identifies them for payment, the contractor has prepared them for marking in the contractually specified way and the 'supervisor' (qv) marks them in that manner.

See also: **Retention of title; Fixtures and fittings; Incorporation.**

Neutral event A term used to describe a cause of delay recognised as giving rise to an entitlement to an extension of time but which does not entitle the contractor to corresponding payment for consequential cost, loss or expense resulting from such delay. It is so called because, in theory at least, the losses resulting from such a delay 'fall where they lie' and are borne equally by each party. Hence, the contractor will be entitled to an extension of time for completion, thereby disentitling the employer to any liquidated or unliquidated damages as a result of the delay and likewise the contractor will recover nothing towards the costs of extended expenditure on his preliminaries, overhead costs and such like. Exceptionally adverse weather conditions, *force majeure* (qv), the occurrence of certain insured or other unforeseeable events (such as war, hostilities, civil disorder etc.) are examples of such neutral events commonly introduced into the extension of time provisions of standard form building contracts such as JCT 98 and WCD 98 (clause 25.4.2), IFC 98 (clause 2.4.2) and ACA 3 (Alternative 2, clause 11.5).

Nominal Less than the actual amount, small or trivial. Generally encountered in relation to money. A nominal sum of money is a sum so small as to be virtually worthless having regard to the circumstances.

A court may award nominal damages to a claimant (qv), even though he has technically proved his case, because it considers that, by his conduct, he deserves no more. Nominal damages may also be awarded for a technical breach of contract. An architect might charge only nominal fees, perhaps because he is hopeful of further commissions from the same client or because the client is a charity which he wishes to support.

See also: **Copyright; Damages.**

Nominated sub-contractors Generally, a sub-contractor (qv) named by the employer. For the purposes of the JCT 98 contract the term refers to any sub-contractor to be engaged for the supply and installation of goods or carrying out of work and whose final selection and/or approval is reserved for the architect by means of either expressly naming the sub-contractor concerned or by use of a prime cost sum (qv) in either:

— The contract bills.
— Any instruction under clause 13.3 for the expenditure of a provisional sum given in the contract bills.
— Any instruction (issued under clause 13.2) requiring a variation (qv).

Provided that, in the case of variations issued under clause 13.2, the work concerned must be additional to that shown on the contract drawings (qv) and/or described in the contract bills (qv) and that: 'any supply and fixing of materials or goods or the execution of work by a Nominated Sub-Contractor in connection with such additional work is of a similar kind to any supply and fixing of materials or the execution of work for which the contract bills provided that the Architect would nominate a sub-contractor'.

Notwithstanding that one or other of those criteria have not been met, it nevertheless may be open to the architect to nominate a sub-contractor, provided he first obtains the contractor's agreement.

Although it may often appear convenient to include sums in the contracts to be expended on nominated sub-contractors, the practice of nomination can give rise to considerable problems, by no means all of which are overcome by even the most comprehensive of provisions for nomination. Such comprehensive and detailed provisions appear in clause 35 of JCT 98, which involves the use of a special series of forms available for tendering and sub-contract purposes comprising: NSC/T parts 1, 2 and 3; NSC/N; NSC/C, NSC/A. Form NSC/W is also available to form a contractual link between employer and nominated sub-contractor, the purpose being to give the employer redress direct against the nominated sub-contractor in certain specified instances. This agreement in no way affects the contractual relationships between the nominated sub-contractor and main contractor[494].

But, notwithstanding even those complex provisions, the process is fraught with possible pitfalls. The nominated sub-contractor is responsible to the contractor for his work and the contractor is responsible to the employer for the whole of the work. Considerable difficulties may arise where a nominated sub-contractor fails and renomination is necessary (clause 35.24) since the employer has a duty to renominate in such circumstances and the contractor has neither the duty nor the right to carry out the work himself[495].

GC/Works/1 (1998), clause 63 also contains provisions for nomination of sub-contractors. However, those are relatively short by comparison with JCT 98 provisions.

Nominated suppliers Provisions for the nomination of suppliers are found in JCT 98, clause 36. A nominated supplier can arise in one of four ways:
— If a prime cost sum is included in the contract bills (qv) and the supplier is named in the bills or by an instruction.
— If a provisional sum is included in the contract bills and in expending it the supply of goods or materials is made the subject of a prime cost sum in an instruction.
— If a provisional sum is included in the contract bills and the supply of goods and materials is from a single supplier by virtue of an instruction, the supply shall be made the subject of a prime cost sum and the supplier is deemed to have been nominated.

[494]*George E. Taylor & Co Ltd* v. *G. Percy Trentham Ltd* (1980) 16 BLR 15.
[495]*North-West Metropolitan Regional Hospital Board* v. *T. A. Bickerton & Son Ltd* [1970] 1 All ER 1039.

— If a variation arises for supply of goods and materials for which there is a single supplier by virtue of an instruction, the supply shall be made the subject of a prime cost sum and the supplier deemed to have been nominated.

The clause provides that (unless otherwise agreed) the architect shall only nominate a supplier who will enter into a contract of sale with the contractor containing the extensive provisions detailed in clause 36.4. A form of tender (TNS/1) is available. If the nominated sub-contractor limits, restricts or excludes his liability to the contractor, it in no way affects the operation of clause 36.4 unless the architect has approved the restriction etc. in writing.

Nomination In general, the naming of a person or firm to undertake a particular task or office. In building contracts, nomination refers to the naming of a person or firm to undertake part of the work or to supply goods. Such nomination is done by the employer. Certain contracts, e.g. MW 98, make no provision for nomination.

See also: **Named sub-contractors; Nominated sub-contractors; Nominated suppliers.**

Notices To give notice to a person means that the matter referred to in that notice has been brought to his attention. A person given notice cannot thereafter deny knowledge of the matter.

Notices may be of three kinds:

Actual Generally, building contracts make express provision to the effect that, any significant notice the parties are, or may be, required to give shall be given, or at least confirmed, in writing (qv). The difficulty about oral notices is, of course, providing proof that they were ever given. A witness to an oral notice would be necessary.

Imputed Where an agent and principal are involved, a notice given to the agent is deemed (qv) to be given to the principal. Thus, a notice given by the contractor to the architect would be deemed to have been given to the employer provided the notice concerned something for which the architect was empowered to act as agent for the employer and provided that there were no express terms in the contract to the contrary. In JCT 98, clause 28.2.1, for example, the contractor is required to give the initial notice of default to the employer. It is thought that it would not be sufficient to give the notice to the architect.

Constructive Notice is deemed to have been given to a party if that party could have been aware of the notice by reasonable enquiry. An example is a notice posted on a site where development is to take place under the Town and Country Planning Act 1990.

The standard form contracts now in common use make express provision for notice(s) to be given in a particular form. Such provisions often go further and in the case of certain notices specify a particular way in which the notice must be given before it will become effective (see, for example, JCT 98, clause 28.1

and also the particular exclusions listed at clause 1.4 of the Supplemental Provisions for Electronic Data Interchange (EDI) (qv) under JCT 98).

GC/Works/1 (1998), clause 1 (3) requires notice to be in writing. Under the NEC Form, the matter is dealt with largely under clause 13 where it is stipulated that each notification required to be given under the contract shall be sent separately from all other communications (clause 13.7) and shall be given 'in a form which can be read, copied and recorded'. JCT contracts deal with the matter differently but are nonetheless specific as to the mode and manner for giving the numerous notices called for under the various versions of the JCT contracts.

Generally, the provisions relating to each type of notice are quite specifically and separately set out and those must be strictly followed. However, under JCT contracts, a 'catch all' provision (e.g. JCT 98, clause 1.7; IFC 98, clause 1.13; MW 98, clause 1.5) also exists whereby: 'if (the) Contract does not specifically state the manner of giving or service of notice ... required or authorised in pursuance of this contract such notice ... shall be given or served by any effective means to any agreed address. If no address has been agreed, then if given or served by being addressed, pre-paid and delivered by post to the addressee's last known principal business address or, where the addressee is a body corporate, to the body's registered or principal office, it shall be treated as having been effectively given or served'.

It is particularly important to note that, where provisions state that special[496] or recorded delivery must be used (e.g. JCT 98, clause 27.1) to direct a notice to the contractor's last known place of abode or business, then subject to evidence to the contrary (such as evidence showing that the notice was incorrectly or inadequately addressed), it will otherwise be deemed (qv) to have been served on the date when, in the ordinary course of business, it would have been delivered. Thus, the contractor will be deemed to have received it on that date, even if it is delayed for a day or two in the post. A similar provision in clause 23.1 of the ACA 3 form deems delivery two working days after pre-paid first-class posting.

It is most important to comply precisely with contractual provisions regarding notices. A party in default, whose notice has expired, may try to plead an irregularity in service if the matter comes before an arbitrator or judge. If the contract requires a notice to be sent by special delivery and it was, for example, delivered by hand, the court might rule that it was improperly served. On the other hand, a number of cases[497] suggest that contractual requirements specifying service of notices in a particular way, e.g. by recorded delivery, are directory and not mandatory. Thus, if the notice is actually received that will amount to valid service. However, such a view should most certainly not be relied upon and wherever possible contractual notice provisions should be rigidly followed in all cases.

---

[496]Note that previous references to 'registered post' are now obsolete with the introduction of what is now termed 'special delivery'.
[497]See, for example, *Goodwin & Sons Ltd* v. *Fawcett* (1965) unreported.

Notional Imaginary or speculative, not known for certain. In a building context, it is generally used with regard to sums of money. An architect, working on a percentage fee basis, may make a calculation of the likely total fee based on a notional figure for the contract sum (qv). If the quantity surveyor knows that he will be delayed in arriving at a final sum to represent loss and/or expense (qv) in a particular case, he may quickly arrive at a notional sum, i.e. what he expects the final sum will be, for the purposes of informing the architect what amount can, with safety, be paid to the contractor as an interim measure.

Novation Commonly (and erroneously) understood by architects and others to mean the concept of a 'consultant switch' (qv); true novation is the substitution of a new contract for an existing one. It can only be done with the consent of all the parties concerned. Unlike assignment (qv) which involves a transfer of rights, novation consists of cancelling an existing obligation and then creating a new obligation in its place.

Clause 22.7 of ACA 3 refers expressly to novation. If the contractor's employment under the contract is terminated and the employer so requires, the contractor agrees and consents to the novation to the employer of the contractor's interest in and under any sub-contracts and to take all necessary steps to make the novation effective. This, of course, requires the sub-contractor's consent. Under JCT 98, where the contractor's employment under the contract may be determined by reason of insolvency, the employer's right (under clause 27.5.1) to withhold any further payments ends in the event that the employer and contractor agree to the novation or conditional novation of the contract.

Nuisance A category of the law of tort (qv). There are three types of nuisance:

Public nuisance An act or omission without lawful justification which causes damage, injury or inconvenience to the public at large. It is a crime as well as a tort. Examples are: obstructing the highway or keeping an immoral house. A private individual has a private remedy for public nuisance only if he suffers damage or inconvenience over and above that being caused to the public at large, e.g. where a builder's skip obstructs the highway *and* the access to private property. Prosecutions for public nuisance are rare.

Private nuisance An unlawful interference or annoyance which causes damage or annoyance to an owner or occupier of land in respect of his enjoyment of his land. Examples are: smell, smoke, noise, encroaching tree-roots, etc. A person wishing to sue for nuisance must prove actual damage. He may adopt self-help and abate the nuisance (see: **Abatement**), e.g. by cutting off the branches of overhanging trees, or he may sue for an injunction or damages or both. An action for nuisance can only be brought by a person with an interest in the land. It is no defence to show that the nuisance existed before the claimant came to his land, but something that was originally a nuisance can be legalised by the passage of time as, for example, where a defendant used some noisy machinery for more than 20 years, but the vibrations caused by it only became a nuisance when the claimant erected a consulting room at the end of his garden near the noise, in which case time would begin to run when the act in fact became a nuisance[498]. A defendant would not, therefore, be able to rely on prescriptive right.

Statutory nuisance Something declared to be a nuisance by statute, such as, for example, where an act specifies any premises 'in such a state as to be prejudicial to health or a nuisance'[499]. The remedy is by way of an abatement notice served by the local authority (qv) on the person responsible. If an abatement notice is not complied with, or the nuisance is likely to recur, the offender can be taken before the magistrates' court which may make a nuisance order and/or impose a fine.

Express contractual provisions relating to the prevention of nuisance generally do not play a significant part in standard form building contracts. However, for obvious reasons such provisions often feature extensively in contracts for operations such as opencast working, landfill and reclamation works and major civil engineering projects. Additional terms are sometimes included in the preliminaries sections of bills of quantities (qv) to deal with noise pollution.

Null Invalid. Devoid of legal effect.
See also: **Void.**

Numbered documents A term which occurs in JCT 98, clause 2.3.5 and which refers to the documents which are to be attached to the nominated sub-contract. The term is also used in connection with DOM/1 (qv) and DOM/2 (qv), likewise to describe the documents attached to the domestic sub-contract documents (qv).

---

[498]*Sturges* v. *Bridgman* (1879) 11 Ch D 582.
[499]See, for example, ss. 91 and 92 of the Public Health Act 1936.

Oaths and affirmations The general rule is that all witnesses must give evidence (qv) on oath or affirmation in proceedings before a court. To reflect the principle of the Welsh Language Act 1993, in civil proceedings where it is possible that the Welsh language may be used by any witness then that witness may elect to take the oath or affirm in English or Welsh as they wish. The practice of requiring evidence to be given on oath or affirmation is also often followed in arbitration proceedings.

Under s. 34 of the Arbitration Act 1996, subject to the right of the parties to agree between themselves any such matter, the arbitrator may decide all procedural and evidential matters, including:

— Whether and to what extent there should be oral or written evidence.
— Whether any and if so what questions should be put to and answered by the respective parties and when and in what form this should be done.
— Whether to apply strict rules of evidence (or any other rules) as to the admissibility, relevance or weight of any material (oral, written or other) sought to be tendered on any matters of fact or opinion, and the time, manner and form in which such material should be exchanged and presented.
— The language or languages to be used in the proceedings.

Under s. 38 (5) of the Arbitration Act 1996, if oral evidence is to be given the arbitrator has power to direct that witnesses shall give their evidence on oath (or affirmation) and the arbitrator may also administer the oath or affirmation as the case may be.

The current general rules about oaths and affirmations are found in the Oaths Act 1978. A false statement on oath or affirmation amounts to the criminal offence of perjury. The usual form of oath in civil proceedings is: 'I swear by Almighty God that the evidence I shall give shall be the truth, the whole truth, and nothing but the truth'. The person taking the oath holds the New Testament or, in the case of a Jew, the Old Testament, in his uplifted hand, and says or repeats this formula after the person administering the oath.

Witnesses not of the Christian or Jewish faith may take the oath with the appropriate ceremonies which are binding on them but, if this would cause delay or inconvenience, they may be required to affirm instead. This also applies to any person who objects to being sworn, e.g. a Quaker. Such people solemnly affirm by repeating after the administrator: 'I, [ABC], do solemnly, sincerely and truly declare and affirm that the evidence I shall give shall be the truth, the whole truth, and nothing but the truth'.

In Scotland the oath is administered in a slightly different way, with uplifted hand but without either Testament, by repeating the words of the oath after the judge or arbitrator, who stands up and holds up his right hand similarly, while saying the words to be repeated. Anyone who wishes to take the oath in the Scottish manner may do so in any part of the UK.

Obiter dictum Part of a judgment (qv) which is not the *ratio decidendi* (qv) or reason for the decision. It is a statement of law made by the judge in the course of a judgment (qv) which is not necessary to the decision or based upon the facts as found. A statement is *obiter* if:

— It is based on facts which were not found to exist or if so found, were not material.
— It is a statement of law which, although it may be based on facts as found, is not material to the decision.

For example, in *Rondel* v. *Worsley*[500] the House of Lords expressed certain opinions that a barrister might be liable for negligence (qv) when acting other than as an advocate and that immunity extended to solicitors when acting as advocates[501]. The case was concerned only with a barrister's liability when acting as advocate[502] and so these opinions were *obiter*.

It is often difficult to decide what is and what is not *obiter dictum* until a later court considers a previous case and isolates the basis of the previous decision. Thus, statements long thought to be part of *the ratio* are sometimes put to one side.

Words said *obiter* may be persuasive in future cases, depending upon the circumstances and the standing of the judge. In the absence of direct authority, they may form the basis of future decisions.

Obscurities Things which are not clear.
See: **Ambiguity.**

Obstruction As distinct from its use in the context of physical impediments to future progress, the term is commonly used in building contracts to connote some act or omission causing interference with the proper administration of the contract. JCT 98, clause 28.2.1.2 specifies interference with or obstruction by the employer of the issue of any certificate due under the contract as a ground on which the contractor may determine his employment under the contract. Other standard form contracts contain similar provisions, e.g. ACA 3, clause 20.2 (b), IFC 98, clause 7.9.1(b), etc.

There is a considerable body of case law on what constitutes interference or obstruction, but for the most part it deals with the contractor's right to recover money without a certificate where the employer has interfered with the independent exercise of the architect's powers as certifier. In such a case the contractor can sue without a certificate[503].

Use of the term obstruction in such clauses is, therefore, generally designed to meet conduct of the employer such as refusing to allow the architect to go on

---

[500][1967] 3 All ER 993.
[501]*Arthur J. S. Hall & Co* v. *Simons*; *Barratt* v. *Ansell & Others (t/a Woolf Seddon)*; *Harris* v. *Schofield Roberts & Hill* [2000] 3 WLR 543 have now changed this position in a landmark judgment in the House of Lords.
[502]See, by way of contrast, the position of a barrister when acting without express authority of his clients discussed in *Connolly-Martin* v. *Davis* TLR 17 August 1998.
[503]*Hickman & Co* v. *Roberts* [1913] AC 229.

site for the purpose of giving his certificate, or directing the architect as to the amount for which he is to give his certificate or as to the decision which he should arrive at on some matter within the sphere of his independent duty. It is unlikely that negligence or omissions by someone who, at the request or with the consent of the architect, is appointed to assist him in arriving at the correct figure to insert in his certificate can amount to interference[504]. It seems, therefore, that in this context, obstruction is used in the sense of impeding. In a different context, obstruction by the employer with the contractor's carrying out of the works etc. amounts to prevention or hindrance which will be a breach of an implied term of the contract.

See also: **Interference; Good faith.**

Occupation This term refers to the actual physical control or use of land. Title to certain personal property (qv) may be acquired by occupation, e.g. taking physical control of it, as is the case with such things as fish, game, etc.

See also: **Adverse possession; Occupier; Occupiers' liability.**

Occupier Someone who owns and occupies land or other premises and who has actual use of that land, etc. An occupier owes a duty of care (qv) to third parties under the Occupiers' Liability Acts 1957 and 1984.

In *Wheat* v. *E. Lacon & Co Ltd*[505], it was said: 'Wherever a person has sufficient degree of control over premises that he ought to realise that any failure on his part to use care may result in injury to a person coming lawfully there, then he is an "occupier" and the person coming lawfully there is his "visitor" and thus is under a duty to his visitor to use reasonable care'[506].

See also: **Dangerous premises; Occupiers' liability.**

Occupiers' liability The Occupiers' Liability Act 1957 s. 2 provides that an occupier of premises owes 'the common duty of care' to his 'visitors', who are those invited or permitted by him to be there, including those who enter under legal authority, e.g. a police officer. The occupier in this context means the person who has physical control or possession of the premises, and may include the landlord[507]. A trespasser is not a 'visitor' for the purposes of the Act, the duty to trespassers being contained in the Occupiers' Liability Act 1984, which replaced the rather complex common law rules. The common duty of care is defined as a duty to take such care as in all the circumstances is reasonable in order to ensure that the visitor will be reasonably safe in using the premises for the purposes for which the occupier invited or permitted him to be there. It does not impose on the occupier any obligation in respect of risks willingly accepted by the visitor as his. The occupier must be prepared for children to be less careful than adults[508], and may expect that a person,

---

[504] *R. B. Burden Ltd* v. *Swansea Corporation* [1957] 3 All ER 243 per Lord Tucker at 253.
[505] [1966] AC 552.
[506] [1966] AC 552 per Lord Denning at 579.
[507] *Wheat* v. *E. Lacon & Co Ltd* [1966] AC 552.
[508] Occupiers' Liability Act 1957 s. 2 (3) (a). *Moloney* v. *Lambeth Borough Council* (1966) 64 LGR 440.

in the exercise of his trade or calling, will appreciate and guard against risks ordinarily incident to it so far as the occupier leaves him free to do so[509].

The duty can be discharged by a reasonable warning of any known danger, but it should be noted that as a result of s. 2 (1) of the Unfair Contract Terms Act 1977 (qv) it is not possible by means of a notice to exclude or restrict liability for death or personal injury resulting from *negligence* (qv).

The obligations imposed by the Act apply to all those occupying or having control over any fixed or moveable structure or any premises or structure, e.g. scaffolding, and so a sub-contractor may be an 'occupier' in respect of his part of the works.

Trespassers are owed a lesser duty under the 1984 Act, which also affords some protection to people exercising rights of access to the countryside or using private rights of way. Section 1 (3) of the 1984 Act says that an occupier owes a duty to a trespasser etc., only if '(a) he is aware of the danger or has reasonable grounds to believe that it exists; (b) he knows or has reasonable grounds to believe that the other is in (or may come into) the vicinity of the danger...; and (c) the risk is one against which, in all the circumstances of the case, he may reasonably be expected to offer the other protection'.

The lesser duty is to take such care as is reasonable in all the circumstances of the case to see that the entrant does not suffer injury (not property damage) on the premises by reason of the danger concerned s. 1 (4). This duty can be excluded altogether by an appropriately worded notice[510].

Offer   An expression by one party of willingness to be bound by some obligation to another. If the offer is accepted, a binding contract results.

An offer may be made in writing or orally or by conduct. It may be made to an individual or group or to the whole world[511]. An offer terminates:
—   If rejected by the offeree.
—   If revoked by the offeror and the offeree has notice (qv) of the revocation before acceptance.
—   If either party dies before acceptance.
—   If a time limit is stipulated and it expires before acceptance.
—   By lapse of time, if not accepted within a reasonable time, and no time limit has been specified.

It is important to note that if one party rejects the offer by another and subsequently decides to accept the offer after all, the offer is no longer available for acceptance unless the offeror agrees. If an offer is made by post, it is only revoked when the offeree receives the revocation. If he has already posted his acceptance (qv) the revocation is of no effect and a full binding contract is formed.

A tender (qv) is an offer. An invitation to tender (qv) is not an offer but what is known as an 'invitation to treat' (qv) or an invitation to make an offer.

---

[509]Occupiers' Liability Act 1957 s. 2 (3) (b).
[510]*Ashdown* v. *Samuel Williams & Sons Ltd* [1957] 1 All ER 35.
[511]*Carlill* v. *Carbolic Smoke Ball Co* [1893] 1 QB 256.

Official Referees See: **Technology and Construction Court (TCC) — Judge of.**
See also: **Courts; Scott Schedule.**

Off-site materials Materials which are intended to be used on the works but
which, for convenience or safety, are not stored on site. Whether the contractor
is to receive payment for such materials will depend upon the express
provisions of the contract. Neither MW 98 nor GC/Works/1 (1998) makes any
such provision for payment for off-site materials. Under the JCT 98 and IFC
98 Forms only those off-site materials, goods or pre-fabricated items (i.e.
'listed items') that the employer has pre-designated by means of a list annexed
to the contract bills (qv), specification (qv) or Schedules of Work (qv) will
qualify for payment and only if the contractor satisfies the various strict
requirements set out in the contract (i.e. JCT 98, clause 30.3.1 to 30.3.5 and
IFC 98, clause 4.2.1(c).1 to .5). These requirements may even extend to the
provision by the contractor of a bond (qv) should the employer so stipulate in
the Appendix (qv) to the contract.

The problems of payment for off-site materials are two-fold. On the one
hand the contractor may be seriously financially embarrassed if he has paid for
large quantities of materials and he may be tempted to bring them on to site
and risk damage to obtain payment; however, the JCT forms of contract do
not permit interim payments for materials brought on to site prematurely. On
the other hand, the employer must be certain that he becomes the owner of
goods for which he has paid and that no other party retains an interest in the
materials or goods (see: **Retention of title**). It is generally not in the employer's
interest to pay for materials off-site because of the difficulty which may be
experienced in proving ownership if, for example, the contractor becomes
insolvent (qv). However, he may in some instances judge it expedient to do so
provided the contract lays down stringent conditions for such payment and
that those stringent conditions are adhered to in all respects.

ACA 3 (Alternative A, clause 16.2A(b)) makes payment subject to whatever
conditions the contract documents (qv) may provide. JCT 98, clause 30.3
provides that in respect of any 'listed items':
— The contractor provides the architect with reasonable proof that
  property in the listed items vests in the contractor.
— The contractor has, if required, provided a bond in the agreed terms.
— The materials are in accordance with the contract.
— They have been set apart and/or are marked to identify the employer
  and the works, and
— Materials must be intended for incorporation.
— Nothing remains to be done to the materials before incorporation.
— The contractor provides reasonable proof as regards insurance against
  loss or damage.
— The contract for supply between contractor and supplier is in writing
  and expressly provides that property (ownership) shall pass to the
  contractor or sub-contractor not later than the time at which they are
  set aside and marked as above.

— Any sub-contractor concerned shall provide similar guarantees.
See also: **Ownership of goods and materials.**

Omissions In the context of building contracts, 'omissions' refers to work or materials which have been priced by the contractor and included in the contract sum, but which the employer no longer requires. The architect issues an instruction by way of a variation (qv) omitting the work or materials and the omitted work is generally valued by means of rates contained in the contract bills (qv). The contract sum is then appropriately adjusted. Under standard JCT contracts unless the contractor otherwise agrees, it will be a breach of the contract if the employer, or the architect on his behalf, omits work included in the contract sum simply in order to have it carried out by others where, for example, the employer subsequently finds he can have the work concerned done more inexpensively by them[512].

An omission may also refer, if so provided by the contract, to the removal of obligations or restrictions imposed by the contract documents on the contractor in respect of working space, working hours etc., e.g. JCT 98, clause 13.1.2. It may also refer to a gap or deficiency in an agreement or document, e.g. bills of quantity (qv).
See also: **Omitted work; Variation.**

Omitted work All commonly used standard form contracts contain provision for the architect to omit work from the contract. Without such a provision, an instruction to omit work would amount to a breach of contract. In general, the value of omitted work is ascertained by reference to the rates in the bills of quantities (qv) or the priced specification (qv) or priced schedule of rates (qv). ACA 3, however, contains provisions (clause 17) for the contractor to submit estimates which are to be agreed. The principal clauses relating to the valuation of omitted work are JCT 98, clause 13.5.2; GC/Works/1 (1998), clause 42; IFC 98, clause 3.7.3; MW 98, clause 3.6.
See also: **Variations; Omissions.**

Operational bills of quantities A system of setting out bills of quantities (qv) with regard to operations rather than the more usual trade bills.

The series of operations are predetermined as is the order in which they are carried out. The materials only are measured by the quantity surveyor; it is the contractor's responsibility when pricing to allow for the labour required for each operation. Individual prices are totalled to obtain the total tender sum. The system has not found great favour, probably because it is not usual for the contractor to be informed of the order in which he is to carry out the work. He may well be able to carry out the total work at a cheaper overall price if left to his own order of working. The system probably works best when the tender is

---

[512]*Carr* v. *J.A. Berriman Pty Ltd* (1953) 27 ALJR 273; *AMEC Building Contractors Ltd* v. *Cadmus Investments Co Ltd* (1997) 13 Const LJ 50.

to be negotiated, so that the contractor can discuss the order of work before bills are prepared.

Order A direction of a court or arbitrator. All directions of a court in any proceedings are termed 'orders' unless they determine a case or issue, when they are usually referred to as judgments. The term may also be applied to secondary or delegated legislation ratified by Parliament but made by ministers under the authority of an Act of Parliament; see, for example, Housing Grants, Construction and Regeneration Act 1996, s. 104 (4).

Order 14 procedure See: **Summary judgment.**

Overheads Generally, the costs of head office administration proportioned to each contract. Included are staff working on the individual contracts and general support staff, rates, electricity, heating, telephones, office equipment, etc.

Although in traditional contracts the cost of head office administration is invariably recovered through an allowance made in the contractor's rates and prices for the measured work, where bills of quantities form an integral part of the contract documentation and those contract bills are, under JCT 98, deemed to be measured in accordance with the principles laid down in the Standard Method of Measurement for Building Works (SMM 7th edition), the contractor will be afforded the opportunity to deal separately with, and to allocate, specific rates and prices to the majority of his project specific overheads such as:
— Site based management and staff.
— Site accommodation.
— Telephone and administration costs.
— Storage facilities.
— Cleaning.
— Water, fuel lighting and power charges.
    Rubbish disposal.
— Safety, health and welfare.

Where the contract provides for work to be carried out and valued on a daywork (qv) basis, it is usual to stipulate the prime cost (qv) rate(s) of labour to be applied or to state the means by which those prime cost rate(s) can be ascertained, and to make further separate provision for the contractor to state the percentage addition he will require to those rates to compensate him for various defined overheads, such as, for example:
— Head office charges.
— Site staff, including site supervision.
— Oncosts of labour, including e.g. allowance for sick pay, fares, travelling and subsistence, third party and other insurances, NI contributions, liability in respect of redundancy payments, etc.
— Profit.

There is often much dispute when a claim arises as to which and/or what proportion of overheads should be allowed. Formulae are sometimes used to arrive at overheads but they are not universally accepted and contractors are now generally expected to keep good and accurate records of the additional overheads expended due to the matters giving rise to the claim.

See also: **Eichleay formula; Emden formula; Formulae; Hudson formula.**

Ownership of goods and materials If goods and materials are unfixed, the employer must take care that he does not pay for them unless he is sure that, on payment, ownership passes to him. The situation was highlighted under a contract made in the JCT 63 Edition where the contractor sub-contracted the roofing to the claimants on the 'blue form'. The blue form provided that the subcontractor should be deemed to 'have knowledge' of the terms of the main contract. JCT 63 provided that ownership of materials was to pass to the employer when their value had been paid to the main contractor by the employer. The main contractor went into liquidation (qv) after the employer had paid him for slates delivered to site by the claimants, but before the main contractor had paid the claimants. It was held that the claimants were entitled to recover the slates and damages or the value of the slates. In effect, the employer was put into the position of paying twice for the same goods because the contractor had no title in the slates to pass on to the employer and there was no privity of contract (qv) between employer and sub-contractor[513].

JCT and NSC contracts have since been amended in an attempt to rectify the situation. However, the position remains complex and the parties' rights may also be significantly affected by the legislation contained within the Sale of Goods and/or the Supply of Goods and Services Acts. Architects should, therefore, check as thoroughly as reasonably possible to establish that good title in materials and goods etc. has passed to the contractor, and they should pay particularly close attention to any duty imposed on them in that regard under the contract (e.g. JCT 98, clause 30.3) before certifying and committing the employer to paying for any unfixed materials.

See also: **Retention of title; Vesting clause; Fixtures; Incorporation.**

---

[513]*Dawber Williamson Roofing Ltd* v. *Humberside County Council* (1979) 14 BLR 70.

P

PC Initials representing prime cost (qv).

Package deal contracts Sometimes known as design and build contracts because they incorporate both elements in one package (hence 'package deal'). However, to amount to a true 'package deal', strictly speaking there should be little or no design input by the employer and so it is most unlikely that the WCD 98 could properly fall to be described a true package deal contract. The main benefit, from the employer's point of view, is that the package deal places all the responsibility for the work, from taking the initial brief to completion of the work, in one place — with the contractor. If something goes wrong or there are defects, the employer is not faced with the usual problem of sorting out design from constructional responsibilities. On the other hand, the employer has no independent advice on which to call if he is in doubt since the contractor, however kindly motivated, will have his own financial interests at heart. An unscrupulous contractor could take advantage of the employer's lack of expertise. It is up to the employer to weigh the pros and cons before deciding which system to adopt or else appoint a professional to supervise the work on his behalf. **Figure 14** compares this type of contract with the traditional form. See also: **Design and build contract; Turnkey contract.**

Parol evidence A term used by lawyers to describe oral and other extrinsic evidence. Once a contract has been reduced to writing, 'verbal evidence is not allowed to be given ... so as to add to or detract from, or in any manner to vary or qualify the written contract'[514]. This basic rule of interpretation is called the parol evidence rule. It covers not only oral evidence but other extrinsic evidence as well; for example, drafts, pre-contract letters etc. are all excluded. It also prevents evidence being given of preliminary negotiations between the contracting parties.

It is subject to exceptions. Thus, it does not apply where misrepresentation (qv) is alleged or where one party claims that there is a collateral contract (qv) or where it is said the written contract does not reflect the actual agreement and so must be rectified. However, it remains a basic rule when interpreting written or printed contracts.

Since most building contracts are in standard form, various optional clauses may be deleted and there may be typewritten or manuscript amendments. Logically the rule would exclude a court or arbitrator from looking at the deletions. In fact the House of Lords has ruled that one is entitled to look at the deleted words 'as part of the surrounding circumstances in the light of which one must construe what [the parties] have chosen to leave in'[515].

---

[514]*Goss* v. *Nugent* (1833) 5 B & Ad 58, 64.
[515]*Mottram Consultants Ltd* v. *Bernard Sunley & Sons Ltd* (1974) 2 BLR 28 per Lord Cross at 47; however, the contrary view was held in *Wates Construction (London) Ltd* v. *Franthom Property Ltd* (1991) 53 BLR 23.

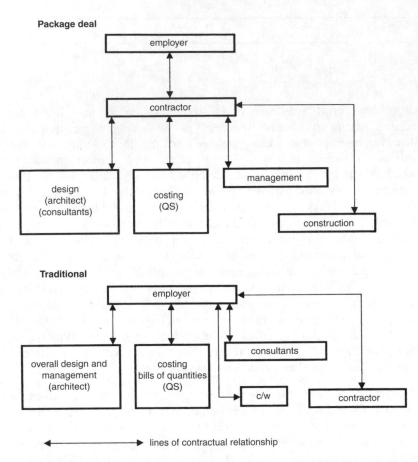

Figure 14 Package deal contract compared to a traditional contract.

'Surrounding circumstances' is an imprecise phrase which can be illustrated but hardly defined. However, in a commercial contract it is certainly right that the court should know the commercial purpose of the contract. This in turn presupposes knowledge of the genesis of the transaction, the background, the context, and the market in which the parties are operating.

Extrinsic evidence will also be admitted to explain the written agreement, and in particular to show the meaning of individual words and phrases used by the parties. The starting point is the ordinary English usage as defined in a standard dictionary, but both courts and arbitrators must give effect to any special technical, trade or customary meaning which the parties intended the word to bear.

See also: **Interpretation of contracts; Rectification.**

Part 20 proceedings The Civil Procedure Rules (CPR) (qv) introduced new terminology in respect of any additional claim(s) brought within the proceedings which do not appear on the claim form (qv) or particulars of claim (qv). Strictly speaking, a Part 20 claim may be a counterclaim (qv) against the claimant or against a wholly new party. For example, an employer might sue the contractor for defective windows and the contractor might counterclaim for payment of monies owed under the contract and might also sue the relevant sub-contractor. This would give rise to two separate Part 20 claims – one against the employer, the other against the sub-contractor.

Part 36 offer/payment The Civil Procedure Rules (CPR) (qv) have modified the rules relating to the settlement of proceedings. The ethos is that where a party makes a genuine attempt to settle proceedings, the other party should face sanctions, usually in costs, if a reasonable settlement is rejected. Part 36 has codified the making of Calderbank offers (qv) and payments into court (qv) in litigation. Both claimants and defendants may take advantage of the Part 36 at any time after the commencement of proceedings[516].

— A claimant may make an offer to settle his claim or part of it or any number of issues within his case for a defined result (e.g. a specified sum of damages or an injunction on certain terms).

— A defendant may make an offer to settle a claimant's claim or part of it or any number of issues within the claim. Where the offer is the payment of damages, that sum must be paid into court[517]. Where the claim is wholly or partly non-monetary, a defendant may make an offer to settle in writing[518].

Once a Part 36 offer or payment has been made, the other party has 21 days in which to accept in writing the offer or payment[519]. Where the offer or payment is accepted, the defendant is generally required to pay the claimant's costs (qv) up to the point of acceptance[520]. Where, on the other hand, the offer or payment is rejected, the matter will continue to trial, where:

— If it was the claimant who made an offer and

 • judgment is given against the defendant for a higher sum (or the other is more advantageous) than the offer, the court may order interest on any sums payable to the claimant at a rate up to 10% above base rate for such period as the court considers just, starting at the date when the defendant could have accepted the offer[521]. Additionally, the court may order the claimant's costs to be paid on an indemnity (qv) basis and/or with interest, not exceeding 10% above base rate[522].

 • judgment is given against the defendant for a sum less than the offer, Part 36 does not specify any sanction, although the making of an offer may be taken into account by the court whilst making any

---

[516]An offer made prior to the commencement of proceedings, although not a Part 36 offer, will be taken into account when the court makes any order as to costs: CPR Rule 36.10.
[517]CPR Rule 36.3.
[518]CPR Rule 36.4.
[519]CPR Rules 36.11 and 36.12.
[520]CPR Rules 36.13 and 36.14.
[521]CPR Rule 36.21 (2).
[522]CPR Rule 36.12 (3).

order as to costs, e.g. if the difference between the offer and the ultimate judgment was small, the court might consider the defendant's failure to accept as being unreasonable.

— If it was the defendant who made a payment or offer (as appropriate) and
  - judgment is entered against the defendant for a sum more than the payment, the claimant will obtain his costs in the usual way.
  - judgment was entered against the defendant for a sum less than the payment, the court will usually award costs against the defendant until the last date upon which the offer could have been accepted and, thereafter, for the claimant to pay the defendant's costs[523].

Partial possession To be distinguished from sectional completion (qv); all the standard forms (except MW 98) make provision for the employer to take possession of part of the works before completion. If partial possession is required under IFC 98, it is no longer necessary to insert a special clause to that effect since clause 2.11 which makes provision for partial possession is now incorporated into the printed form. Under GC/Works/1 (1998), clause 37 the PM can instruct the contractor to allow partial possession. In other forms, like JCT 98 and IFC 98, the contractor's agreement (which he cannot unreasonably withhold) must first be obtained. On partial possession:

— The architect must issue a statement giving particulars of the relevant part of the works that has been taken into possession and the relevant date when it was taken; from which time the defects liability (qv) period for that part will then begin to run.
— The statement should also provide the value of the work taken over. (This is for the purposes of the contract only.)
— Half the retention in respect of the part taken over must be released.
— The amount of liquidated damages must be reduced in proportion to the value of the work taken over to the total contract sum.
— Of considerable importance is the fact that, from the date that partial possession is taken, the contractor's liability, if any, for the insurance of that part of the works is removed.

For partial possession to have contractual significance, it requires a formal act such as the issuing of a statement to that effect in the particular terms (such as those above) prescribed by the contract. Partial possession will not be implied if the employer simply moves items of furniture into the part in question[524].

Where is it known at the outset that completion is required in sections the contract must be amended accordingly. JCT 98 has a special supplement for the purpose.

Particulars of claim This formal document sets out the basis of the claimant's case against the defendant. It must contain a concise statement of the facts

---

[523]CPR Rule 36.20.
[524]*English Industrial Estates Corporation* v. *George Wimpey & Co Ltd* (1973) 7 BLR 122.

upon which the claimant relies[525]. The Practice Direction to Part 16 requires a claimant to set out certain matters, including any allegation or fraud, misrepresentation or breach of trust and details of mitigation of damage[526]. See: **Statement of case; Defence.**

Partnering Partnering was one of the recommendations in the Latham Report (qv). Contrary to common misconception it is not a scheme to replace use of traditional standard form building contracts, sub-contracts and consultant's appointments or the like. It should not be confused with partnership, joint venture agreement, management contracting, construction management or other such arrangements. Partnering agreements are an 'add on' intended to supplement those traditional contractual relationships to create incentives for all parties involved in the construction process to co-operate towards the overall success of the project. Very often the parties will prepare a 'charter' which will set out their joint aspirations. It is rare for the charter to define a legal relationship although it may sometimes overlap into that territory. Hence, there is still a necessity for legally binding contracts.

Partnering may be a long-term objective, (sometimes referred to as 'strategic partnering') or it may be project specific (i.e. 'project partnering'). In either case, the means by which it is achieved will vary according to the individual circumstances of the parties and/or projects concerned. However, the aims are always broadly the same: to create a spirit of teamwork that will ultimately provide mutual benefits for all those involved in the project(s), such as:

— A co-operative management approach.
— Improved efficiency of design and construction.
— Minimised and shared risk.
— Reduced costs, maintenance of expected profit levels and certainty of build costs.
— Defect free construction.
— Fast, efficient and quality construction.
— Timeous and reliable design.
— Improved communication between all parties from employer and design team through to suppliers and sub-contractors.
— Early recognition of problem areas and potential disputes and quick and efficient means of resolving disputes before they can escalate.

The means by which these incentives are achieved and the degree of success offered to all those involved in the process varies considerably[527]. A growing trend appears to be towards some form of profit (or 'gainshare') scheme or perhaps an 'open book' policy whereby the chosen team of consultant(s), contractor, sub-contractor(s) and supplier(s) all agree on an open book policy where obtaining a tender price for the project is of less importance than simply agreeing on a fixed level of overheads and profits to be charged on the auditable cost.

---

[525]CPR Rule 16.4 (1) (a).
[526]Paragraph 10 of the Practice Direction to CPR Part 16.
[527]See, for example, the results of the study undertaken by the Economic and Social Research Council carried out during 1996/1997.

However they are achieved, such agreements must be carefully drawn up to avoid the potential for certain crucial pitfalls, such as:

— The fraud or negligence of one or more of those who are a party to the partnering agreement.

— The implication of a true partnership (and not a 'partnering') agreement; as to which, see **Partnership.**

— Effective dispute resolution procedure for the partnership that will not conflict with or be overridden by those in the various individual main contract, sub-contract(s), supply contract(s) or consultant appointments.

— In the case of public sector or public utilities works that the partnering agreement does not breach EC or UK law aimed at ensuring free and open competition.

Partnership A way of carrying on business which is governed primarily by the Partnership Act 1890 under which it is defined as 'The relation which subsists between persons carrying on business in common with a view of profit'. It has been, and for the moment remains the most common type of professional business arrangement, particularly suited to small professional practices where partners of the same profession work closely together. The characteristics of a traditional partnership are that the partners share profits and losses (not necessarily equally) and they carry on the business together. Each partner carries unlimited liability for partnership debts. It is known as 'joint and several liability' because they are liable together and independently. Thus, a creditor may sue the partnership or an individual partner to recover a debt. For example, if a partnership runs into debt which the assets of the firm will not cover and one partner removes himself from the jurisdiction (qv), the other partner or partners will be liable for the whole of the debt to the full extent of their personal assets. If a partner dies or becomes bankrupt (qv), the partnership comes to an end. It also ends when a partner retires or a new partner is taken into the firm. Although there is no legal maximum, the number of partners is normally no more than 20, but certain professional partnerships may have an unlimited number of partners, e.g. architects, surveyors, estate agents, solicitors, accountants, etc.

Each partner has the power to bind the others in regard to any matter concerning the partnership. Partners may be bound, even though a partner acts beyond his authority, if the general public has reason to believe that he is acting on behalf of the partnership. For this reason, all partners must show the utmost good faith (qv) in their dealings with one another. That means revealing all matters to one another which may affect the partnership. It is not necessary, although desirable, to draw up a deed of partnership; a simple written or oral agreement will suffice. Whether or not a partnership exists is a matter of fact. A court will look at all the circumstances. Sharing of profit and loss suggests a partnership, but the situation, particularly in small firms, may be confused.

Frequently, employees share in the profits by way of bonuses. Therefore, some indication that a person has a more fundamental interest is required, e.g. involvement in the making of policy decisions. A person may be deemed to be a partner if he is represented on office stationery as such, whatever the internal arrangements of the firm may be. Where some members of a firm are named as 'associates' or 'executives', it is desirable that their names are separated from the names of the partners on the firm's notepaper or they may well find themselves becoming liable in the event of the firm becoming insolvent (qv).

Growing concerns over the increasing size and number of negligence actions against personally liable professionals who, until now, have had unlimited liability, coupled with the modern trend towards ever larger and more diverse professional practices where, in reality, partners may have little or no effective control over the conduct of their co-partners, has led the UK government to promote a new concept of Limited Liability Partnership. Following publication of a series of consultation papers and draft bills and regulations[528], the Limited Liability Partnerships Act 2000 received Royal Assent on 20 July 2000. As its title suggests, a key feature of this radical legislation is the introduction of a new status of partnership – the Limited Liability Partnership (or LLP). An LLP properly formed by incorporation under the Act and registered as such will be a separate corporate legal entity, distinct from its members (owners). It has unlimited capacity, in its own right, to contract, hold property and generally to undertake the full range of business activities which a traditional partnership may undertake. As such it more closely resembles a company than a partnership and except to the extent otherwise provided by that or any other Act, the law relating to partnerships does not apply to an LLP.

Like a company, clients will generally contract with the LLP and not with a partner contracting as principal and on behalf of other partners. In that case, where, within an LLP, a partner is negligent in his conduct for a client, the client's rights of action in contract will be solely against the LLP. The likely success of any separate action in tort against the individual partner of the LLP concerned will generally be considerably less than in the case of a traditional partnership since such individual liability will depend, among other things, on a finding that the individual member of the LLP had assumed personal responsibility for the advice he had given, that the client had relied on that assumption of responsibility and that such reliance was reasonable[529].

In addition to its provisions concerning the legal status of the LLP, the means by which it must be registered and incorporated, the past, present and future membership and relationship between members, the Act also makes provisions:

— Generally ensuring the preservation of an income tax and capital gains regime taxation applicable to members of the LLP as if they were partners carrying on business in a traditional partnership, notwithstanding that the LLP is, in fact, a separate body corporate.
— Ensuring that present corporate insolvency and winding up procedures, such as company voluntary arrangements, winding up, receivership and administration etc., are adapted and extended to encompass LLPs.

---

[528] URN/97/597; URN 98/874; URN 99/1025; URN 00/617 and URN 00/865 published between February 1997 and May 2000.
[529] *Williams and Another* v. *Natural Life Health Foods Ltd and Richard Mistlin* [1998] 1 WLR 830.

— To make certain that non-conformance with regulations made by or under the Act can give rise to criminal liability.

See also: **Liability; Limited Liability Partnership Act 2000.**

Party wall Under the common law the term has not been precisely defined and is capable of a number of meanings that amount to a technical term used to describe a particular type of wall between properties and falling broadly into one of three categories:

— A wall divided vertically, the whole wall being subject to reciprocal easements (qv).
— A wall divided vertically into strips, one belonging to each adjoining owner.
— A wall belonging entirely to one owner, subject to his neighbour's rights to have it maintained as a dividing wall[530]. See **Figure 15.**

In London, in one form or another since 1844, and throughout the remainder of England and Wales since 1 July 1997, the common law rules relating to party walls have been considerably modified by statute. The term 'party wall' is now statutorily defined as a wall standing on land of different owners not taking account of projecting artificial foundations, which is part of a building; or that part of a wall which separates buildings belonging to different owners[531]. Hence, the term excludes such dividing walls that are not part of a building. For a full consideration of the term as defined by statute, see **Party Wall Act 1996.**

Party Wall Act 1996 There are now special procedures for party walls under the Party Wall Act 1996 which came into force on 1 July 1997. It currently applies only to England and Wales. If anything is to be done to a party wall as defined by the Act, notice is to be given in certain forms. A 'party wall' is defined as a wall, standing on land of different owners not taking account of projecting foundations, which is part of a building; or that part of a wall which separates buildings belonging to different owners. A 'party structure' is a party wall, floor or other structure separating parts approached by separate entrances, while a 'party fence wall' is a wall, standing on land of different owners not taking account of projecting foundations, which is not part of a building, but separates adjoining lands.

If the two adjoining owners do not agree (and it is often unwise to agree in advance), each party must appoint a surveyor to whom certain powers are given by the Act to determine the difference and to decide, subject to the provisions of the Act, what contribution each party is to make to the cost of the works. Both building and adjoining owners have statutory rights which they can exercise under the Act and those rights can never be overlooked or set aside.

There are three basic situations covered by the Act:

— Building a new party wall.

---

[530]*Watson* v. *Gray* (1880) 42 LT 294.
[531]Section 20, Party Wall Act 1996.

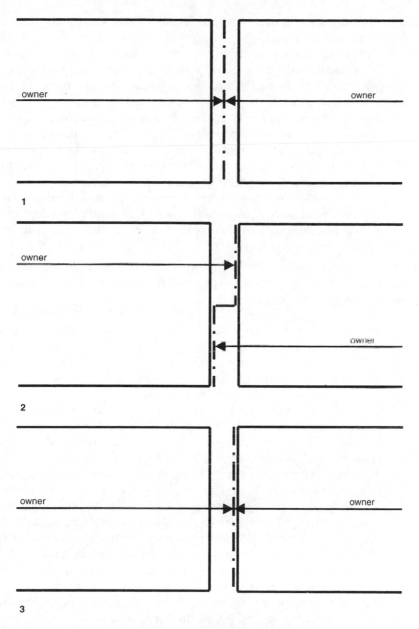

Figure 15 Party wall categories.

— Work to existing party walls.
— Adjacent excavations and constructions.

Building a new party wall This applies where adjoining land is not built on at the line of a junction or only built as a boundary wall (i.e. not a party fence

wall or the external wall of a building). There are two situations:
(1)   If the wall is intended to straddle the boundary, one month notice of
      wish to start work must be given. The notice must indicate desire to
      build and describe the intended wall. If notice of consent is received, the
      wall must be built half and half or as agreed, the cost borne by each in
      proportion to use.
(2)   If the wall is wholly on the applicant's own land, one month notice of a
      wish to start work must be given. The notice must indicate a desire to
      build and describe the intended wall as before, but the building owner
      has the right to project foundations, if necessary, under adjacent land
      any time within 12 months from expiry of the notice, but the work must
      be at the building owner's own expense and the adjoining owner or
      occupier must be compensated for damage caused by building the wall
      or the foundations. This also applies where the adjoining owner refuses
      consent to a party or party fence wall.

Work to existing walls A building owner has certain rights in respect of
existing walls. The scope is very broad and the following is a summary. The
building owner has the right:
—   To underpin, thicken or raise, but if it is not due to defect or lack of
    repair, he must make good all damage to adjoining premises, internal
    furnishings and decorations and if it is a party structure or external
    wall, he must carry up any adjoining owner's flues and chimneys which
    rest on or form part of the party structure or external wall as may be
    agreed or settled by the disputes process.
—   To repair or demolish and rebuild a party structure or party fence wall
    if the work is necessary because of defects or lack of repair.
—   To demolish a partition which does not conform with statutory
    requirements and build a party wall which does conform.
—   To demolish structures over public ways or passages belonging to other
    persons and rebuild them so as to conform to statutory requirements.
—   To demolish a party structure and rebuild it so as to make it of
    sufficient strength or height for any intended building of the building
    owner or to rebuild it to lesser thickness or height provided it is still
    sufficient for any adjoining owner, but he must make good all damage
    to adjoining premises, internal furnishings and decorations and if it is a
    party structure or external wall, he must carry up any adjoining owner's
    flues and chimneys which rest on or form part of the party structure as
    may be agreed or settled by the disputes process.
—   To cut into a party structure or away from a party, party fence, external
    or boundary wall any foundation, chimney breast or other projection
    over the building owner's land or take away or demolish overhanging
    parts of wall or building of adjoining owner to the extent necessary to
    enable a vertical wall to be erected or raised against the wall or building
    of an adjoining owner, but all damage to adjoining premises, internal
    furnishings and decorations must be made good.

— To cut into an adjoining owner's wall to carry out weatherproofing of a new wall erected against it, but he must make good all damage to the wall.
— To carry out other necessary works incidental to the connection of a party structure with the premises adjoining.
— To raise a party fence wall or to raise it for use as a party wall or to demolish it and rebuild it as a party fence or party wall.
— To reduce or to demolish and rebuild a party or party fence wall either to not less than 2 metres if not used by the adjoining owner other than as a boundary wall or to a height currently enclosed by the building of an adjoining owner, but the building owner must reconstruct or replace any existing parapet or construct one if needed.
— To expose a party wall or structure, but adequate weathering must be provided.

A building owner may exercise these rights with the written consent of the adjoining owner. If adjoining land is built on at the line of a junction as a party or party fence wall or the external wall of a building, the building owner must give a two months 'party structure notice' of the date when work will start before exercising any right under the Act. The notice must state the name and address of the building owner, particulars of the proposed work, whether special foundations are intended and plans, sections and details including the loads to be carried. The notice ceases to have effect if the work is not begun within 12 months of the date the notice is served or if it is not continued with due diligence. There is provision for the adjoining owner to serve a counter notice. If no consent is received within 14 days of the date of service of party structure or counter notices, dissent is deemed and a dispute is deemed to have arisen.

Adjacent excavations and constructions There are two situations:
— Where a building owner proposes to excavate and erect a structure any part of which is within 3 metres horizontally from any part of the structure of an adjoining owner and which extends to a lower level than the level of the bottom of the foundations of the adjoining structure.
— Where a building owner proposes to excavate and erect a structure any part of which is within 6 metres horizontally from any part of the structure belonging to an adjoining owner and which extends to a lower level than a point measured at 45° from the point of intersection of the external face of the adjoining structure and the bottom of the foundation.

The owners of such structures are deemed to be adjoining owners for the purposes of this section even though the property is not touching the boundary.

The building owner must give one month's notice of the date when work will start. The notice must set out the proposals and whether underpinning or other strengthening or protection is proposed. Plans and section must show the site and the depth of any excavation proposed and if the erection of a building is proposed, its site. The notice ceases to have effect if work is not begun within 12 months of the date the notice is served or if the work is not continued with due diligence. The building owner may at own expense strengthen the foundations of the adjoining structure or may be required to do so by the adjoining owner. If there is no consent within 14 days of the date of service of notice, dissent is deemed and a dispute is deemed to have arisen.

There are various other provisions in relation to matters such as disputes, the appointment of surveyors and access.

Passing of risk Goods and materials are said to be at someone's risk when he is liable for the accidental loss of or damage to them.

The basic presumption in sale of goods as to the transfer of risk initially is that risk will move to the buyer at the same time as ownership is transferred to him (Sale of Goods Act 1979, s. 20), but the parties to a contract can always provide otherwise. This is almost invariably the case under the standard forms of building contract in common use where, although ownership may pass to the employer, e.g. on incorporation (qv), into the works or on payment, the risk remains with the contractor. For example, under JCT 98, the contractor's obligation (clause 2.1) is to 'carry out and complete the Works...'. He does not fulfil that obligation until the architect certifies practical completion. The passing of property does not, in this instance, transfer the risk. ACA 3, clause 6.2 is an explicit clause to the same effect: 'The risk of loss or damage to any Section or to the Works shall remain with the contractor until the taking-over of such Section or of the Works, as the case may be.'

See also: **Sale of goods.**

Patent A Crown grant of sole rights with regard to an invention. The grant is normally valid for, and gives the patentee a monopoly over his invention for, a period of 20 years from the date when the specification is filed. Payment in respect of patent rights (i.e. the right to use a patented article or process belonging to another) is generally the responsibility of the contractor. JCT 98, clause 9 deems such sums to have been included in the contract sum (qv) and provides that the contractor shall indemnify the employer against any claims arising from the infringement of patent rights by the contractor (clause 9.1). The contractor is not liable to indemnify the employer and monies payable shall be added to the contract sum where the contractor is complying with architect's instructions (clause 9.2). GC/Works/1 (1998), clause 12 is a broadly similar provision.

See also: **Indemnity clauses.**

Patent defect A defect which is discoverable by reasonable inspection. In the context of the building contract, the term embraces all the items which the architect or the clerk of works might be expected to find and bring to the contractor's attention so that remedial work can be carried out. Patent defects are plain to see, or at least that is the theory. Whether the architect could or should have seen defects on site during site visits has exercised more than one judicial mind. Where the final certificate (qv) is conclusive or partially so, its issue may preclude the employer from bringing any proceedings against the contractor for defects, whether patent or latent.

See also: **Certificates; Inspector; Latent defects; Supervision of works.**

Payment into court Now more properly referred to as a Part 36 payment[532] (qv). In any action for debt or damages the defendant may pay money into court in satisfaction of all or any of the claimant's claims. Before such a payment will properly qualify as a payment into court it must be made according to certain detailed procedures and requirements laid down in the Civil Procedure Rules (qv). In particular, the rules (Part 36) require:

(1)   Where payment is made in respect of proceedings in the County Court or District Registry, the paying party must lodge with the court concerned a notice of payment in the prescribed form. If payment is made by way of a cheque (qv), it should be made payable to Her Majesty's Paymaster General.

(2)   Where the court concerned is the Royal Courts of Justice, cheques are made payable, instead, to the Accountant General of the Supreme Court and a special form accompanying it should be sent to the court's Funds Office. In addition to completing and lodging the other forms prescribed in the rules, a sealed copy of the claim form must also be sent to the court.

In all cases it is also necessary to serve the claimant with notice of the payment and that notice, too, must be given in a prescribed form. Payment in will not properly be made until such notice has been served.

Under the Civil Procedure Rules, in the event that the paying party wishes to withdraw or increase an amount previously paid in or the other party wishes to accept the amount concerned, other very detailed provisions and restrictions apply and those, too, must be strictly and fully complied with.

In general, if the claimant wishes to accept the amount paid he should do so within 21 days of the payment being made, in which case he will not first require the permission of the court before doing so. By accepting the payment within that time the action (or those issues subject to the payment in) will be stayed and the claimant will also almost invariably be entitled to his costs of the proceedings up until the date when he served an appropriate notice accepting the payment. After 21 days have expired the payment may still be taken, but in that case the defendant's agreement is required or the court will have to give permission and will consider the cost consequences.

Alternatively, the claimant may leave the money in court and continue with his action, in which case, the fact that a payment into court (i.e. a Part 36 payment) has been made will not, except in the most exceptional circumstances[533], be communicated to the trial judge until all questions of liability and the amount of money to be awarded have been decided. If, in the event, the claimant recovers less or no more than the amount paid in then in the majority of cases, unless the court considers it would be unjust to do so, the claimant will be ordered to pay any costs incurred by the defendant after the latest date on which the payment or offer could have been accepted without needing the court's permission.

See also: **Costs; Sealed offer; Calderbank offer; Part 36 offer.**

---

[532]Being a reference to Part 36 of the Civil Procedure Rules.
[533]Disclosure of a payment into court will usually require the trial judge to withdraw. See: **Bias.**

Payments Under all commonly used standard form building contracts the custom has been that provision is made for payment to be made by the employer to the contractor in instalments as the works progress. That custom now has statutory backing so that all building contracts qualifying as a 'construction contract' (qv) within the meaning of the Housing Grants, Construction and Regeneration Act 1996[534] (qv) must now make express provision for periodic payment to the contractor during the course of the project. Non-qualifying contracts or those that are for short-term projects (currently set at a predicted duration of less than 45 days) remain unaffected, so that in those cases payment may still be made in a single lump sum on completion of the project.

In addition to the basic contractual provisions concerning periodic payment, all of the significant standard form building contracts currently in use also provide for a process of interim certification (qv) by the architect (or other contract administrator) as a part of the procedure leading to payment being made. See also: **Advances on account; Certificates; Due date; Performance; Stage payments.**

Penalty Sums of money inserted in a contract which is extravagant and unconscionable, the purpose being to coerce a party to performance. A 'penalty clause' is invalid and the sum is irrecoverable in contrast to liquidated damages (qv).

Although many contractors think otherwise, sums inserted in the usual 'liquidated and ascertained damages' clauses in standard form building contracts are usually moderate, and there appears to be no reported English case in which a sum has been disallowed as a 'penalty' merely because of its amount under a building contract. It is wrong to speak of liquidated damages clauses as 'penalty clauses'.

Peremptory order An order which may be made by an arbitrator, or judge, usually restating a previously ignored order or direction and prescribing a specific period of time within which the defaulting party must comply with the order concerned. Once a peremptory order is made, if still not complied with there are then several options open to the tribunal. For example, where the peremptory order requires the claimant to provide security for costs, if ignored then an award may be made dismissing the claim. Where there is non-compliance with peremptory orders on other matters such as, for example, the provision of certain documents or service of statement of claim, defence or other pleading, statement or the like, the arbitrator or judge may:
— Ignore the statements or other documents etc. and go on to make his award without further reference to them.
— In making his award, draw his own conclusions regarding the non-conformance with his order.
— Order that the defaulting party shall be liable for the costs of non-compliance in any event and irrespective of the overall outcome of the proceedings.

---

[534]For a list of those contracts that qualify as construction contracts for the purposes of the Act, see s. 104 to s. 107 at Part II of the Housing Grants, Construction and Regeneration Act 1996.

Moreover, subject to any contrary agreement the parties themselves may otherwise have made, the arbitrator, or a party to the arbitration, may begin specialist court proceedings seeking an order requiring the defaulting party's compliance with a peremptory order.

Performance The carrying out of an obligation imposed by contract or statute. In building contracts, *complete performance*, where the contractor carries out the whole of the works in accordance with the contract documents and the employer pays the contract sum (qv), will discharge the contract. *Partial performance* by one party may be sufficient evidence of his intention to be bound by the terms of a contract if he has not made formal acceptance. Whether or not performance is complete is a matter for the courts to decide in each particular case. The point is particularly important where payment depends upon the whole of the work being completed. The courts will, however, grant relief to the contractor who can show, in such a case, that he has achieved *substantial performance,* i.e. the work is complete save for some minor omissions or defects. 'Where a contract provides for a specific sum to be paid on completion of specified work, the courts lean against a construction of the contract which would deprive the contractor of any payment at all simply because there are some defects or omissions'[535].

See also: **Entire contract; Specific performance.**

Performance bonds See: **Bonds.**

Performance specification An alternative to the specification (qv) as traditionally understood. Instead of describing precisely all the work and all materials required in a building, the performance specification sets out criteria which must be met by the contractor. The idea is to give the contractor maximum scope for initiative and price competition. For example, a traditional specification might describe an external wall in terms of type of brick, number of courses to a given height, thickness of wall, size of cavity, material for the internal leaf, insulation type and thickness, wall ties, damp proof course, etc. A performance specification would require that the wall would last a given number of years, be waterproof, have a given U-value, have a certain colour range, have certain maintenance characteristics, etc. The criteria may be very precise or very broad and commonly contain the overall requirement of compliance with Building Regulations (qv) and British Standards. A performance specification always carries a design (qv) requirement.

The writing of a performance specification is a skilled task and may take longer than a traditional specification. It is a mistake, therefore, to use a performance specification to attempt to overcome pressing deadlines. It is important to make a clear distinction, in the specification, between those criteria which are mandatory and those which are at the contractor's discretion. Outline dimensioned drawings are usually provided with the specification and form

---

[535]*Hoenig* v. *Isaacs* [1952] 2 All ER 176 per Denning LJ at 181.

part of the contract. Where WCD 98 is used, the other essential part of the contract documentation is the contractor's proposals.

Very often, the architect will prepare a performance specification for work for which he intends to invite tenders with a view to nomination (qv). A lift installation is a good example of work which requires a performance specification in order that a proper comparison of prices can be made.

See also: **Design and build contracts.**

Performance specified work A term particular to and defined in JCT 98, clause 42. It describes work that must meet stipulated requirements and which, when completed, must achieve the standards of performance described in the bills of quantities.

Where sufficiently described in the bills, the contractor will be expected to price for the work concerned as part of his tender. However, if at time of tender the employer can provide only limited information – such as simply the location and performance objectives of the work concerned – and the contractor can do little more than assess its effect on his programme and costs of preliminaries, then the performance specified work should be priced simply by way of a provisional sum (qv).

Within the contractually stipulated time, or if none is stipulated then within a reasonable time before starting the performance specified work, the contractor must provide the architect with his sufficiently detailed proposals for carrying out the work and in doing so must ensure that those proposals are also given prior approval by the planning supervisor (qv) for the project. The architect has 14 days within which to consider the contractor's proposals and in that time, if he considers them to be in any way deficient, then on written notice he may require the contractor to make good that deficiency.

Beyond the general provisions of the contract, further specific provisions are also made in relation to performance specified works to deal with: variations instructions, (clauses 42.11 and 42.12); instructions for integration within the overall design of the works (clause 42.14), corrections of errors and omissions in information provided by the employer concerning performance objectives and the like.

Period of suspension A term used in the JCT 98 (Appendix) to refer to the continuous period of time during which the works may be suspended due to causes not of the contractor's making, before the contractor then has an express entitlement to determine his employment under the contract (JCT 98, clause 28).

The contract makes provision for three periods of suspension:

— Suspension for a continuous period (usually one month or more) will entitle the contractor to determine his employment where that suspension is the result of the architect's failure to provide information at the times required under the contract or where the suspension is the result of an architect's instruction, delay by the employer's own contractors engaged on the works or the employer not affording the contractor proper access to or from the site (clause 28.2.2).

— Suspension for a continuous period (usually three months or more) will entitle the contractor to determine his employment where it is the result of *force majeure* (qv), loss or damage caused by any one of the clause 22 specified perils or civil commotion (clause 28A.1.1.1 to clause 28A.1.1.3).

— Suspension for a continuous period (usually one month or more) will entitle the contractor to determine his employment where it is the result of architect's instructions given as a consequence of default of a statutory undertaker acting in that capacity, hostilities (qv) or terrorist activity (clause 28A.1.1.4 to 28A.1.1.6).

See also: **Determination; Frustration.**

Persistent neglect A term previously used in JCT 80 (clause 27.1.3) whereby, if the contractor refused or *persistently* neglected to comply with a written notice from the architect requiring removal of defective work or improper materials or goods and the works were thereby materially affected (qv), the employer would have grounds for determination of the contractor's employment. The express requirement for the neglect to be '*persistent*' was removed by Amendment 4 to the JCT 80 published in July 1987, since when all subsequent reprints of the JCT 80, and now JCT 98 (clause 27.2.1.3) provide simply that the contractor's employment may be determined if he neglects to comply with a written instruction ordering removal of non-conforming work etc. The position under the IFC contract is now also essentially the same, with the requirement for persistent inactivity being removed in July 1988, by Amendment 3.

Nevertheless, it remains that any such notice given under JCT 98 or IFC 98 for the purposes of finally determining the contractor's employment must not be given unreasonably or vexatiously. In that case, for the contractor's neglect to react to the architect's instruction to come within the terms of JCT 98 (clause 27.2.1.3) or IFC 98 (clause 7.1), there will still be a necessity for a reasonable degree of persistent neglect to act on the part of the contractor.

Personal injury See: **Injury to persons.**

Personal property Also called personalty. All forms of property other than freehold estates and interests in land. It is contrasted with real property (qv) and covers everything (other than freehold estates and interests) which is capable of being owned. Some things are incapable of being owned, e.g. the air or running water. Such an item is known as *res nullius* – a thing belonging to nobody. Personal property is not confined to tangible objects, which are known as chattels, but includes intangible rights such as debts and copyright. Rights of this sort are called choses in action (qv) as opposed to choses in possession (qv). For example, a lender of money has a present right to repayment from the borrower – a chose in action. That right is a property right enforceable by means of legal action and may, subject to conditions[536], be transferred to a third party by assignment (qv). Leasehold interests are, for historical reasons, classified as personal property and are called chattels real in contrast to chattels personal. **Figure 17** (see: **Property**) shows the position in diagram form. See also: **Assignment.**

Personal representative An executor or administrator of the estate of a deceased person. He is a trustee and stands in the shoes of the deceased. In contractual terms, a personal representative is a named person acting as agent for one of the parties with full authority. The architect is not the personal representative of the employer. See also: **Agency; Representative; Trust.**

Person-in-charge Clause 10 of JCT 98 and WCD 98 and clause 3.4 of IFC 98 each provide for the contractor to keep a competent person-in-charge constantly upon the works. The person concerned is clearly intended to be the site agent or foreman and is to be of sufficient seniority to be capable of receiving instructions from the architect, such instructions being deemed (qv) to have been given to the contractor himself. A similar term is also mentioned in MW 98, in clause 3.3. The difference is, however, that under that contract he need not be constantly upon the works. He need only attend at all reasonable times. What is reasonable will depend upon the size and complexity of the work and, if known, the dates of the architect's visits. See also: **Competent; Site manager.**

PFI See: **Private Finance Initiative.**

Plaintiff See: **Claimant.**

Planning consent Statutory control over the development and use of land in England and Wales is now principally consolidated within five Acts: The Planning and Compensation Act 1991; The Town and Country Planning Act 1990; The Planning (Listed Buildings and Conservation Areas) Act 1990; The Planning (Hazardous Substances) Act 1990; and The Planning (Consequential Provisions) Act 1990. Those in turn are modified and expanded upon by numerous supplementary Acts, General Development Orders and Regulations along with various other statutory instruments and governmental policy statements and guidance notes. Controls are exercised by local planning authorities. Except in the case of a 'permitted development' of the type listed in schedule 2 of the Town and Country Planning (General Permitted Development) Order 1995 or one of the other very few clearly defined instances, where it is proposed to carry out any 'development' (as defined by s. 55 of the Town and Country Planning Act 1990) an application must be made to the planning authority for

---

[536]See *Trendtex Trading Group* v. *Credit Suisse* [1980] 3 All ER 721.

permission to do so. A decision whether to grant planning consent will be largely influenced by the development plan covering the area within which the development is proposed to take place.

It is often prudent to obtain outline permission before making a detailed application. That will involve making the application in the prescribed form issued by the local planning authority accompanied by an appropriate certificate relating to the nature of the ownership of the land to be developed and submission of a minimum of drawings and information. The authority will then either give or refuse consent to the principle of the development, for example, whether an office block would be permitted in a particular area. Obtaining detailed permission will similarly require completion of the appropriate form issued by the planning authority, accompanied by an appropriate certificate relating to the nature of the ownership of the land to be developed and full details of the proposed development including, for example, design, external appearance, proposals for external features and parking facilities and the like. Depending on the size and nature of the development, other more far reaching information such as the likely impact of the development on existing traffic flows and so on may also be necessary. Development must begin within five years of the date of the full planning permission or normally the permission lapses. If outline permission has been obtained, application must normally be made for detailed permission within three years. The authority has wide powers to make conditions on the permission and to reserve matters for further approval. Appeal may be made to the Secretary of State for the Environment against refusal of planning permission or against conditions attached to the consent.

Planning regulations are exceedingly complex and the advice of the local planning officer should always be sought when any development is contemplated.

See also: **Building line; Notice.**

Planning supervisor A competent individual, company or partnership appointed by the person for whom the project is carried out, or if he opts to do so by his agent, to have overall responsibility for co-ordinating the health and safety aspects of the design and planning phase of the project in accordance with the Construction (Design and Management) Regulations 1994 (qv).

His responsibilities include ensuring proper and timely preparation of the health and safety plan (qv) and health and safety file (qv) — for which he remains responsible during the project — the giving of advice to the client or his appointed agent concerning the allocation of appropriate and sufficient resources for health and safety, overseeing health and safety aspects of the design and, if appropriate, the giving of relevant written notification of the project to the health and safety executive, giving them such details as:

— A signed declaration of his and of the principal contractor's appointments.
— The address of the construction site and type of project concerned.
— The names and addresses of the client, planning supervisor, principal contractor and any chosen contractor(s).

— The planned date for the start of the construction phase and its planned duration.
— The estimated maximum number of workers and expected number of contractors who will be on the site.

A competent planning supervisor will be expected to have membership of a relevant professional body, be familiar with and have a good knowledge of construction practices generally and of those particularly relevant to the specific project for which he is appointed, and to have knowledge of and be familiar with design functions, health and safety and fire safety. He should also have ability to work along with and co-ordinate the different design and construction activities.

Appointment of a planning supervisor must be made as soon as there is sufficient known about the project to allow the client to make reasonable enquiries and to be satisfied that the appointee has sufficient competence, knowledge, ability and resources briefly outlined above to cope with the role and to fulfil his responsibilities under the CDM Regulations in relation to the particular project concerned. Once appointed, the competence of the planning supervisor to cope with the project or any changes in the nature of it must be kept under review but even in the event that it should become necessary to terminate, change or renew the appointment the position must remain competently filled at all times until the end of the construction phase.

Plans A very general word of imprecise meaning. It is usually taken to mean the drawings to a small scale showing work to be carried out. A 'plan' is, strictly, a horizontal section through a portion or the whole of the work to any scale as opposed to a vertical section or cut. In broad terms, it may refer to any idea or scheme of action.

Plant A rather broad term referring to the equipment used by the contractor. Its meaning may be restricted by the wording of the contract. Thus, JCT 98, clause 27.6.1 refers to 'all temporary buildings, plant, tools, equipment and Site Materials ...'. In that instance it is clear that plant refers to something other than equipment, tools or temporary buildings. Mechanical diggers, mixers and vehicles are indicated. Guidance on what might or might not fall within the category of plant may sometimes be obtained by reviewing whether and if so how any specific item in question may be categorised for the purposes of any Standard Method of Measurement (qv) used in preparing bills of quantities (qv) for the project or, for example, by reviewing any schedules of basic plant charges such as may be issued by the Royal Institution of Chartered Surveyors or other professional body. In general terms, however, 'plant' might be used to describe any kind of mechanical or non-mechanical equipment, including scaffolding and huts. Plant can be either temporary – such as dumpers, cranes and the like – or permanent and built into the works – such as boilers, fans and the like.

Pleadings See: **Statement of case.**

Points of law Sometimes referred to as questions of law; in civil cases they are questions which concern the proper interpretation and construction of the parties' legal rights, obligations, remedies and the like as opposed to questions of fact or opinion. The term is usually found in the context of arbitration (qv) where, if a question of law that will significantly affect a party's rights should arise in the course of the proceedings, then one or other of them may, with the arbitrator's consent, apply to the court to give judgment on that question[537]. Questions of English or Welsh law must be referred to the English or Welsh courts. Those concerning the interpretation and effect of Northern Irish law will be heard and decided by the courts of Northern Ireland[538]. Where, however, in the course of his decision and in his award an arbitrator has himself decided a question of law, only in the most carefully regulated and limited circumstances will the parties then be able to appeal against that decision to the courts[539].

Possession In the absence of an express term (qv) in the building contract, a term will be implied that the contractor must have possession of the site in sufficient time to allow him to complete the works by the contract completion date (qv)[540]. Failure by the employer to give sufficient possession will amount to a breach of contract entitling the contractor to damages[541].

Most standard forms state the date on which possession must be given (JCT 98, clause 23.1.1; IFC 98, clause 2.1; ACA 3, clause 11.1 read together with the Time Schedule) and JCT 98 and IFC 98 empower the employer to defer the giving of possession for a stated period. MW 98 does not give a date for possession. It gives a date on which the works may be commenced (clause 2.1) but, in practice and in law, this must also be the latest date for possession. GC/Works/1 (1998), clause 34 makes provision for the employer to give possession within the period or periods specified in the Abstract of Particulars (qv). Where no period is specified, the order must be given within a reasonable time after acceptance of the contractor's tender. The contract period, and hence the date for completion, is to be stated in the Abstract of Particulars (qv).

Under the NEC (qv), the contract data provide for possession and commencement dates to be stipulated separately, with further provision being made for numerous separate possession dates to be stipulated where, for example, it is proposed to give the contractor possession of different parts of the site at different times.

If the employer fails to give possession on the due date, then unless the contract entitles him to defer the giving of possession (such as he may do for up to six weeks under JCT 98, clause 23.1.2 and under IFC 98 under clause 2.2), the contractor will be entitled to sue for damages. Similarly, unless the contract makes express provision for the date for completion to be extended by reason

---

[537]Section 45 Arbitration Act 1996.
[538]Section 82 (1) Arbitration Act 1996.
[539]Section 69 Arbitration Act 1996.
[540]*Freeman & Son* v. *Hensler* (1900) 64 JP 260.
[541]*Rapid Building Group Ltd* v. *Ealing Family Housing Association Ltd* (1984) 1 Con LR 1.

of the deferment by the employer of giving possession (e.g. under JCT 98, clause 25.4.13 and IFC 98, clause 2.4.14), then the date for completion may become at large and so invalidate any liquidated damages clause(s) (see: **Time at large**). It is not possible simply to overcome the problem by issuing an instruction to postpone the work and, subsequently, awarding an extension of time to cover postponement.

Possession of the site and the carrying out of work are different things. The contractor will need to do a number of things which are not strictly carrying out the works and for which he must have possession of the site and so, even where the deferment is sanctioned as provided under JCT 98 and IFC 98, such deferment of possession may in any event entitle the contractor to claim for reimbursement of any consequent direct loss and/or expense.

Generally, only when the architect has certified practical completion (qv) will the contractor's licence (qv) to occupy the site come to an end and the employer will then have the right to take, or re-take, possession of the site. However, with the exception of MW 98, all commonly used JCT contracts make express provision (e.g. under JCT 98, clause 18; IFC 98, clause 2.11; MC 98, clause 2.8; WCD 98, clause 17) for the employer to take possession of any part or parts of the works before that time, provided he first obtains the contractor's agreement. Where partial possession is taken, it will have the effect of transferring responsibility to the employer for putting in place and maintaining appropriate insurance cover for the relevant part or parts, and the rate of liquidated damages (qv) will be reduced proportionally. The part or parts concerned will also, for all the purposes of the contract, be deemed to have reached practical completion.

Possessory title Title to land acquired by occupying it for 12 years without paying rent or otherwise acknowledging the rights of the true owner. The period is 30 years in the case of Crown land.

See also: **Adverse possession.**

Postponement All standard form contracts in common use allow the architect to postpone the execution of any work to be done under the contract. There is usually an express term to that effect (e.g. JCT 98, clause 23.2; IFC 98, clause 3.15; MC 98, clause 3.5) and in the case of WCD 98 the power to postpone extends not only to the execution of construction works but also to the preparation of any design(s) for which the contractor is responsible under the contract. Postponement is a serious step. It will entitle to the contractor to:
— An extension of time (e.g. JCT 98 and WCD 98, clauses 25.4.5.1; IFC 98, clause 2.4.5).
— Financial reimbursement for loss and expense (e.g. JCT 98, clause 26.2.5; WCD 98, clause 26.2.4; IFC 98, clause 4.12.5).
— Determine his employment if the works are postponed for a longer period than allowed by the contract or, if no period is stipulated, a reasonable period (JCT 98 and WCD 98, clause 28.2.2.2; IFC 98, clause 7.9.2(b)).

GC/Works/1 (1998) allows suspension of all, or any part of the works under clause 40 (2) (g). Consequential extensions of time are provided for under clause 36 (2), provided the suspension is not caused by the contractor's default. Provision for the contractor to determine his employment is given under clause 58 where, under sub-clause (3) (g) determination may arise specifically in the event of excessive suspension which is not the result of the contractor's own default.

ACA 3 allows suspension of the works under clause 11.8. Consequential extension of time and resulting alteration of the date for take-over of the works is provided for under clause 11.9 and additional payment in respect of any postponement is dealt with under clause 11.8 and, where and to the extent provided, clause 17.1. Clause 11.8 is essentially a provision for postponing the date of taking-over, however, some degree of suspension is implied. It is likely that in addition the architect also has wider powers of suspension under clause 8.1 (f), which enable him to issue instructions 'on any matter connected with the Works'. In that case, an extension of time would be awarded under clause 11.6 and, where appropriate, clause 17. There is no provision for termination of the contractor's employment due to suspension of the works unless the delay lasts for 60 consecutive days or more and is due to:

— *Force majeure* (qv).
— Clause 6.4 contingencies.
— War, etc.

Thus, if suspension is ordered by the architect the contractor is thrown upon his common law rights if the resulting delay lasts for an unreasonable period (which in the circumstances might well in any event be considered to be 60 consecutive days). MW 98 probably gives the architect power to postpone under clause 3.5; extension of time would fall under clause 2.2 but there is no provision for reimbursement of loss and/or expense. Any such payment would have to be agreed by the employer or would become the subject of an action by the contractor at common law. The contract expressly allows the contractor to determine his employment if the employer suspends the carrying out of all, or substantially all, of the works for a continuous period of at least one month.

It has been held that an instruction to postpone will be implied if the architect issues an instruction to the contractor which necessarily entails postponement of the work, even though the instruction is not issued under the appropriate clause and does not specifically mention postponement[542].

The employer, in the absence of an express term (qv) has no implied right to postpone the work. There is an implied term (qv) in every building contract that the employer will allow the contractor to begin work on the date fixed for commencement and to continue working so as to complete the works by the contract completion date (qv). Without the express term, therefore, the contractor may be able to treat postponement as an act of repudiation (qv) on the part of the employer and sue for damages. In general, the wording of

---

[542]*M. Harrison & Co (Leeds) Ltd* v. *Leeds City Council* (1980) 14 BLR 118 and *Holland, Hannen & Cubitts (Northern) Ltd* v. *Welsh Health Technical Services Organisation and Others* (1981) 18 BLR 80.

the common suspension or postponement clauses will not extend to an express right to defer the giving of possession (qv) of the site.

See also: **Suspension.**

Practicable steps A phrase found in JCT 98, clause 25.4.7 where it refers to 'delay on the part of' nominated sub-contractors or nominated suppliers which the contractor has taken all 'practicable steps to avoid or reduce'. The steps which the contractor is required to take are those which, in practice, he can take, i.e. those which are feasible. In order to decide whether the contractor has taken all practicable steps, consideration must be given to all of the circumstances. That is to say, the circumstances surrounding the matter which the practicable steps are designed to reduce or avoid along with other circumstances relevant to the particular project and particular contractor concerned. An obligation to take all practicable steps imposes a stricter standard than might otherwise be the case where, for example, the obligation is expressed to be one to take measures that are 'reasonably practicable'. In that case a more subjective test involving questions of cost, common practice and the like may apply. Where the obligation is simply to take all practicable steps, then what will be 'practicable' should not be measured by what may or may not be 'reasonable' or 'equitable'. Questions of cost will also be less significant if not altogether irrelevant[543]. The test in that case is, therefore, more objective. Instead of taking account of the particular financial and/or other particular constraints affecting the particular contractor concerned, the practicability of the steps he has or has not taken should be measured, in the circumstances, against what should be expected of the average contractor.

Practical completion A phrase found in JCT 98, principally in clause 17. It marks the date at which:

— The defects liability period (qv) begins (clause 17.2).
— The contractor's liability for insurance (qv) under clause 22A ends.
— Liability for liquidated damages (qv) under clause 24 ends.
— Liability for damage caused by frost occurring thereafter ends (clause 17.2).
— The employer's right to deduct full retention ends. Half the retention percentage becomes due for release (clause 30.4.1.3).
— Regular interim certificates (qv) cease to be issued (clause 30.1.3).
— The period for the architect's final review begins under clause 25.3.3.

Despite the enormous importance of the date, the contract does not define 'practical completion'. Under clause 17.1 the architect must issue a certificate forthwith (qv) when:

— in the architect's opinion practical completion is achieved, and
— when the contractor has complied sufficiently with his obligations under the CDM Regulations (qv), and

---

[543] *Adsett* v. *K. L. Steelfounders & Engineers Ltd* [1953] 1 All ER 97 and *Hammond* v. *Haigh Castle Co Ltd* [1973] 2 All ER 289.

— when the contractor has provided any 'as built' drawings specified in the contract to be provided in connection with any performance specified work.

Other JCT contracts have clauses to similar effect.

It is generally agreed that practical completion does not mean 'nearly complete'. Some commentators refer to it as complete for all the practical purposes of the contract, but it is not the same thing as substantial completion (qv).

Some architects insist that all work is totally complete before issuing a certificate of practical completion. Others certify when a considerable amount of work is complete and only that which can be finished without inconveniencing the employer remains. It would appear to be going too far to insist on total completion before issuing a certificate, otherwise the contract could have referred merely to 'completion'. The addition of the word 'practical' must have some relevance. From a legal point of view, the phrase is ambiguous. The question is whether it covers the situation where the works are substantially finished but there are defects. This is an important matter since the architect's power to order the remedying of defects during the defect's liability period is limited to defects 'which shall appear' during that period.

There is conflicting case law. In *J. Jarvis & Sons Ltd* v. *Westminster Corporation*[544] the court took the view that 'practical completion' means that there must be no defects apparent in the works at the date on which the architect issues the certificate. 'The defects liability period is provided in order to enable defects not apparent at the date of practical completion to be remedied. If they had been apparent, no such certificate would have been issued.' In other words, the architect can issue his certificate even if he knows that some latent defects (qv) are present. In contrast, in *P. & M. Kaye Ltd* v. *Hosier & Dickinson Ltd*[545] it was suggested that the architect could withhold his certificate until all known defects, except trifling ones, were corrected. In *H. W. Neville (Sunblest) Ltd* v. *Wm Press & Son Ltd*[546], the High Court favoured the view expressed in *Jarvis*. The judge said, 'I think that the word "practically" gave the architect a discretion to certify that (the contractor) had fulfilled its obligation ... where very minor *de minimis* works had not been carried out, but if there were any patent defects in what (the contractor) had done the architect could not have given a certificate of practical completion'.

It seems, on balance, that the architect is justified in issuing his certificate if he is reasonably satisfied that the works accord with the contract, notwithstanding that there are very minor defects which can be remedied during the defects liability period (qv).

ACA 3 uses the phrase 'fit and ready for Taking-Over' (clause 12). 'Taking-Over' may be considered to be loosely equivalent to practical completion but there are some important differences. The contractor must notify the architect when he considers that the works are or will be fit and ready for taking-over.

---

[544](1970) 7 BLR 64 per Viscount Dilhorne at 75.
[545][1972] 1 All ER 121.
[546](1981) 20 BLR 78.

Practical completion

Certificate of
Practical Completion

Issued by: E W Pugin
address: Gothic Buildings
Birmingham

Employer: The Duke of Omnium
address: Trollope House
Belstead

Contractor: Gerrybuilders Ltd
address: The Crescent
Belstead

Works: Mansion
situated at: Belstead Gardens, Belstead

Contract dated: 1 February 2001

JCT 98 / IFC 98

Job reference: EWP / ABC

Certificate no: 1

Issue date: 5 July 2001

Under the terms of the above-mentioned Contract,

I/we hereby certify that in my/our opinion

Practical Completion of

* the Works

Delete as appropriate

* ~~Section no xxxxxxxxxxxxxxx of the Works~~

has been achieved

* and the Contractor has complied with the contractual requirements in respect of information for the health and safety file

This item applies to JCT 98 only.

* and the Contractor has supplied the specified drawings and information relating to Performance Specified Work

on   4 July   20 01

To be signed by or for the issuer named above

Signed *E W Pugin*

Distribution					
[X] Employer	[ ] Structural Engineer	[X] Planning Supervisor	[ ]
[X] Contractor	[ ] M&E Consultant	[ ]	[ ]
[X] Quantity Surveyor	[X] Clerk of Works	[ ]	[X] File

F853A/B   for JCT 98 / IFC 98   © RIBA Publications 1999

Figure 16   Certificate of practical completion.

The architect is expressly given discretion to issue a taking-over certificate upon receipt of the contractor's written undertaking to complete with all due diligence (qv) any work contained in the architect's or contractor's list. If the architect opts to wait until all the outstanding listed items are complete, he must then issue his certificate forthwith (qv) when the items are completed. His certificate marks the date at which:

— The contractor's liability for loss or damage to the works and to goods intended for the works ends (clause 6.2).
— The contractor's liability for insurance under clause 6.4, alternative 1, ends.
— Liability for liquidated or unliquidated damages, under clause 11.3, alternatives 1 or 2 respectively, ends.
— The maintenance period begins (clause 12.2).
— The regular calculation of fluctuations (if applicable) on interim certificates ends (clause 18.1).
— Any reference to arbitration can be opened under clause 25D.9.
— The period for review of extensions of time granted begins (clause 11.7).

GC/Works/1 (1998) makes reference, at clause 34 (1), to the works (or any relevant section thereof) being 'completed' in accordance with the contract by the date or dates for completion. The PM is specifically required, by clause 39.1, to issue a certificate stating the date when the works or any section of them is to that extent complete. The wording indicates that this is equivalent to 'practical completion' and his certificate marks the date at which:

— The contractor's liability for liquidated damages ends.
— The contractor is entitled to receive the estimated final sum less one half of the reserve.
— Reference to arbitration can be opened.
— The maintenance period begins. At the end of the longest relevant maintenance period the PM must issue a further certificate when the works are in a 'satisfactory state'.

See also: **Completion.**

Precedent See: **Judicial precedent.**

Preliminaries That part of the bills of quantities (qv) which describes the works in general terms and lists the contractor's general obligations, the restrictions imposed by the employer and the contractual terms.

Prescription The vesting of a right by reason of lapse of time. Prescription is the most important method of acquiring easements (qv) over property such as rights of light and rights of way. It is based on long enjoyment as of right. At common law it was necessary to prove that the right had been enjoyed since 1189 – being 'time immemorial' or the beginning of legal memory – but because of the difficulty of proving enjoyment for so long a period, evidence of use for a period of 20 years raised a presumption that the right had existed in 1189.

A prescriptive claim could be defeated by showing that the right must have arisen at a later date, and to make matters easier the courts evolved the doctrine of 'lost modern grant', under which if user could be proved for 20 years, a lawful grant would be presumed. That presumption could be defeated by proof that during the period when the grant could have been made there was nobody who could lawfully have made it.

ᐧ The Prescription Act 1832 was passed to simplify these difficulties so that claims to easements generally cannot be defeated by showing that the user commenced after 1189 if 20 years' uninterrupted enjoyment as of right is shown. If 40 years of enjoyment without interruption is proved, the right becomes absolute unless it has been enjoyed by written agreement or consent[547]. In the case of right of light (qv) there is only one period, 20 years[548], and the actual enjoyment, rather than user as of right, suffices[549].

The Act makes no change in the common law requirements as to prescription itself. The right claimed must have been exercised *nec vi, nec clam, nec precario* − it must not be exercised forcibly (*vi*), secretly (*clam*) or with consent (*precario*).

See also: **Right of light.**

Presumption A conclusion or inference which may or must be drawn from other established facts. Presumptions are important in the law of evidence (qv).

Pre-tender design Under the BPF System (qv) this is the term used to refer to the design and specification carried out by the design leader (qv) and consultants (qv) before tenders are invited.

Pre-tender information The information, in the form of drawings, schedules or reports, which the employer or his architect provides for the contractor to consider when preparing his tender. Some information will also be provided and sent to the contractor. Much if not all of that pre-tender information will generally be incorporated into and will become part of the contract documentation when the contract is concluded. Other pieces of information may be retained by the architect, and the contractor will be notified of their existence and availability and when the contract is made it will often be an express provision of the contract that, even if he has not in fact taken steps to inspect that information, the contractor will, nevertheless, be deemed (qv) to have done so.

It is important that all pre-tender information be accurate. Inaccurate or misleading information can lead to an action for damages or the contract being set aside on the grounds of misrepresentation (qv).

---

[547]Oral consent, even if evidenced by annual payments, is insufficient: *Plasterers' Co* v. *Parish Clerk's Co* (1851) 6 Exch 630.
[548]Section 3 Prescription Act 1832.
[549]*Colls* v. *Home and Colonial Stores Ltd* [1904] AC 179, 205.

Price The monetary value of something. The price at which a builder is prepared to carry out work. It will include the cost of labour, materials and overheads together with an addition for profit. 'Prices' is a word often used to refer to the sums which the builder inserts against the items in bills of quantities. It connotes the total (extended) price for the total item or number of items described or for the total quantity given as opposed to the unit rate from which a total price is calculated. Notably, JCT 98, clause 13.5.1.2 and GC/Works/1 (1998), clause 42(5) each make a clear distinction between the terms 'rates' and 'prices' in connection with the rules for valuation of variations.

See also: **Schedule of Prices; Priced statement.**

Priced Activity Schedule/Activity Schedule An optional attachment to main contracts JCT 98 and IFC 98 and sub-contract DOM/1 which, if used, must be attached to the contract and the Appendix completed accordingly. The schedule itself may take whatever form the parties agree and may be as detailed or general as they deem appropriate in the particular circumstances of the project concerned. However, in essence it must describe some or all of the construction activities involved in the project and against each description should be set the value which the parties agree represents that part of the contract sum properly attributable to each of those activities concerned.

Although it need follow no particular format and may be as comprehensive or general as the parties wish, it should be remembered that as regards the items that it describes the schedule must in future be used as the basis for determining all or some of the amounts due to the contractor on interim certificates (JCT 98, clause 30.2.1.1 and IFC 98, clause 4.2.1 (a)). Consequently, it must be drawn up with sufficient clarity and accuracy so that the quantity surveyor can properly relate and apportion it to the work done on each activity and thereby neither overvalue nor undervalue the amounts payable to the contractor.

Historically, a similar schedule − termed 'schedule of activities' − has also been a feature of the BPF system. However, unlike the schedule anticipated by the JCT and DOM/1 contracts, which is not intended to replace the use of traditional bills of quantities, under the BPF system the priced schedule is prepared by the contractor to replace the bills and to set out his design and management intentions and construction activities and to be used as part of the tender documents, for managing construction work, monitoring progress and for payment of the contractor.

Priced programme Under the BPF System (qv) the design leader must produce a priced programme. This consists of a schedule of his design activities and a programme showing when the activities will be carried out. Where separate consultants (qv) are appointed, they must undertake a similar exercise. Priced programmes become part of the BPF master programme (qv) and master cost plan (qv).

The priced programme is used as a plan of work and a basis for reporting progress. Payment to the design leader and consultants is based on

completed activities shown in the priced programme.
See also: **Schedule of Activities.**

Priced statement One of the procedures available under JCT 98 (clause 13.4.1.2), IFC 98 (clause 3.7.1.2) and WCD 98 (clause 12.4.2) for the valuation of work instructed by way of variation or of work approximately quantified in the contract documents. Upon receipt of a relevant instruction or at commencement of approximately measured work, the contractor may choose to provide the quantity surveyor with a statement of his price for undertaking the work concerned. Provided the contractor has sufficient information to enable him to do so, his statement must be prepared and submitted to the quantity surveyor within 21 days. Otherwise it must be sent as soon as practicable after such sufficient information is made available to him. Thereafter the priced statement must be wholly or partly accepted or rejected in writing by the quantity surveyor within a further 21 days, following consultation with the architect. Failure by the quantity surveyor to give any such response will be deemed to constitute a rejection of the contractor's proposals thereby entitling the contractor to assume that the contents of the statement are formally disputed. If and to the extent expressly rejected, the quantity surveyor must give his detailed reasons for doing so and must put forward alternative proposals which the contractor then in turn has 14 days to wholly or partly accept or reject.

It may be argued that the procedure is otiose since in arriving at the value of his priced statement the contractor is bound in any event to follow the traditional valuation procedures and to adopt, where possible, bill rates and prices as a basis for the statement. Nevertheless, used in a proper and timely manner the process has distinct advantages. Not least it allows issues or disputes over the correct valuation of the work to be addressed, in detail, at the time the work is undertaken, thereby avoiding prolonged uncertainty over its cost to the employer, and corresponding value to the contractor. In addition, where the priced statement procedure is adopted the contractor, should he choose to do so, may separately but at the same time provide the quantity surveyor with pre-estimates of any direct loss and expense and/or any extension of time that he considers would be acceptable to him in lieu of awards that might otherwise be made to him at a later date under the appropriate extension of time and loss and expense provisions of the contract.

Prime contracting A procurement system whose principles were proposed by the Egan Report (qv). The Construction Supply Network Project (CSNP) has been set up to develop and promote the new approach. Government has published a document called the Building Down Barriers Approach which identifies and integrates certain techniques. CSNP identifies five key phases in prime contracting:

(1) Inception: During this phase, the client team establishes the client's needs, carries out option studies, drafts the strategic brief and appoints an adviser.

(2) Definition and qualification: The client team selects the prime contractor who in turn drafts the project programme and identifies key supply chain partners (qv).
(3) Concept design: The prime contractor continues the principal role by exploring the client's functional requirements, drafting the project brief, involving the supply chain, developing and appraising potential solutions and providing initial GMP based on optimum whole life cost.
(4) Detailed design and construction: In this phase, the prime contractor completes the design, constructs the building 'right first time', optimises whole life costs, develops a compliance plan and hands over the building.
(5) Post handover: The prime contractor monitors and maintains the facility until proof of compliance.

Further information can be obtained from *The Prime Contractor Handbook of Supply Chain Management* which is being developed by Richard Holti, Davide Nicolini and Mark Smalley, supported by the Defence Estate Organisation, Department of the Environment, Transport and the Regions, AMEC Construction and John Laing.
See also: **Continuous improvement; Supply clusters.**

Prime cost The actual cost to the contractor of undertaking work, e.g. the wages paid, the cost of supervision, the price of materials and of sub-contract work. In contracts let on the basis of reimbursing the contractor his prime cost, it is important to have a precise definition of what prime cost is to be reimbursed.

Where variations are incapable of being properly valued by measurement, the work concerned will often be valued, in the first instance, by reference to its prime cost. With the exception of the MW 98, the substantial majority of the other commonly used JCT contracts provide in that case that the prime cost is to be arrived at in accordance with the definition of prime cost of daywork issued by the Royal Institution of Chartered Surveyors.

Prime cost (PC) sums A term found in many standard forms of contract. Its meaning is subject to some variation, depending upon the contract or the person using the phrase. It is often confused with the term 'provisional sums' (qv) and the phrases 'PC sums' and 'provisional sums' are used indiscriminately. A prime cost sum is a sum of money included in a contract, usually by means of an item in the bills of quantities (qv), to be expended on materials or goods from suppliers or on work to be carried out by subcontractors nominated by the employer. The contractor has to add his required profit to this sum at tender stage. By definition a prime cost sum should be a specific and accurately known amount and should be obtained as a result of a direct quotation or tender from the supplier or sub-contractor concerned. The reason for confusion with a provisional sum becomes clear when it is appreciated that, in practice, a PC sum is seldom put in the bills as a precise amount. Thus, a figure of £468.50 is obtained from the supplier and a figure of £500.00 is put in the bills 'to allow for increases for various reasons'. The additional £31.50 is, in effect, a small contingency sum. Alternatively, a PC sum is inserted before quotations have been invited. The contractor's profit is calculated on the bill sum (i.e. £500.00) and must be adjusted when the final supply sum is known. Where bills of quantities are based upon the Standard Method of Measurement (qv), 7th edition, 'prime cost sum' is referred to but nowhere defined.

GC/Works/1 (1998) mentions PC items in clause 63 and in clause 63 (3), the sum to be paid to the contractor for those items is the actual cost to the contractor of any payments properly due to the relevant nominated sub-contractor or supplier and includes the actual cost to the contractor of any other incidental packing, carriage or delivery to site. GC/Works/1 (1998), clause 63 (4) entitles the contractor then to add a pro rata allowance for profit. JCT 98 mentions PC sums principally in clause 30.6.2.

Principal contractor A competent firm, individual or partnership appointed under and for the purposes of the Construction (Design and Management) Regulations 1994 and whose trade or business involves carrying out or managing, or arranging for others to carry out or manage, construction work on the project to which he is appointed. Commonly this will be the main contractor engaged for the building works, but of overriding importance is the need to ensure that the appointee has sufficient knowledge, ability and resources to fulfil the duties ascribed to the principal contractor under the CDM Regulations. Those duties include:

— Development, from the earliest possible opportunity, of the health and safety plan (qv) so that it incorporates, for example:
- the approaches to be adopted for managing health and safety during construction.
- assessments made under other relevant legislation such as the Management of Health and Safety at Work Regulations.
- common arrangements for emergency and welfare procedures
- arrangements for fulfilling those other duties of the principal contractor under regulations 16 and 18 of the CDM regulations
- reasonable, flexible arrangements for monitoring compliance with and managing health and safety.

— Taking reasonable steps to ensure co-operation between all contractors to enable compliance with relevant statutory provisions.

— Ensuring that all contractors and employees comply with the health and safety plan (qv).

— Ensuring that unauthorised persons are excluded from the construction site.

— Ensuring that notification given under Regulation 7 is available to be read by all interested parties.

— Promptly providing the planning supervisor with information that he does not already have but which the principal contractor has and which he might reasonably expect should, by virtue of Regulation 14, be included in the health and safety file.

In addition to any contractual provisions empowering him to do so, the principal contractor also has power, under the regulations, to give reasonable directions to any contractor engaged on the project as may be necessary to enable him to fulfil those and his other duties. He may also include in the health and safety plan such rules for the management of the construction work as are reasonably necessary to ensure health and safety.

Appointment of principal contractor should be made as soon as practicable before management of the construction and the construction phase of the project begins and, critically, sufficiently early to enable timely and proper development of the health and safety plan in conformance with Regulation 15 (4) of the Regulations. However, notwithstanding that urgency, no appointment should be made unless and until the client has made sufficient checks and/ or has sought sufficient advice to satisfy himself that the proposed principal contractor has sufficient competence to properly fulfil the duties required of him. Such checks should include establishing that the proposed appointee has, for example:

— Sufficient numbers of properly skilled, knowledgeable, experienced and trained people to carry out or manage the works.
— Allocated sufficient time to carry out and complete the various stages of construction work without risk to health and safety, including fire safety.
— An employment policy that conforms with health and safety law.
— Sufficient technical and managerial expertise to deal with the risks to health and safety specified in the health and safety plan.
— Competent procedures in place to deal with high risk areas identified by the designers and/or planning supervisor.
— Procedures in place for ongoing monitoring of compliance with health and safety legislation.

Once appointed, the position of principal contractor must remain filled continuously until such time as the construction phase is completed.

Priority of documents Standard form contracts often contain an express term dealing with the priority to be given to the various contract documents. In the absence of such a term, where there is a contract in printed form with handwritten or typewritten insertions, additions or amendments which are inconsistent with the printed words, the written words prevail[550].

This sensible rule can be, and in JCT 98 and in IFC 98 is, overridden. In JCT 98, clause 2.2.1 and IFC 98, clause 1.3, it is stated that 'nothing contained in the Contract Bills, (or, in the case of IFC 98, the Specification/Schedules of Work/Contract Bills), shall override or modify the application or interpretation of that which is contained in the Articles of Agreement, the Conditions, (in the case of IFC 98, the Supplemental Conditions), or the Appendix'. This clear wording means that specially written clauses in the bills, e.g. dealing with insurance, will not prevail if they conflict with the wording of the printed form[551]. It is to be noted, however, that the written clause must conflict with the

---

[550]*Robertson* v. *French* (1803) 102 ER 779.
[551]See, for example, *M.J. Gleeson (Contractors) Ltd* v. *London Borough of Hillingdon* (1970) 215 EG 165.

Conditions or Articles etc. and not merely amount to an explanation of, or extension to, them. By way of example; a provision in the bills of quantities requiring compliance with an agreed programme may not be seen as an attempt to override, modify or otherwise affect the printed terms of the contract[552].

GC/Works/1 (1998), clause 2 governs the priority of contract documents and also accords them an artificial order of precedence.

More sensibly, ACA 3, clause 1.3 provides that 'the provisions of this Form of Agreement shall prevail over the provisions contained in any other of the Contract Documents save only the following provisions which shall prevail over anything contained in this Form of Agreement'. The parties can then expressly afford the documents whatever alternative order of precedence they wish.

See also: **Interpretation of contracts.**

Private Finance Initiative (PFI) Introduced in 1992, PFI is not a procurement system. It is a procedure based upon the idea that the private sector should be involved in providing and operating various assets which might otherwise never have been started but with the eventual aim being that the project concerned will return to the public sector. The idea has much to commend it, but despite its obvious advantages there are also many complications.

On the plus side, private finance invested in the public sector introduces a high level of technical, managerial and financial skills and experience whilst at the same time a construction company willing to engage in such a project might actually be in the position of creating its own workload. However, it is usual for a special purpose vehicle (SPV) to be set up for the express purpose of obtaining finance and carrying out the project. More often than not it is a joint venture company between the finance providers and the building contractor. To ensure the SPV secures a satisfactory return on investment the agreements with central or local government are normally for periods of as much as 25 years. While that apparently ensures sufficient time to make a substantial profit, the long time period places a high level of risk on the SPV which more often than not must enter into several undertakings about the services to be provided. The system is not yet fully proven and there are still misgivings among some construction companies who have indicated that they have had to bear most of the losses.

Projects that the government has said would be suitable for PFI schemes include hospitals, prisons, public sector offices, types of housing, roads and railways. It is a development that certainly has a future and architects must have a thorough grasp of the implications[553].

Privilege A rule of evidence (qv) whereby a witness may refuse to answer certain questions or the parties may legitimately refuse to produce certain classes of document.

---

[552]*Moody* v. *Ellis* (1983) 26 BLR 39 CA.
[553]For useful further information, see: *The Private Initiative: the Essential Guide* (1996) RICS Business Services.

In the law of defamation (qv) privilege refers to a defence, e.g. statements made by witnesses in judicial proceedings are absolutely privileged so that the person making them is not liable in defamation. Other statements may be privileged to a lesser extent (generally termed 'qualified privilege'), provided that the contents are honestly believed to be true by the writer and there is an absence of malice. Communications between client and professional adviser may also be privileged and it is generally accepted that communications between a client and his solicitor enjoy absolute privilege for obvious reasons. If a defendant (qv) seeks to use some part of his own privileged document in evidence, he will be deemed to have waived his privilege in the whole document so far as it relates to the same subject matter[554].

See also: **Without prejudice.**

Privity of contract A rule of English law which means that only the actual parties to a contract can acquire rights and liabilities under it[555]. The rule applies even though the contract itself provides that a third-party shall be entitled to sue[556].

The doctrine of privity of contract is subject to several common law exceptions. For example:
— The covenants (qv) in a lease are normally binding not only on the original parties but on their successors in title.
— A husband who insures his life in favour of his wife or children may, under statute, create enforceable rights in them.
— Agency (qv).

Privity of contract is an inconvenient notion in modern commercial practice. Commonly a contractual relationship between two parties involves a series of other linked transactions, yet the law generally treats each link as an entirely separate relationship. The reality of that situation is now recognised with the enactment of the Contracts (Rights of Third Parties) Act 1999 (qv) which came into force in November 1999. The doctrine of privity, so far as it relates to England, Wales and Northern Ireland, is now significantly modified to the extent that, subject to the provisions of that Act, a third party who is not a party to a contract may in his own right enforce a term of the contract if the contract makes express provision to that effect or if one or more terms of the contract purport to confer such a benefit.

In building contracts, unless the Contracts (Rights of Third Parties) Act 1999 has effect to alter the position, the practical consequences of the doctrine of privity are two-fold:
— The main contractor carries responsibility for a sub-contractor's work etc. so far as the employer is concerned.
— The employer cannot sue the sub-contractor directly in contract, unless there is a separate direct contract between them, e.g. as where JCT Agreement NSC/2 is signed.

---

[554]*Great Atlantic Insurance Co* v. *Home Insurance Co* (1981) 2 All ER 485.
[555]*Dunlop* v. *Selfridge* [1915] AC 847.
[556]*Tweddle* v. *Atkinson* (1861) 1 B & S 393.

As there is no direct contractual relationship between sub-contractor and employer, neither can sue the other in contract, although a breach of the sub-contract may, at the same time, amount to a tort (qv), in which case the employer may be able to sue the sub-contractor in tort, e.g. for negligence. However, recent cases have severely restricted if not altogether eliminated that possibility.

Procurable The normal meaning is 'obtainable'. Clause 8.1 of the JCT 98 form provides an important qualification of the contractor's obligations under 2.1. It provides that materials, goods and workmanship must be of the respective kinds and standards described in the contract bills (qv) only so far as procurable. Therefore, if he is unable to obtain goods etc. of the required kinds or standards, his obligation appears to end. It should be noted that the materials etc. must be unobtainable and not simply more difficult or costly to obtain than previously, or reasonably, expected. Thus, unless expressly qualified as such, an obligation to procure appropriate materials and goods is not to be taken as meaning simply that one must use one's reasonable — or even one's best — endeavours. The goods or materials in question must be unavailable on the open market.

On what is to happen then, the contract is silent. In practical terms, much will depend upon circumstances. If, for example, the contractor cannot obtain the kind of materials for which he tendered because he was late in placing his order, the onus is probably on him to offer an alternative of at least equal standard. Similarly, if the materials were unobtainable at the time he offered to provide them. If, however, the materials become unobtainable through no fault of the contractor, it is suggested that the architect will be obliged to issue an instruction (qv) varying the materials, with consequent adjustment to the contract sum (qv).

Architects will be prudent to make full enquiries before accepting that materials are not procurable. It is not unknown for a contractor to plead this clause because the materials are either more expensive than he anticipated or more difficult to obtain. Neither situation falls under clause 8.1.

Productivity payments Sometimes known as 'bonus payments' or 'incentive schemes'. They are paid to operatives by contractors to encourage rapid completion of work. In practice, every operative on site expects to receive a bonus and haggling over payments is a major source of grievance. Many contractors agree special bonus schemes with their workpeople (qv), but schemes which are in accordance with the rules of the Construction Industry Joint Council or other wage fixing body rank for inclusion in fluctuations payments under JCT 98, clause 38.1.1.4.

Programme A schedule or chart showing stages in a scheme of work. JCT 98, clause 5.3.1.2 makes reference to a master programme (qv) but it is not a contract document (qv) and footnote [r] to clause 5.3.1.2 provides that this clause should be deleted if a master programme is not required. If the clause is retained and a master programme is required, then as and when from time to time it is subject to change following the granting of an extension of time or where the date for completion is changed by a clause 13A quotation, it should be updated.

The contractor may produce many subsidiary programmes during the course of a contract to assist him to plan the work efficiently. The architect will also produce programmes, particularly at the commencement of the design stage, to help him organise the design team. Project planning and programming are being developed in ever more sophisticated ways using computer software and modelling. Such programming is now used as an aide not only in organising and monitoring the efficiency of the basic construction operations but also as an aide in many other aspects of successful project management, such as cost/value and cash flow forecasting and monitoring, substantiation of financial claims, the distribution and allocation of off-site plant and overhead resources, tendering and health and safety management, extensions of time etc.

Among the more popular forms of programme are:
— Network analysis and critical path.
— Precedence diagrams.
— Bar (Gantt) charts.
— Advancing fronts.

Each method has its own particular advantages depending upon the type of job and the people for whom it is intended.

GC/Works/1 (1998), clause 33 is very specific in its requirements. The contractor cannot merely provide some theoretical programme which he later may choose to ignore or manipulate depending on his future intentions, actual progress or claims potential. Under this form of contract the contractor 'warrants' that his programme:
— Shows the sequence of operations that he does, in fact, propose to follow.
— Gives details of any temporary work, methods of work, labour and plant that he intends using.
— Shows all critical activities.
— Is in all respects achievable.
— Conforms to the contract.
— Provides reasonable time for the ongoing preparation and release of information by the design team.
— Is drafted in such a way as to allow for effective monitoring of the progress of the works.

Similarly, under the NEC (clause 31), provision of a detailed programme by a stipulated date is an express requirement and, as a minimum requirement, any such programme submitted for acceptance by the project manager must contain details of:
— Start date.
— Method statements and details of the resources necessary for each operation.
— Operations the contractor proposes to undertake and their order and timing.

— Any operations the employer or others employed by him will undertake and their timing.
— Dates when the contractor requires to be provided with anything to be provided by the employer.
— Proposed date(s) for completion of sections or phases of the work to facilitate work by the employer

and provision for and details of:

— Float.
— Time risk allowances.
— Health and safety requirements.
— Any particular procedures set out in the contract.
— All other information specifically referred to in the works information as being required to be shown.

This programme must also be revised from time to time throughout the project, as and when necessary in accordance with the contract.

See also: **Master programme; Accepted programme.**

Progress meeting A meeting required under the provisions of GC/Works/1 (1998), clause 35. The contractor's agent must attend regular progress meetings to assess the progress of the Works. Subject to contrary instructions, they must be held at intervals of not less than one month. The time and place is to be specified by the PM. Not less than five days before each meeting, the contractor must submit a written detailed progress report which must include details of requests for information, causes of delays, extensions of time requested and setting out proposals to ensure prompt completion. Clauses 3 (a) to (e) list the specific matters that must be contained in the report.

Following the meeting, the project manager must – within seven days – provide the contractor with a written statement giving his opinion on whether the works are delayed, early or on time, and setting out also:

— Matters he considers will or may in future delay completion or cause extra cost.
— Estimates of the likely extra cost, if any, that may be caused.
— The steps considered necessary to avoid or reduce such delay or extra cost.
— The situation regarding any claims for extension of time and any awards made in that respect.
— Responses to any outstanding requests for information, drawings and the like.

Progress meetings are commonly held no matter what form of contract is used but, unless properly structured and chaired, it is doubtful whether they serve any really useful purpose. They tie up key personnel for anything up to half a day and generally produce nothing which could not be produced by other less labour intensive means.

Project management An extremely popular but nevertheless very loose term referring to the management of a building project.

A project manager may be appointed by the employer to co-ordinate the entire job from inception to completion. His relationship with the other professionals must be clearly set out and their respective powers and responsibilities established. Since practice varies from contract to contract, it is impossible to define the role of the project manager precisely. He may be appointed to take over the whole of the architect's traditional management and co-ordinating functions together with those of the main contractor. The concept is still in the process of evolution.

The supporters of project management suggest that it provides an efficient and cost effective method of producing a building. Opponents believe that it fragments existing responsibilities and fails to achieve any improvement in timing and cost. Project managers can be architects, engineers, quantity surveyors, surveyors or managers specialising in the building field.

See also: **Client's representative.**

Project manager Referred to specifically in GC/Works/1 (1998), where clause 1 (1) defines the project manager (PM) as 'the person employed in that capacity named in the abstract of particulars and appointed by the employer to act on his behalf in carrying out those duties described in the contract (subject to the exclusions set out in the abstract of particulars) or such other person as may be appointed in that capacity for the time being by the employer'. Hence, under GC/Works/1 (1998), subject to any specific exclusions set out in the abstract of particulars, extensive powers of delegation are conferred on the PM by clause 4. Even in the case of excluded matters, all decisions that are to be communicated to the contractor will nevertheless come through the project manager. The project manager also has wide discretion to delegate his powers to other named representatives provided he does so expressly and in writing to the contractor.

The appointment of a project manager is also a feature of the NEC. Again he is given extensive powers and responsibilities in relation to, among other things, the acceptance or rejection of the contractor's proposed programme, receiving and replying to communications under the contract, consideration and acceptance or rejection of claims for extension of time or additional compensation, instructions in relation to stopping, starting or acceleration of any work, testing and inspection of the works, assessment of sums due to the contractor by way of interim and final payments, the issue of payment and other certificates and the issuing of instructions changing the works information etc.

Prolongation claim A claim made by the contractor for financial reimbursement because the contract period has been extended as a result of the default of the employer. It is expressly mentioned in clause 46 of GC/Works/1 (1998) but not in other standard forms. Contractors commonly refer to all claims for loss and/or expense (qv) as 'prolongation claims.' This is misleading. It implies that either every extension of the contract period carries an automatic claim for reimbursement or that a financial claim cannot be made unless an overrun of the contract period has occurred. Both of these implications are wrong.

See also: **Claims.**

Proof of evidence A written statement of what a witness of fact will say. It is produced mainly for the benefit of counsel who will use it to examine a witness before a court or arbitration hearing and to assist in cross-examining witnesses for the other party. In Scotland it is referred to as a 'Precognition'.

The proof is usually written after discussion with the solicitor with responsibility for the conduct of the case concerned since it is he, after consultation with counsel, who will decide what is and what is not important to be included in the statement. Where the case involves opinion evidence from an expert witness the expert will receive formal instructions concerning the matters on which he is required to give evidence. He will then be expected to prepare his statement and report entirely independently so that the opinions that he holds are most accurately represented and expressed in his own way.

See: **Expert witness.**

Property In legal terms, 'property' denotes something capable of being owned. Property is divided into two sorts (see **Figure 17**): real and personal, very roughly land and moveable goods respectively.

See also: **Bailment; Chattels; Corporeal property; Hire; Incorporeal hereditaments; Lien; Personal property; Real property.**

Provisional quantities In otherwise accurately measured bills of quantities (qv) it is common to find some quantities noted as 'provisional'. They usually refer to items which are unknown or uncertain in extent at the billing stage and should not be confused with bills of quantities or items that describe 'approximate quantities'. Provisional quantities may, for example, be given with regard to items of work such as substructure or drainage where the extent of the work that will have to be done simply cannot be properly or even reasonably accurately measured.

It is not uncommon for the quantity surveyor to include items in bills of quantities for the excavation of rock or running sand or for the necessity to excavate below the water table. The quantity is only given as an estimate. As the work proceeds it is re-measured at the rate(s) the contractor has inserted against the item in the bill of quantities. Provisional quantities are also taken for such things as cutting holes through walls and floors for plumbing and other services. They are often taken from a schedule supplied by the specialist concerned and are commonly referred to as 'builder's work'.

See also: **Approximate quantities.**

Provisional sum A term used to denote a sum of money included in the contract by the employer, normally as an amount in the bills of quantities (qv). It is provided to cover the cost of something which cannot be entirely foreseen or detailed accurately at the time tenders are invited. For example, the architect may know that he requires a retaining wall to be constructed, but does not know accurate dimensions or details. He may ask the quantity surveyor to make an estimate (qv) of the likely cost and insert that sum in the bills at tender

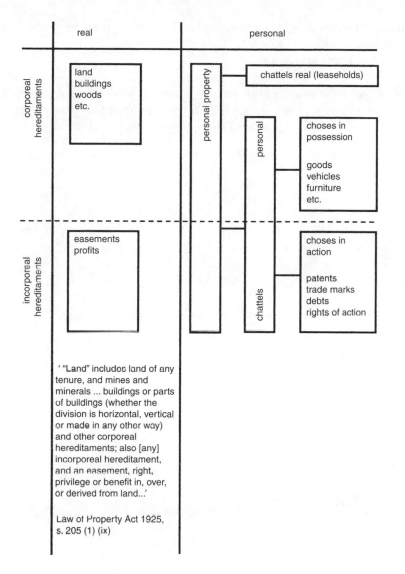

Figure 17 Division of property.

stage. During the progress of the contract, the architect may issue an instruction, together with full details of the wall, to the contractor. At final account (qv) stage, the quantity surveyor deducts the provisional sum from the contract sum (qv) and adds back the value of the retaining wall ascertained in accordance with the contract provisions for the valuation of variations (qv). JCT 80, clause 1.3 has since 1988 contained a definition of 'provisional sum' as one including a sum provided for work whether or not identified as being for defined or undefined work. Where bills of quantities are based upon the

Standard Method of Measurement (qv), 7th edition, the term 'provisional sum' is defined in General Rule 10.
See also: **Prime cost sums.**

Proxy A person authorised to act on behalf of someone else, e.g. a lawfully appointed agent, or in company law a person appointed to represent and vote for another at meetings as well as the formal document of appointment.
See: **Agency.**

Quality of work The standard or degree of excellence. A term used in IFC 98, clause 1.2 and in JCT 98, clause 14.1 where the quality of work for which the contractor is deemed to have made allowance in the contract sum is that described in the contract bills or, where no bills have been prepared, as described in the specification or schedules of work. Where neither bills, specifications or schedules of work are prepared as contract documents, the quality for which allowance is made in the contract price is to be discerned from a review of all of the relevant contract documents, including any relevant contract drawings. IFC 98 provides that, in the case of inconsistency in description between such documents, the contract drawings will prevail.

The quality of work is the subject of much argument on site and it is notoriously difficult to specify a quality to any fine degree. The use of British Standards, Codes of Practices, standard specification clauses and the definition of tolerances is all helpful, but the contractor will usually base his pricing on his knowledge of the architect and employer. Where quality is specified precisely, the contractor is bound to provide materials and workmanship to that quality, but not, it should be noted, above the quality described. GC/Works/1 (1998), clause 3.1 is extensive and quite specific in its terms and, *inter alia*, requires the contractor to undertake a form of quality testing and to warn the PM of any goods or materials which he considers are unsuitable for incorporation. Under JCT contracts it is expressly provided that, where the question whether the quality adopted by the contractor is suitable is a matter of opinion to be decided by the architect, that opinion will be a reasonable one.

Quantity surveyor A professional person whose expertise lies mainly in the fields of the measurement and valuation of building and civil engineering work and cost advice. Most standard forms of building contract make reference to the quantity surveyor, generally with regard to the valuation of work in progress, variations, financial claims and the preparation of the final account. Under the JCT 98 the quantity surveyor's powers include receiving and consulting with and advising the architect upon any priced statement (qv) or quotation given by the contractor in relation to variation works pursuant to alternative A of clause 13.4 or clause 13A of the contract.

ACA 3 contains an optional provision (in clause 15.2) allowing for the appointment of a quantity surveyor. Under the rather unusual provisions of MW 98, the appointment of a quantity surveyor is expressly dealt with but once appointed the contract makes no express reference to the specific duties the quantity surveyor will then be expected to undertake.

Long before a contract is placed, the quantity surveyor will be involved in advising the employer and the architect on probable costs of the completed building. He can produce a cost plan which is a highly sophisticated method of controlling costs throughout the design and development stage. If the work is of sufficient size, he will also produce bills of quantities (qv) or such other document for pricing which he will advise is necessary in the particular circumstances. He will generally carry out all the negotiations with the contractor which have cost implications. The quantity surveyor is normally appointed by the employer, sometimes on the advice of the architect, and he will usually be a member of the Royal Institution of Chartered Surveyors, which issues a code of conduct and recommended fee scales for its members.

Although in practice the quantity surveyor will often take responsibility for the ascertainment of any amounts due to the contractor in respect of claims for loss and expense, under the terms of the JCT 98 and IFC 98 that duty falls, in the first place, to the architect. It will only properly become the responsibility of the quantity surveyor to carry out such ascertainment as and when he is instructed to do so by the architect. It is to be noted that even if so instructed, it is not a matter for the quantity surveyor to decide whether the contractor has an entitlement. His function is simply that of ascertaining the amount of loss and expense due.

Quantity surveyors are now also employed within contractors' firms where they may specialise in estimating, claims preparation and measurement of work in progress.

Quantum meruit 'As much as he has deserved' – a reasonable sum. This Latin phrase is often used as a synonym for *quantum valebat* (qv) which means 'as much as it is worth'. More properly it refers to the doctrine whereby the contractor becomes entitled to payment of 'as much as he has earned' by virtue of an implied obligation on the part of the employer to pay a fair remuneration according to the extent and quality of the work done. It is the measure of payment where the contract has not fixed a price or where, for some reason or another, the contract price is no longer applicable. At common law there are four situations in which a *quantum meruit* claim may apply:

— Where work has been done under a contract without any express agreement as to price.
— Where there is an express agreement to pay a 'reasonable price' or a 'reasonable sum'.
— Where work is done under a contract which both parties believed to be valid at the time but which is in fact void (qv).
— Where work is done at the request of one party but without an express contract, e.g. work done pursuant to a letter of intent (qv). This is a claim in quasi-contract (qv) or restitution since, 'In most cases where work is done pursuant to a request contained in a letter of intent, it will not matter whether a contract did or did not come into existence; because if the party who has acted on the request is simply claiming payment, his claim will usually be based upon a *quantum meruit*, and it will make no difference whether that claim is contractual or quasi-contractual. A *quantum meruit* claim ... straddles the boundaries of what we now call contract and restitution ...'[557].

---

[557]*British Steel Corporation* v. *Cleveland Bridge & Engineering Co Ltd* (1981) 24 BLR 94.

If extra work is done completely outside the contract then payment on a *quantum meruit* may be implied[558] but this is very rare. Many contractors erroneously assume that they are entitled to claim on a *quantum meruit* basis merely because they are losing money but in fact such a claim will only lie, if at all, where what the contractor does is substantially different from what he undertook to do. *Quantum meruit* is usually taken as a 'fair commercial rate'[559].

Where a valid contract exists for the supply of goods or the supply of services but the parties have neglected to determine the price or charge and have failed to provide any means by which the price or charge may be arrived at, a statutory right to a reasonable price or charge is provided for under s. 8.-(2) of the Sale of Goods Act 1979 and s. 15.-(1) of the Supply of Goods and Services Acts respectively. Strictly speaking, this is not to be taken as one and the same as *quantum meruit* but for all practical purposes they may amount to one and the same.

Quantum valebat A concept related to *quantum meruit* (qv), which requires the payment of a fair price for goods or materials supplied rather than for work done. Although technically different, it is usually subsumed within a *quantum meruit* claim and no distinction is drawn.

Quasi-contract A term used where parties' conduct gives rise to an implied contract such as, for example, where there is an implied promise by one party to pay the other a *quantum meruit/quantum valebat* (qvv) in return for something done or given by the other or, where money is received for the use of someone else. The underlying principle is that by finding the existence of a quasi (or implied) contract, unjust enrichment or unjust benefit can be avoided. The aim is 'to prevent a man from retaining the money of, or some benefit derived from, another which it is against conscience that he should keep'[560].

In the context of the construction industry, the most common instance is that of a *quantum meruit* (qv) claim for work done or services rendered. Under the provisions of the Law Reform (Frustrated Contracts) Act 1943, where a contract is frustrated money paid under the contract may be recovered, subject to a claim for set-off (qv) for expenses incurred by the recipient of the payment. If a partnership (qv) is determined prematurely, the court has power to order the full or partial return of any premium paid by a partner for admission to the firm. See also: **Frustration.**

Queen's (King's) Counsel Senior barristers (qv) who have been appointed counsel to Her Majesty on the recommendation of the Lord Chancellor. Also called 'leading counsel' or 'silks' because they wear silk gowns.

Queen's (King's) enemies A traditional term used in contracts to refer to enemies of the state. It was formerly found in GC/Works/1, where 'King's

---

[558] *Sir Lindsay Parkinson & Co Ltd* v. *Commissioner of Works* [1950] 1 All ER 208.
[559] *Laserbore* v. *Morrison Biggs Wall* (1993) CILL 896.
[560] *Fibrosa Spolka Akcyjna* v. *Fairbairn Lawson Combe Barbour Ltd* [1943] 2 All ER 122.

enemy risks' was listed among the 'accepted risks' for insurance purposes, defining that expression by reference to s. 15.-(1) (a) of the War Risks Insurance Act 1939 where the definition is 'risks arising from the action taken by an enemy or in repelling an imagined attack by an enemy, as the Board of Trade may by order define'. The term has fallen out of use and no longer appears in that contract.

Quotation A price given usually in the form of an offer (qv) for the carrying out of work or the supply of materials or both. It is normally expected to be a precise figure, capable of acceptance (qv) so as to form a binding contract. See also: **Estimate; Offer; Tender.**

Rates A local tax assessed on and made payable by a business in respect of its occupation of land or buildings. JCT 98, clause 6.2 requires the contractor to pay 'any fees or charges (including any rates or taxes) legally demandable' and these are to be added to the contract sum. The substantive law on the matter of rating is to be found in the Local Government and Finance Act 1988. Whether or not a contractor's temporary site accommodation, stores, welfare facilities and the like are to be treated as attracting business rates will be a question of interpretation in each particular case and in that regard the local valuation office with responsibility for the area concerned inevitably has to exercise a degree of discretion. In doing so a number of factors will be considered. Among them will be matters such as whether there is likely to be a sufficient degree of permanency of the proposed accommodation, whether the huts, stores and the like are in actual possession of the contractor, whether they are of benefit and value to the contractor and whether they are for the exclusive use of the contractor.

Ratio decidendi The principle of law on which a judicial decision is based. It is the reason or ground for the decision and makes a precedent (qv) for the future. For the purpose of the doctrine of precedent it is the *ratio* which is the vital element in the decision. Not every statement of law made by a judge in the course of his judgment is part of the *ratio*. It must be distinguished from *obiter dictum* (qv) which is a statement made 'by the way' and which is not necessary for the decision. In general, the *ratio* of a case will be the statement of the principles of law which apply to the legal problem disclosed by the facts before the court. The area is fraught with difficulty because:
— A judge does not usually state that a particular statement is the *ratio*.
— A judge may give what may appear to be alternative *rationes decidendi*.
— A later court may distinguish the precedent.
— Even if the facts found in an earlier case appear identical with those in a later case the judge in the later case may draw a different inference from them[561].
— Where the court consists of more than one judge, i.e. the Court of Appeal or House of Lords, the result may be agreed unanimously but each judge may have differing reasons for arriving at any particular conclusion.

Real property Most legal systems recognise a distinction between land, which is immovable and as a general rule indestructible, and other pieces of property such as cars, books or clothes. In England, for historical reasons ownership may exist in respect of both real and personal property (qv). Real property

---

[561] *Qualcast (Wolverhampton) Ltd* v. *Haynes* [1959] 2 All ER 38.

(reality) is, broadly speaking, a freehold estate or interest in land. In law, 'land' has a very wide definition; it includes not only the actual soil itself but all the things growing upon it or permanently attached to it, as well as rights over it. This has important consequences in building contracts because as goods and materials are incorporated into the building they generally cease to be personal property and become part of the land.

Real property is a term which is applied solely to interest in land. Interests under leases – leaseholds – are 'interests in land' in one sense, but for historical reasons are classed as personal property (qv). They occupy an anomalous position and are technically known as chattels real (qv). **Figure 17** shows the position in diagram form.

Reasonable Although a term widely used in a variety of contract clauses – where, for example, work done or material provided must be to the architect's 'reasonable satisfaction', instructions must be complied with, defects rectified or breaches remedied within a 'reasonable time' or valuations must be based on rates and prices that are 'fair and reasonable' – the meaning is virtually impossible to define satisfactorily. What is reasonable in one case will almost certainly be considered unreasonable in another. However, that is not to say that reasonableness is to be measured wholly subjectively. Whether some action or opinion is reasonable must be measured against what the man in the street (or the man on the top of the Clapham omnibus) would think of as reasonable. Thus, when assessing whether some act or opinion is reasonable, the circumstances must also be considered with a degree of objectivity. In the context of building contracts, the question whether something is or is not reasonable is probably best answered, therefore, by considering what the average architect, engineer, project manager, builder or the like would consider reasonable having regard to the particular facts and circumstances under which the parties found themselves in the specific case in point. As a general rule where the contractor has to satisfy clearly specified standards and is also expressly required to supply work and materials that are to the reasonable satisfaction of the architect, the architect's insistence upon full compliance with the express terms of the specification will not have been construed as unreasonable.

Where a party is compelled to carry out some act or perform a task to the extent that it is 'reasonably practicable' to do so, then the addition of the qualification of reasonableness may allow matters such as financial considerations to be taken into account where they might otherwise not be relevant[562].

Further guidelines concerning what may be considered reasonable are set out in the Unfair Contract Terms Act 1977 (qv). Although those are valid for that Act alone, they may be a useful indicator of the statutory position in a particular case. However, in the context of construction contracts the question whether the action taken or opinion formed etc. is or is not reasonable ultimately will be one for the court or arbitrator to decide in all the circumstances.

---

[562]*Jordan* v. *Norfolk County Council and Another* [1994] 4 All ER 218.

Reasonable time What is a reasonable time will depend upon the circumstances of each particular case. It is a favourite expression in contracts when it is impossible to set down exactly how much time is intended. It might well be equated with 'appropriate time' in some cases. JCT 98, clause 17.2 lays down that defects etc. shall be made good by the contractor 'within a reasonable time after receipt of such schedule.' In fact a reasonable time could be a week, in the case of a very small job with few defects, or many months if the job is large. The question what is or is not a reasonable time essentially must therefore be concerned with the particular circumstances of each case. It is not a question of what – taken in the abstract – would be the 'ordinary time' to be expected for compliance or performance. Rather, the test is one which should have regard to all of the prevailing circumstances that existed at the time compliance or performance was to take place, although circumstances caused by or contributed to by the party whose performance is under review must, for obvious reasons, be ignored. What is clearly intended is that the contractor will organise himself so as to make steady progress in completing the work. He should not, for example, start work to rectify a defect and then stop for a week, before starting again, etc.

ACA 3 attempts to clarify the matter by stating precise times as often as possible. It does not state that the times are reasonable and, indeed, in some cases they appear to be unreasonable, e.g. five working days to agree to the contractor's estimates (clause 17.3).
See also: **Reasonable.**

Receiver A person appointed under a security, generally held by way of a fixed or floating charge over a debtor's assets, when the debtor defaults on payment of a debt. By way of example: where an overdraft facility is given, a receiver may be appointed by the lending bank when the borrower later fails to repay the amount borrowed under the overdraft. Although appointment of a receiver is a procedure aimed primarily at the enforcement of a security and involves the receiver in obtaining the best possible price for the charged assets for the principal benefit of the charge holder, receivership can sometimes result in the ailing business being rescued where, by selling the business as a going concern, the recovery to the lender is improved.

The term Official Receiver refers to a receiver appointed by the court and he is usually an officer of the court. An official receiver is a civil servant appointed as an interim measure, in bankruptcy, until a trustee in bankruptcy has been appointed.
See also: **Bankruptcy; Insolvency; Liquidation.**

Recitals The introductory paragraphs or statements in a contract which recite (or describe briefly) the background to the agreement and briefly set out in general terms its aims and purposes. They set the agreement in context. It is against the background of those 'narrative recitals' and in the context of those purposes outlined in the 'introductory recitals' that the subsequent operative clauses of the agreement should then be read and construed. The operative clauses prevail over the recitals, but if the operative clauses of the deed are ambiguous the recitals may be an aid to interpretation of the operative clauses. Recitals usually begin 'Whereas'.

In standard form building contracts they may be of great importance, particularly as regards description of the works or the site (qv). For example, in JCT 98 and the Agreement for Minor Building Works (MW 98), the recitals are the only place in the contract where the exact nature of the work to be undertaken by the contractor is specified.

Rectification A discretionary remedy (qv) whereby the court – or, if expressly so empowered, an arbitrator (e.g. JCT 98, clause 41B.2) – can order the correction of errors in a written contract. It is a remedy rarely granted.

The House of Lords described the remedy as one available 'where parties to a contract, intending to reproduce in a more formal document the terms of an agreement upon which they are already *ad idem*, use, in that document, words which are inapt to record the true agreement reached between them. The formal document may then be rectified so as to conform with the true agreement which it was intended to reproduce, and enforced in its rectified form.'[563]

Rectification will only be ordered where the written document fails to represent what the parties agreed. It will not be ordered where the document fails to represent what they intended to agree. It must be shown that the parties were in complete agreement on the terms of the contract, but by an error wrote them down wrongly.

See also: **Clerical errors; Errors.**

Re-examination The final stage in the examination of witnesses in judicial or arbitral proceedings. Following cross-examination (qv) the witness may be re-examined by or on behalf of the party calling him with the object of reinstating any of the witness's testimony that has been shaken in cross-examination. Leading questions may not be asked and new matters cannot generally be raised.

See also: **Examination-in-chief; Witness.**

Reference The proceedings before an arbitrator (qv) and so the 'costs of the reference' mean the costs incurred by the parties in the conduct of the proceedings as opposed to the costs of the award (qv), which are the arbitrator's fees and expenses. The same term is used for a written testimonial about someone's character and abilities.

Referral A document prepared by the party initiating adjudication proceedings, whereby he sets out his case and the remedies sought in relation to the matters on which he requires the adjudicator's decision. It should provide the adjudicator with a brief, but nevertheless clear understanding of the background to the contract generally and, more specifically, must clearly and

---

[563] *American Airlines Inc* v. *Hope* [1974] 2 Lloyds Rep 301.

succinctly set out all of the matters that the referring party wishes the adjudicator to consider in support of his claims. In particular it should provide details of the material facts and terms of the contract specifically relied upon to demonstrate the breach(es) and entitlement(s) being contended for. It, and any further documents relied upon to support those allegations and claims, should be received by the adjudicator and simultaneously copied to the other party within seven days of the initial notice requiring the dispute to be adjudicated.

Given the relatively short time within which it must be prepared, and given the inevitably quick and somewhat informal nature of adjudication, little or no opportunity exists to perfect an incomplete or ill prepared referral. Hence, careful attention should be given at the outset to its drafting and in the majority of cases appropriate professional advice and assistance should be sought to ensure that it is properly and comprehensively prepared and supported.

Registered office Every company, private or public, must have an office registered with the Registrar of Companies. The office need not be, and often is not, the normal place at which the company does business. It is often the address of the company's solicitors or accountants, but whatever the address the important thing is that members of the public must have somewhere, not subject to overnight change, to which correspondence may be sent, or where notice of commencement of legal proceedings and the like may be served.

Invariably, where the parties to a building contract are corporate bodies each will give its registered address in the introduction to the agreement. Unless the operative clauses of the contract expressly provide to the contrary, service of any notice, document or the like at that registered office will generally then be taken to be good service on the company concerned. However, most standard form contracts currently in use recognise that the registered office given in the agreement is unlikely to be the company's everyday trading address. In that case, it is now common for contracts such as JCT 98 (clause 1.7), IFC 98 (clause 1.13), GC/Works/1 (1998) (clause 1(3)) and NEC (clause 13.2) to provide for the parties to agree on an alternative convenient business address to be specified for the purposes of receiving such notices and the like. Provided to do so does not contradict any other express notice provision elsewhere in the contract, service to that alternative agreed address will be deemed to be good service.

Under JCT 98, where the contract makes no specific provisions for service, or where the parties have simply neglected to make any agreement over a suitable alternative address for that purpose, the fall back position will generally be that all notice and documents will be effectively served if directed to 'the addressee's last known principal business address or, where the addressee is a body corporate, to the body's registered or principal office' (JCT 98, clause 1.7).

Registrar An officer of the High Court or County Court who is defined as such within the Insolvency Rules 1986[564].
See also: **Courts.**

---

[564]SI 1986/952.

Regular progress A term used in many standard forms of contract to indicate the way in which the work is to be carried out. To amount to 'regular progress' the progress of the work must bear a relationship to the contractual completion date (qv). What is regular progress will depend upon the precise terms and circumstances of the contract.

See also: **Regularly and diligently.**

Regularly and diligently The phrase used in JCT 98 (clause 23.1.1) to describe the contractor's obligation as to progress. Breach of this obligation is a ground for determination under JCT 98 (clause 27.2.1.2) and under IFC 98 (clause 7.2.1 (b)). This phrase probably means more than an express restatement of the contractor's common law obligation as to progress, i.e. it must bear some relationship to the specified date of completion and his progress must be constant, systematic and industrious. Whether or not the contractual standard is achieved is probably best judged objectively having regard to such matters as:

— The number of men retained on site compared to the number one would expect to be needed by a competent and experienced contractor faced with the same relevant commitments.
— The plant and other equipment employed on the work in relation to the type and volume of work to be undertaken in the time available for completion.
— The progress actually being made on site relative to the volume and complexity of the work still to do to achieve contractual completion.

A slow rate of progress judged against the performance of other contractors is an indicator that the contractor is not proceeding 'regularly and diligently' although low productivity on site may well be explained by other factors which are outside the contractor's control. Hence, account must also be taken of any particular extenuating circumstances that are outside the contractor's control which are hindering, or which may in due course hinder or prevent, the desired rate of progress.

There is also at least one line of authority making it clear that merely going slowly is not, in itself, a breach of contract and it will always be a difficult question to decide whether a contractor's progress amounts, in fact, to a failure to proceed regularly and diligently. However, the Court of Appeal has now given some useful guidance on the matter. 'What particularly is supplied by the word "regularly"; is not least a requirement to attend for work on a regular daily basis with sufficient in the way of men, materials and plant to have the physical capacity to progress the work substantially in accordance with the contractual obligations. What in particular the word diligently contributes to the concept is the need to apply that physical capacity industriously and efficiently towards the same end. Taken together the obligation upon the contractor is essentially to proceed continuously, industriously and efficiently with appropriate physical resources so as to progress the works steadily towards completion substantially in accordance with the contractual requirements as to time, sequence and quality of work.'[565]

That view usefully expands on previous judicial opinion which acknowledges that, whereas the words generally convey a sense of activity, of orderly progress, of industry and perseverance, 'such words provide little help on the question of how much activity, progress and so on is to be expected ...'[566].

In this context, a comparison of contracts is notoriously difficult. In GC/Works/1 (1998) clause 34 (1), the contractor's progress obligation is to 'proceed with diligence and in accordance with the Programme (qv) or as may be instructed by the PM', while ACA 3, clause 11 requires him to proceed 'regularly and diligently and in accordance with the Time Schedule'.

The contractor's obligation as to progress is important in relation to claims for extension of time (qv) as well as determination (qv) of employment.

Regulations The term may be used widely to refer to privately imposed restrictions and/or directions or to those restrictions, directions and/or rules imposed by law such as, for example, where they are the result of UK and/or EU legislation. In the former sense, the word is found in GC/Works/1 (1998) clause 22, which compels the contractor to comply with 'the occupier's rules and regulations which have been provided to him or made available to him for inspection, both in respect of the Site and in respect of any larger premises of which the Site forms part'.

In the latter sense, JCT 98, clause 6.1.1 provides that the contractor shall 'comply with and give all notices required by ... any regulation or by law ... of any local authority or any statutory undertaker which has any jurisdiction with regard to the Works ...'. In this sense the regulation(s) envisaged by the contract will amount to a form of delegated legislation which the local authority or other statutory body is empowered by Act of Parliament to make. Thus, albeit indirectly, the regulation concerned will nevertheless have statutory force. Likewise, under EU law certain institutions have similar power to make regulations which, although not necessarily requiring specific implementation by national authorities may nevertheless be regarded as having legal effect and will be binding in the UK.

Reinstatement A word used normally in connection with the insurance provisions of the standard forms (see, for example, JCT 98, clause 22A.1). Reinstatement means the putting back of materials or workmanship in the same state and to the same standard as they were before the need for reinstatement arose. The reinstatement value may well be greater than the straightforward value of works because reinstatement will include all necessary demolition and ancillary work. It is, therefore, important that insurance covers the full cost of all work, including a percentage for professional fees.

---

[565] *West Faulkner Associates* v. *London Borough of Newham* (1995) 11 Const. LJ 157 per Simon Brown LJ at 161.
[566] In *London Borough of Hounslow* v. *Twickenham Garden Developments Ltd* (1970) 7 BLR 81.

Relevant event A term unique to the JCT 98 form of contract and defined (by clause 1.3) as the events specified in clause 25 which provide grounds for the awarding of an extension of time (qv). There are eighteen such relevant events listed in clause 25.4.

Remedies See: **Rights and remedies**.

Remoteness of damage A contract breaker is not liable for all the damage which ensues from his breach of contract, nor is a tortfeasor (qv) responsible for all the damage which flows from his wrongful act. Some damage is said to be too remote and is therefore irrecoverable. In a contract the basic rule was stated in *Hadley* v. *Baxendale* (1854):

> 'Where two parties have made a contract which one of them has broken, the damages which the other party ought to receive in respect of such breach of contract should be such as may fairly and reasonably be considered *either* as arising naturally, i.e. according to the usual course of things, from such breach of contract itself, *or* such as may reasonably be supposed to have been in the contemplation of both parties, at the time they made the contract, as the probable result of the breach of it.'[567]

There are two branches of this rule (indicated by the italicised either/or in the quotation above) and it should be noted that under the second rule the contract breaker is only liable if he knew of the special circumstances at the time the contract was made[568]. A similar test is applied in tort where the phrase 'reasonably foreseeable' is used as opposed to 'reasonably contemplated'.
See also: **Damages; Knowledge.**

Removal of defective work JCT 98, clause 8.4.1 gives the architect power to order removal from site of work, materials or goods which are not in accordance with the contract. By clause 8.4.2, provided he first obtains the agreement of the employer and has also consulted with the contractor, the architect may instead allow the defective work or materials to remain and make an appropriate deduction to the contract price. It appears from the strict wording of clause 8.4 that those are the architect's only options, i.e. entire removal or complete acceptance of the defect. Nothing in the clause expressly empowers the architect to, nor should he instruct the contractor to, carry out corrective measures that would suffice to reduce the defects to an acceptable level.

If the defective work or materials etc. are not 'acceptable' under clause 8.4.2 then seemingly the architect is not empowered merely to order that the defective work should be corrected. He must order its total or partial removal from site in order for the instruction under clause 8 to be valid. In *Holland, Hannen & Cubitt (Northern) Ltd* v. *Welsh Health Technical Services Organisation and Others* (1981), a case on JCT 63 (in which clause 6 (4) had essentially the same wording as clause 8.4.1 of JCT 98), Judge John Newey QC said:

> 'In my opinion, an architect's power under clause 6 (4) is simply to instruct the removal of work or materials from the site on the ground that they are not in

---

[567] *Hadley* v. *Baxendale* (1854) 9 Ex 341 per Alderson B at 354.
[568] *Victoria Laundry (Windsor) Ltd* v. *Newman Industries Ltd* [1949] 1 All ER 997.

accordance with the contract. A notice which does not require removal of anything at all is not a valid notice under clause 6 (4).'[569]

Renomination Many standard forms of building contract provide a mechanism whereby the architect, on behalf of the employer, may nominate specialists as sub-contractors. Problems may arise where the nominated sub-contractor defaults or fails. This problem was considered by the House of Lords under JCT 63[570], where a subcontractor nominated under a PC sum (qv) went into liquidation. The liquidator (qv) refused to complete the sub-contract. It was held that in these circumstances the employer, through the architect, was bound to make a fresh nomination. The main contractor was neither bound nor entitled to take over the nominated sub-contractor's work. That principle appears to be of general application in the sense that where the original contract provides for work to be done by a nominated sub-contractor, if the nominated sub-contractor defaults or otherwise fails, the employer must provide a substitute. In the absence of an express term to the contrary the main contractor is neither bound nor entitled to do the nominated sub-contract work himself. This general position may, of course, be affected, or indeed reinforced, by the particular wording of the contract.

Under the current JCT 98 Standard Form, the specific problem has been overcome so that by clause 35 the contract now makes express provision (at clause 35.24), detailing the circumstances and timing of the architect's duty to renominate as a consequence of the original sub-contractor failing. ACA 3, clause 9.7, dealing with any 'named sub-contractor' or any sub-contractor 'named in any instruction' provides that in such circumstances the main contractor 'shall select another person to carry out and complete the execution of the work ...'. Under GC/Works/1 (1998), clause 63 (7), the employer may (but is not bound to) renominate.

Any immediate loss to the contractor arising from the nominated subcontractor's withdrawal or failure falls on the main contractor[571], since the nominated sub-contractor's failure is not a default or breach of contract on the part of the employer. However, since the architect is bound to renominate, the employer is responsible for any loss arising from any delay in him doing so. The architect has a reasonable time (qv) in which to make the renomination (JCT 98, clause 35.24.10) which time runs from the date of receipt of the contractor's request for a renomination instruction. But, it should be noted that:

— An apparent delay in renomination does not of itself make the period of time involved unreasonable unless the delay is caused by the fault of the architect or employer.
— The architect is entitled to have regard to the interest of the employer by seeking lump sum tenders from proposed renominees. However, it has also been suggested that, unless the contract wording provides

---

[569] *Holland, Hannen & Cubitts (Northern) Ltd* v. *Welsh Health Technical Services Organisation and Others* (1981) 18 BLR 80 per Judge Newey at 120.
[570] *North-West Metropolitan Regional Hospital Board* v. *T. A. Bickerton & Son Ltd* [1970] 1 All ER 1039.
[571] *Percy Bilton Ltd* v. *Greater London Council* (1982) 20 BLR 1.

otherwise, the main contractor is not bound to remedy defects in the work of the original nominated sub-contractor where these arise before completion of the sub-contract works, and such loss falls on the employer. To be valid, the renomination must cover both existing defective sub-contract work and completion work, otherwise the main contractor is entitled to reject the renomination[572].

Repair The word is found in the insurance clauses of contracts. It has its ordinary meaning – to restore to the same condition as obtained before the event which necessitated a repair being carried out; in other words there must be 'disrepair'.

Representative One who stands in the place of another. JCT 98 makes provision (at clause 11) for not only the architect but also his representative to be allowed access to the site. Similarly, JCT 98, clause 13.5.4 provides that, in respect of work to be valued on a daywork basis, vouchers specifying the various details of time and resources spent daily upon the work should be given weekly to the architect or to his authorised representative. The rules of agency (qv) govern a representative. It is, therefore, important that the architect specifies who is to be appointed as his authorised representative and the extent of the representative's authority. The architect must put that information in writing and must communicate it to anyone who may have dealings with his representative. The contractor should do likewise in respect of his representatives. Such information is commonly exchanged and minuted in the first contract meeting. See: **Employer's representative.**

Repudiation This is the term used to describe those breaches of contract which consist of one party clearly indicating, at a time before the contract has been fully performed, that he no longer intends to fulfil his contractual obligations. In general, the innocent party is not bound to accept the repudiation; he may affirm the contract if he wishes. If he accepts the repudiation, the contract is discharged and the innocent party may sue for damages.

Although the concept of repudiation is simple in theory, there are considerable difficulties in practice. It is not always clear whether there has been a wrongful repudiation and it is for this reason that most standard form building contracts contain clauses entitling one party to terminate on the happening of specified events.

See also: **Anticipatory breach of contract; Breach of contract; Damages; Determination.**

Rescission The termination or abrogation of a contract by one of the parties. A contract may be rescinded on grounds of misrepresentation, mistake, or fraud (qv).

Rescission is effected by taking proceedings to have the contract set aside by the court (as in the case of misrepresentation) or by giving notice to the other

---

[572]*Fairclough Building Ltd* v. *Rhuddlan Borough Council* (1985) 30 BLR 26.

party of one's intention to treat the contract as at an end. *Restitutio in integrum* (qv) is an essential precondition to the right to rescind. If it is impossible, the parties are left to their other remedies, e.g. damages. In practical terms, the defendant must indemnify the claimant against the obligations created by the contract[573].

Resident architect If frequent or constant inspection of the works is considered necessary, then in line with the provisions of the Standard Form of Agreement for the Appointment of an Architect (SFA/99) and in Conditions of Engagement for the Appointment of an Architect (CE/99) the task of resident site inspector most commonly falls to a clerk of works (qv) who, according to paragraph 3.10 of those agreements, will be appointed and paid directly by the employer to act solely for that purpose under the direction of the architect. However, in the circumstances of a large or complex project it may be desirable for the resident inspector to act beyond the limited powers of a clerk of works and to have perhaps all, or at least some, of the powers and authority given to the architect named under the contract. In that case, the architect may himself take up residence on site or may appoint another – generally less senior – member of the practice to act as resident architect. A resident architect should have his authority and powers clearly and specifically defined and must be distinguished from the clerk of works who has no power to issue instruction etc. but only the duty to inspect the work.

Resitutio in integrum Restoration to the original position. Before a contract can be rescinded (see: **Rescission**) this principle must be satisfied. 'The principle of *resitutio in integrum* does not require that a person should be put back into the same position as before; it means that he should be put into as good a position as before, e.g. if property has been delivered, it must be restored, and the party seeking rescission must be compensated for the money etc. which he has expended as a result of obligations imposed on him by the contract. The court must do what is practically just, even though it cannot restore the parties *precisely* to the state they were in before the contract'[574].

Respondent The person against whom an appeal (qv) is brought (in litigation) or against whom a claim (qv) is made in arbitration (qv) proceedings.
See also: **Arbitration; Claimant.**

Restitution An obligation on one party to restore goods, property or money to another. It arises in situations where goods, etc. have been transferred by virtue of mistake (qv), illegality or other lack of legal authority, and is intended to avert injustice.
See also: **Letter of intent; Quantum meruit/valebat.**

---

[573] *Boyd & Forrest* v. *Glasgow Railway* [1912] SC (HL) 49.
[574] *Erlanger* v. *New Sombrero Phosphate Co* (1878) 46 LJ Ch. 425.

Restraint of trade (1) In the context of employment law, it refers to a means by which an employer seeks to impose on an employee an agreement that, on leaving his employment he will not then move to another job in the same trade or profession and in doing so take with him useful experience gained with his present employer. When sought to be enforced, it is a practice that gains little favour with the courts who, on policy grounds, will generally tend towards narrowly construing the effect of such agreements so as not to preclude individuals from their fundamental right to earn a living.

(2) In a corporate context the term refers to attempts by companies, contrary to law, to prevent free trade by means such as creating monopolies, fixing prices or otherwise restricting the opportunity for open competition.

Restrictive covenant A negative obligation affecting freehold land and restraining the doing of some act on or in relation to the land in question, e.g. a prohibition in the title deeds against using the premises other than as a private dwelling. A restrictive covenant is enforceable not merely between the original parties to the agreement but also between the successors in title to both parties. A restrictive covenant 'runs with the land' provided that it exists for the benefit or protection of the land. The burden or liability to be sued on a restrictive covenant binds a subsequent purchaser[575].

Restrictive covenants are registerable as land charges under s. 10 of the Land Charges Act 1925 and registration amounts to actual notice (qv) of the existence of the covenant to every prospective purchaser. Such covenants remain enforceable indefinitely and may in practice hinder conversion and development. In some cases outmoded restrictive covenants may be modified by the Lands Tribunal (qv).
See also: **Covenant.**

Restrictive tendering procedure One of three public procurement tendering procedures sanctioned under EU law; the others being the 'open' and 'negotiated' procedures. It is the commonly preferred method where the prospective employer wishes to save time in future assessing the technical and financial merits of candidates, particularly where it is proposed to let a series of similar contracts. Under the restricted procedure, expressions of interest are invited from contractors wishing to be considered as suitable to submit a tender for a project or series of projects. Those interested may, in appropriate circumstances, be required by the prospective employer to complete a pre-qualification questionnaire designed to assess exact requirements in relation to the particular contract(s) concerned. They will be assessed on the basis of information provided on those questionnaires, if used, and generally on the basis of their technical expertise, proven track record and financial status before then being selected or rejected for inclusion on a list of approved candidates.

Subject to ensuring genuine competition, where the restricted procedure is used the prospective employer may prescribe a maximum and minimum

---

[575]*Tulk* v. *Moxhay* (1848) 2 Ph 774.

number of tenderers (with a maximum range of between 5 and 20) for inclusion on the list.

Retention fund; Retention monies A sum or sums of money held by the employer as safeguard against defective or non-conforming work or materials provided by the contractor. It is a safeguard for the employer against latent defects or defects which may subsequently develop and the contractor's possible failure to complete the contract. It is provided for the general protection of the employer[576]. The fund is a percentage (normally 5%) of the work properly executed by the contractor. It is built up by deducting the appropriate percentage from the quantity surveyor's valuation of work in progress at each certificate.

JCT 98, clause 30.5.1, IFC, clause 4.4 (if the employer is not a local authority) and ACA 3, clause 16.4 state the employer's interest as trustee (without obligation to invest). ACA 3 and the private edition of JCT 98 provide for the employer to set the money aside in a separate bank account. This reflects a requirement of the general law whenever the contract provides that the retention money is to be held in trust[577]. The purpose is to safeguard the contractor's money in the event of the employer becoming insolvent and an employer who neglects to set the money aside may be required by mandatory injunction to do so. An employer who uses the retention fund for his own ends, for example by applying it to capitalise his further business, as distinct from ensuring it is held safe for the contractor's benefit, would be in breach of his trust[578].
See also: **Fiduciary.**

Retention of title Many supply contracts contain a clause whereby the seller retains title in the goods until he has been paid for them. The right to retain title is recognised by statute in s. 19 (1) of the Sale of Goods Act 1979 and clauses to that effect have also become increasingly common in contracts made in the building industry.

The purpose of such provisions is to protect the seller in case of the buyer's insolvency (qv). A typical retention of title clause provides that the seller retains ownership of the goods sold, notwithstanding delivery, until the goods have been paid for, or sometimes until all debts due by the buyer to the seller have been paid. A retention of title clause may become of no real value once the goods have been incorporated into the building, because ownership passes to the employer as soon as they are actually built into the works[579]. It seems that the clause is also worthless where the materials have been admixed with other materials to form a new material, e.g. sand mixed with cement and water.

---

[576]*Townsend* v. *Stone Toms & Partners* (1985) 27 BLR 26.
[577]*Wates Construction (London) Ltd* v. *Franthom Property Ltd* (1991) 53 BLR 23 and *Rayack Construction Ltd* v. *Lampeter Meat Co Ltd* (1979) 12 BLR 30.
[578]*Wates Construction (London) Ltd* v. *Franthom Property Ltd* 53 BLR 23 and *Rayack Construction Ltd* v. *Lampeter Meat Co Ltd* (1979) 12 BLR 30.
[579]*Reynolds* v. *Ashby & Son* [1904] AC 466.

The effectiveness of such clauses otherwise was upheld by the Court of Appeal in the *Romalpa* case (1976)[580] and again since then has been reinforced in *Clough Mill Ltd* v. *Geoffrey Martin* (1984)[581].

The latter case concerned the supply of yarn on credit terms, the contract of sale providing that the ownership of the yarn was to remain with the sellers, who reserved the right to dispose of it 'until payment in full for all the [yarn] has been received . . . in accordance with the terms of this contract or until such time as the buyer sells the [yarn] to its customers by way of bona fide resale'. Payment was stated to become due immediately on the buyer's insolvency, and various other rights were reserved by the sellers, all of which were upheld by the Court of Appeal. The clause did not require to be registered as a charge under section 95 of the Companies Act 1948.

Similarly, in *Archivent Sales & Developments Ltd* v. *Strathclyde Regional Council*[582], builders' merchants supplied ventilators to a contractor and delivered them to site. The sale was on terms that 'until payment of the price in full is received by the company the property and the goods supplied by the company shall not pass to the customer'. A Scottish judge upheld the validity of the clause.

Following an earlier decision of the Joints Contracts Tribunal not to alter the wording of its 1963 or 1980 editions of the standard form contracts[583], in 1985 the JCT decided to amend JCT 80 so that, where any work was sub-let the contractor should undertake to ensure that the relevant sub-contract includes terms whereby property in materials on site and those partially or wholly incorporated into the works will transfer to the employer when paid for under the main contract. Similar provision is now made in JCT 98, clause 19.4.2.2, and IFC 98, clause 3.2.2, which provide that it shall be a condition in any sub-contract the contractor may enter into that; 'where, in accordance with clause 30.2 of JCT 98 (and clause 4.2.1 (b) of IFC 98) the value of any . . . materials or goods shall be included in any interim certificate under which the amount properly due to the Contractor shall have been paid by the Employer to the Contractor, such materials or goods shall be and become the property of the Employer and the sub-contractor shall not deny that such materials or goods are and have become the property of the Employer'.

Notwithstanding those provisions, the architect is unwise to include goods brought on site in interim certificates without proof of ownership or, at the very least, before satisfying himself that the sub-contract does, in fact, include such express provisions. If he does not do so, the employer may later be faced with the situation of having paid the main contractor for such materials only to then find, if and when the main contractor becomes insolvent, that the main contractor has neglected to incorporate the appropriate provisions in the sub-contract. On its worst case the sub-contract may in fact be made in the sub-contractor's standard terms containing an express retention of title clause.

---

[580] *Aluminium Industrie Vaassen BV* v. *Romalpa Aluminium Ltd* [1976] 2 All ER 552.
[581] [1985] 3 All ER 982.
[582] (1985) 27 BLR 98.
[583] Joint Contracts Tribunal formal notice of 1978, titled *Retention of title (ownership) by suppliers of building materials and goods.*

Clause 16.2 (A) of ACA 3 enables the value of goods and materials intended for but not incorporated in the works to be included in interim certificates where the contract documents expressly so provide, but this is to exclude 'the value of any such goods and materials where the Architect is not satisfied ... that the property in such goods is vested in the contractor'. GC/Works/1 makes no specific provision for the effect of retention of title clauses.
See also: **Ownership of goods and materials; Incorporation; Fixtures.**

Revocation The withdrawal of an act already done or promised. For example, the revocation of an offer may be made at any time before acceptance (qv) or the revocation of a will. It may be by individual or company or it may occur through the operation of law or by death or by order of the court.

RIBA contracts An incorrect and outdated method of referring to the standard forms of contract published by the Joint Contracts Tribunal (qv). The title was correct until 1977, when the Royal Institute of British Architects withdrew its name from the documents which are now correctly referred to as the JCT Forms. The 1998 editions of these contracts are prepared and issued by the Joint Contracts Tribunal and published by RIBA Publications, a division of RIBA Companies Ltd. Copyright (qv) in all of the 1998 editions of the JCT, IFC, MW, MC and WCD forms now vests in The Joint Contracts Tribunal Limited.

Since January 2000, the Private and Local Authorities Editions of JCT 98, WCD 98, IFC 98, MW 98 and the Nominated and Named Sub-Contract Forms have also been made available on CD-ROM. This software package provides subscribers with the opportunity to access and interrogate the forms using a word search or clause search facility. It also allows each of the forms to be edited, text added and clauses deleted or altered to suit the user's particular requirements and amendments. The software is also available in a 'Small Works Service' providing subscribers with similar facilities but limited to the IFC 98, Named Sub-Contract and MW 98 Forms.

The current JCT 98 derives from a form agreed as long ago as 1893 with further subsequent editions being issued in 1909, 1931, 1939 and 1963. All were known as 'the RIBA Contract' and are so referred to in the law reports (qv) and textbooks.

Right first time A term used in connection with prime contracting (qv). The idea appears to be that with appropriate management of the supply chain through detailed design and involving value engineering and risk management all construction activities will be right first time and effort and money is not wasted on correcting problems whether of design or construction.
See also: **Continuous improvement; Supply chain partners; Supply clusters.**

Right of light A negative easement (qv) which entitles one owner to prevent his neighbour building so as to obstruct the flow of light through particular windows. The property enjoys the privilege of 'ancient lights'. In determining whether there has been an actionable interference with a flow of light the test is: how much light is left, and is that sufficient for the comfortable use and enjoyment of the house according to the ordinary requirements of mankind? It is a right to receive a reasonable amount of light and nothing more[584].

The test most commonly applied is the 'forty-five degree' test, i.e. the interference will not be considered a nuisance (qv) if the light can still flow to the window at an angle of 45° from the horizontal.

Under the Rights of Light Act 1959 s. 2, the owner of land over which a right of light might be acquired by user may now register as a land charge a notice identifying the properties and specifying the size and position of a notional screen. This prevents any right of light being acquired by the adjoining property and circumvents the cumbersome common law necessity of erecting an actual screen[585]. While the notice is in force the other party may seek cancellation or variation of the registration[586].

See also: **Prescription.**

Right of way (1) The right to pass across land belonging to another. The right may be public, in which case any member of the public has the right to use it, or private, when it is an easement (qv) for the benefit of adjoining property (qv) or land. In the latter case, only the owner of the land and such people as he permits may use it.

(2) A public right of way as usually created by Act of Parliament or by custom (qv) as access from one public place to another. By amendments to the existing law made by the Rights of Way Act 1990, whereas unlawful disturbance of the surface of certain footpaths, bridleways and highways which causes inconvenience to the public rights of way over them is an offence, the Act provides that, where a footpath or bridleway passes through agricultural land and it would be unreasonable for the occupier of such land to do otherwise, the surface of the footpath or bridleway may be disturbed (for example, during ploughing or sowing crops etc.), provided that, within a stipulated period (i.e. 24 hours generally and in the case of crop sowing within 14 days), the surface must be reinstated to its full width and the position of the footpath or bridleway must be clearly indicated for those wishing to use it. Failure to meet those requirements will be an offence. **Figure 18** shows examples of public and private rights of way.

See also: **Highway; Prescription; Access to neighbouring land; Countryside and Rights of Way Act 2000.**

Rights and remedies A phrase found, for example, in JCT 98, clauses 26.6, 27.8 and 28.5 where it is stipulated that the rights and obligations expressed in those clauses are without prejudice (qv) to any other *rights and remedies* which the contractor (or employer) may possess. That is to say, the parties' common law

---

[584]*Colls* v. *Home & Colonial Stores* [1904] AC 179.
[585]Section 3 (1)
[586]Section 3 (3)

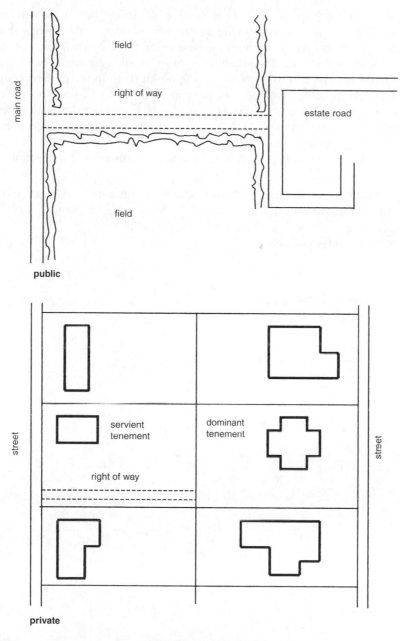

Figure 18 Rights of way.

rights are unaffected. In the absence of anything to the contrary, that would be the position anyway if the sub-clause were left out (see also: **Unfair Contract Terms Act 1977**). The rights are the parties' rights at common law; the remedies are the remedies available to satisfy those rights.

Riot A term that often appears in relation to contracts of insurance. In that context (and for most other practical purpose) the term applies when three or more people carry out a common purpose using force or violence, not merely in or about the common purpose but displayed in such a manner as to alarm a reasonable person, against persons who opposed them in the execution of that common purpose[587]. People taking part in a riot are guilty of an offence at common law. The term is used in JCT 98, clause 1.3, under the general head of 'specified perils' (qv) and in ACA 3, clause 11.5 (c), alternative 2, as a ground for awarding an extension of time.
See also: **Civil commotion; Civil war; Commotion; Disorder; Insurrection.**

Romalpa clause A retention of title (qv) clause is commonly so called after the case of *Aluminium Industrie Vaassen BV* v. *Romalpa Aluminium Ltd* (1976)[588] in which the effectiveness of the device was upheld.
See also: **Retention of title.**

---

[587]*The Andreas Lemos* [1983] 1 All ER 590 and *Field* v. *Metropolitan Police Receiver* [1904–7] All ER Rep 435.
[588][1976] 2 All ER 552.

Sale of goods Comprehensive statutory provisions regulating sale of goods are contained in the Sale of Goods Act 1979, which applies throughout the UK.

The Act applies only to sales of goods. As such it has no direct application to building contracts which are contracts for work and materials[589]. However, sales of building materials to the contractor and similar transactions are within the scope of the Act, which implies certain conditions and warranties as to fitness for purpose (qv), satisfactory quality (qv), etc.

Under the Act, property in the goods passes at the time when the parties intend it to pass. Subject to any apparent contrary intention, s. 18 sets out certain presumptions concerning when the parties intend property in the goods should pass from seller to buyer. It should be noted that the fact that the price may remain unpaid does not affect the position regarding transfer of ownership unless the contract provides that property is to pass when the price is paid. In practice, it is now common for the seller to contract on terms which include a retention of title (qv) clause. Similarly, parties may agree that property may pass before the goods are delivered and in the case of 'specific goods', the presumption is that the parties intend property to pass immediately at the time the contract is made.

In general, a non-owner cannot transfer title of goods, and nobody can give a better title than he himself possesses[590]. In the past, this principle has caused practical problems under building contracts, especially with regard to off-site goods and materials paid for under the contact by the employer where the contractor is not, however, the owner of the goods concerned[591]. By amendment to JCT 80 and introduction of a new clause 19.4.2 (which is now largely reproduced in JCT 98, clause 19.4.2), the Joint Contracts Tribunal has attempted to overcome such problems. Further safeguards protecting the employer's interest in relation to the transfer of title in off-site materials and goods are now contained in JCT 98, clause 30.3, whereby the architect's discretion to include the value of off-site materials within interim certificates is removed. Only such off-site items, if any, as the employer may list at the time the contract is made will fall to be paid for under interim certificates and then only provided:

— the listed items are set aside or are uniquely identified as being destined for the particular project and employer concerned, and
— the contractor provides the architect with reasonable proof of satisfactory insurance of the listed items whilst they are held off site and until delivered to the project, and

---

[589]The relevant Act governing building contracts, which are generally mixed contracts for supply of goods (e.g. concrete) and services (e.g. building a wall) is the Supply of Goods and Services Act 1982, which applies similar constraints and principles to building contracts.
[590]*Bishopsgate Motor Finance Co Ltd* v. *Transport Brakes Ltd* [1949] 1 KB 322.
[591]*Dawber Williamson Roofing Ltd* v. *Humberside County Council* (1979) 14 BLR 70.

— the contractor provides the architect with reasonable proof of ownership, and

— the contractor, if so required, has provided a bond in favour of the employer.

See also: **Ownership of goods and materials; Retention of title.**

Sanction Generally, a reaction which indicates approval or disapproval of something – usually conduct – tending to induce conformity with required standards. The word may be used in the sense of authorisation or alternatively for the penalty laid down for contravention of some legal requirement.
See also: **Approval.**

Satisfactory quality Goods sold in the course of business must be of satisfactory quality[592]. Goods are defined by s. 14 (2A) as being of satisfactory quality 'if they meet the standard that a reasonable person would regard as satisfactory, taking into account any description of the goods, the price (if relevant) and all other relevant circumstances'. There is little judicial guidance as to the differences between this test and the former requirement of goods being of merchantable quality (qv).

Schedule contract A contract based upon a Schedule of Prices (qv).

Schedule of Activities See: **Activity Schedule**.

Schedule of Basic Prices of Materials Under JCT 98, if fluctuations clause 39 is introduced into the contract the contractor will be required to submit a schedule of basic rates of materials, goods electricity and, if specifically required, fuel, so that, in the event of any market increases adjustment of those prices can be carried out. The schedule of the 'basic prices' is annexed to the contract bills of quantities (qv) and the prices listed on it will be those on which the prices in the contract bills (qv) are calculated. They are deemed to be the market or 'basic price' of the relevant materials and goods at the base date (qv) in which case the contractor should be asked to present substantiation that they properly represent the rates and prices used in the bills of quantities. Careful checking is necessary because certain materials have standard prices and prices vary with the amount required.

Schedule of Prices Where time is very short or where for some other reason it is not possible or desirable to prepare bills of quantities or bills of approximate quantities (qv), a schedule may be prepared giving comprehensive descriptions of the work to be carried out and the materials to be used. The employer may put prices against each item and require tenderers to state what percentage

---

[592]Sections 14 (2)-(2C) of the Sale of Goods Act 1979, as introduced by s. 1 of the Sale and Supply of Goods Act 1994.

above or below those prices the contractor will require to carry out the work. Alternatively, tenderers may be asked to put their own prices against the items.

It is extremely difficult to prepare a tender by this method or to compare tenders received, because contractors normally balance their rates according to the amount of work and materials required. The system is most commonly used for small contracts or for contracts for maintenance work and is also referred to as a Schedule of Rates.

See also: **Term contract.**

Schedule of Rates Contracts which do not include bills of quantities (qv), but rely on drawings and specifications, require the contractor to submit a schedule of his rates used to arrive at the tender figure in order that variations can be accurately and fairly valued. Although MW 98, clause 3.6 provides that the architect should have regard, where relevant, to the 'Specification/schedules/ schedules of rates' when fairly valuing variations, surprisingly no mention is made of a schedule of rates in the contract documents listed in the first recital to the contract. The ACA 3 contract makes provision of a schedule of rates an option (see Recital C).

Schedule of Work On projects of a relatively simple nature, where formal bills of quantities are either inappropriate or otherwise not used, the architect may prepare a Schedule of Work more fully describing, item by item, the work shown and described briefly on the contract drawings. Such a schedule, which may be prepared in varying degrees of detail and complexity, may then be provided to prospective contractors to assist them for tendering purposes. Under IFC 98 (Second Recital A) the use of a Schedule of Work in lieu of bills of quantities is expressly provided for and once priced by the contractor the schedule then becomes a contract document. As such it will be used, where relevant, for the purposes of valuing variations under clause 3.7.

Scheme for Construction Contracts See: **Housing Grants, Construction and Regeneration Act 1996.**

Scott Schedule A formal document sometimes used in litigation or arbitration, which sets out the issues in dispute and the contentions of the opposing parties in tabular form. There is no set form prescribed but the object is to present the issues in dispute as clearly as possible. It is common for some of the issues to be resolved at this stage, thus simplifying and shortening the hearing. It is good practice to agree the headings for the various columns at the earliest possible stage in the case management process. Where the issues are being referred to arbitration as opposed to litigation the form and contents of the schedule may be raised at the preliminary meeting before the arbitrator.

The Scott Schedule was invented by Mr G. A. Scott QC about 60 years ago, and is most suited to cases involving a multiplicity of claims where each party is required to set out his case positively item by item and to answer each other's case in the same way. The schedule can be extended to claims between defendants and to third and subsequent parties. From the completed schedule representative items are selected for trial, so avoiding the necessity of trying each and every one. Various examples of a Scott Schedule are set out in *Keating on Building Contracts,* 6th edn, pp. 481–485 and *Powell-Smith and Sims' Building Contract Claims,* 3rd edn, p. 351.

Scott v. Avery clause a term written into a contract whereby the parties agree that the award of an arbitrator is a condition precedent to the commencement of legal proceedings. The validity of such a clause was upheld by the House of Lords in the case of the same name in 1856, where it was held that in the face of such a clause a party had no right to sue until arbitration had taken place and that such a condition was not contrary to public policy. The usual wording is: 'Arbitration shall be a condition precedent to the commencement of any action at law' or 'The obligation shall be to pay such a sum as may be awarded upon arbitration under this clause'. Other words may have the same effect. However, in the context of the building industry it should be noted that no standard form of contract currently contains such a Scott *v.* Avery type of clause[593]. Where parties have agreed to arbitrate their disputes or differences, on an application from the defendant the courts will now (except in the most extraordinary circumstances where that agreement is held to be null and void, inoperative or incapable of being performed), stay the court action[594]. The claimant will therefore be prevented from litigating the issues and will be bound to refer the matter to arbitration as agreed. In the exceptional event that the action is not stayed and litigation is allowed to proceed, s. 9-(5) of the Arbitration Act 1996 provides that any clause in the contract which has effect as Scott *v.* Avery clause will be of no effect in relation to those particular proceedings.
See also: **Arbitration; Stay of proceedings.**

Seal Technically, a device affixed on wax or impressed on a wafer as a mark of authentication. Since early times it has been essential for parties to a deed (qv) to each affix their seal to it before the deed would be held to be valid. Consequently every deed had to be executed 'under seal'. Contracts executed as a deed in this way were known as specialty contracts and differed from simple contracts in three primary respects:
  — Under the Limitation Act 1980 the limitation period (see: **Limitation of actions**) is 12 years as opposed to six years.
  — Consideration (qv) is not necessary to support promises made under seal.
  — In theory the parties cannot deny statements of fact contained in a deed, including its recitals (qv).
The requirements for a deed to be validated by sealing are radically altered by s. 36A (4) of the Company's Act 1985 as amended by s. 130 (1) of the

---

[593] *Scott* v. *Avery* (1856) 25 LJ Ex 308.
[594] Section 9 (4) Arbitration Act 1996.

Companies Act 1989 and s. 1 (1) (b) of the Law of Property (Miscellaneous Provisions) Act 1989, which have, in effect, abolished the general requirement for sealing by individuals or on companies incorporated under the Companies Act. In those cases, the Acts introduce alternative requirements for the valid execution of a deed depending upon whether the party executing it is an individual or a corporate body.

In the case of an individual the law now requires that:

— The contract must make clear on the face of the document that it is intended by the individual concerned to be a deed.

— The deed must be signed by the individual concerned in the presence of a witness who attests the signature, or at his direction and in his presence and in the presence of two witnesses who each attest (qv) the signature (Law of Property (Miscellaneous Provisions) Act 1989 – s. 1 (3)).

— The deed must be delivered as a deed by the individual or by a person authorised by him to do so.

In the case of incorporated companies, the law now requires[595] that the document must:

— Be expressed to be executed by the company.

— Make it clear on its face that it is intended by those executing it to be executed as a deed.

— Be signed by a director and company secretary or by two directors of the company.

By way of marginal notes, the attestation provisions in the IFC 98, Private Editions of JCT 98 and WCD 98 all give clear guidance on how to ensure compliance with the requirements outlined above. No such facility is given in the standard form MW 98.

Although the use of a seal in the context of deeds is now largely (although not altogether) abolished, the need to affix a seal still has relevance in other contexts. It is still commonplace for a seal to be affixed to certificates of admission to various professional institutes and other such bodies, and documents such as share certificates, a company's memorandum of association and probate of a will also require sealing.

See also: **Deed; Attestation; Locus sigilli.**

Sealed offer In proceedings before the Lands Tribunal (qv) about compensation claims for compulsory purchase, the acquiring authority may make an unconditional offer of compensation in a sealed envelope. If the sum eventually awarded is the same or less than the amount of the offer, then the claimant does not get his costs. It is the equivalent of a payment into court (qv). The existence of the sealed offer is not disclosed to the Lands Tribunal until it has given its decision.

Sealed offers are sometimes, albeit rarely, used in arbitration (qv). The practice was approved by Donaldson J in *Tramounta Armadora SA* v. *Atlantic*

---

[595]By s. 36A (4) of the Companies Act 1985.

Shipping Co SA (1978)[596] where it was said that:

'A sealed offer is the arbitral equivalent of making a payment into court in settlement of the litigation or of particular causes of action in that litigation. Neither the fact, nor the amount, of such a payment into court can be revealed to the judge trying the case until he has given judgment on all matters other than costs. As it is customary for an award to deal at one and the same time both with the parties' claims and with the question of costs, the existence of a sealed offer has to be brought to the attention of the arbitrator before he has reached a decision. However, it should remain sealed at that stage and it would be wholly improper for the arbitrator to look at it before he has reached a final decision on the matters in dispute other than as to costs, or to revise that decision in the light of the terms of the sealed offer when he sees them.'

There are, in fact, substantial objections to the practice, which are usefully summarised in the Commercial Court Committee Report on Arbitration 1978 (Cmnd 7284), paras 62–65, the main objection being that 'the arbitrator, unlike a judge, will know that some offer of settlement has been made, although he will not know how much'.

An alternative course is to make an *open offer* – generally referred to as a Calderbank offer (qv) – on terms that its existence and contents must not be disclosed to the arbitrator until he has reached a final decision on liability, when it will be drawn to his attention. The offer should include an offer to pay costs up to the date of acceptance if accepted within 21 days and should address the matter of what interest, if any, is included within it. The letter of offer should state that it is intended to have the effect of a payment into court. At the end of the hearing before the arbitrator he should be asked to make an interim award on liability and amount and, without the existence of the offer being disclosed, he should be asked to defer consideration of costs until he has made his interim award on the other issues. This procedure is commonly used and is suggested in *Keating on Building Contracts*, 6th edn, pp. 448–449, Butterworths.

An alternative procedure is to make a 'without prejudice' (qv) offer, backed up by a deposit of money in the joint names of the parties or their solicitors, and once again to ask the arbitrator to make an interim award on liability and amount. 'If the claimant in the end has achieved no more than he would have achieved by accepting the offer, the continuance of the arbitration after that date has been a waste of time and money. *Prima facie,* the claimant should recover his costs up to the date of the offer and should be ordered to pay the respondent's costs after that date.'[597]

See also: **Calderbank offer; Payment into court; Part 36 offer/payment.**

Sectional completion Completion of the works in sections or parts. JCT contracts, except MW 98, have supplements which enable the employer to stipulate that the works should be completed in accordance with different completion dates. The supplement also allows the employer to stipulate different dates for possession (qv) and different amounts of liquidated damages (qv) for each section. MC 98 has a supplement called 'phased completion' which amounts to

---

[596][1978] 2 All ER 870 per Donaldson J at 876–877.

[597]*Tramounta Armadora SA* v. *Atlantic Shipping Co SA* [1978] 2 All ER 870 per Donaldson J at 877–878.

much the same thing. Sectional completion must be distinguished from partial possession (qv).

Seizure and vesting See: **Vesting and seizure.**

Serial contract If it is desired to carry out a number of contracts in succession, this type of contract may be employed. On the basis of the successful tender for the first contract, further contracts are negotiated. To operate properly, all the projects must be similar in construction and type so that negotiation for future contracts on the basis of the original contract is feasible. It is usual for the employer to make some sort of limited commitment to the successful tenderer for the whole series. However, it is not something which can be legally enforced since it is always subject to the successful outcome of negotiations. The advantage is that one set of tendering takes place and the contractor can use the experience gained on the first contract to improve efficiency thereafter. For maximum benefit for the employer, the basic terms for succeeding contracts in the series should be established when calling for the initial tender. An intended programme for all contracts in the series should be set down at the outset if the contractor is to be able to calculate the potential benefits to the full. The system should produce savings for both parties but it is often difficult to operate in practice. In the case of public sector procurement, care must be taken to ensure that any attempt to create serial contracts based on a process of negotiation does not contravene EU legislation designed to ensure equal competition throughout the EU.

Service of notices, etc. All the standard forms of contract require the service of notices, certificates, etc. to follow certain procedures and time limits. In order to preserve the effect of such notices, employer, architect and contractor must carefully observe the procedure laid down. A general provision concerning the service of notices and other documents is given in JCT 98, clause 1.7. However, that clause is made subject to the overriding proviso that it will be relevant only if and to the extent that the contract does not elsewhere specifically state the manner in which the particular notice etc. must be given. For example, under clause 27.1, for the employer to instigate a valid determination of the contractor's employment a notice under that clause must be given in writing and must be served by actual delivery (by hand), or by special or recorded delivery. JCT 98 deals with certificates (qv) in clause 5.8 and throughout the contract under the appropriate clauses, and numerous other notices etc. are specifically dealt with under other clauses as they arise.

ACA 3 deals with notices in clause 23.1 and throughout the contract particularly as far as timing is concerned. GC/Works/1 (1998) deals with certificates under clauses 39 and 50 and with notices throughout the contract. IFC 98 deals with notices generally in clause 1.13 and with certificates in clause 1.9. Throughout the contract, notices are also dealt with as they arise.

Set-off a defence to a claim used to reduce or extinguish a claim, whereas a counterclaim (qv) may give rise to an award for damages. Set-off acts in a similar way to abatement (qv), but is of general application. In *Hanak* v. *Green*[598] it was considered where a' set-off might arise:

Legal The mutual set-off of liquidated debts.

Equitable The test is whether a cross-claim is so closely connected with the main claim as to render it manifestly unjust to determine one claim without the other[599].

Insolvent There are specific rules for the setting-off of mutual dealings (qv) under s. 323 of the Insolvency Act 1986 and rule 4.90 of the Insolvency Rules 1986[600].

Additionally, parties may agree to extend or limit the rights of set-off arising out of any agreement between them.

An important change to the rights of set-off has been introduced by s. 111 of the Housing Grants, Construction and Regeneration Act 1996 which excludes the right of set-off from building contracts (qv) where the requisite withholding notice has not been served at the proper time. It appears that the court, in absence of such a notice, will ignore any set-off or abatement otherwise allowable at law or in equity[601].
See: **Abatement; Counterclaim; Mutual dealings; Housing Grants, Construction and Regeneration Act 1996.**

By virtue of statutory provisions (first introduced by the Supreme Court of Judicature Act 1873 s. 24 (3), latterly by s. 49 (2) of the Supreme Court Act 1981 and refined by s. 41 of the Supreme Court of Judicature Act 1925) set-off may also be pleaded as a defence in litigation or arbitration.
See also: **Counterclaim.**

Setting out The procedure whereby the dimensions of a structure are transferred to the site by means of measuring tapes, theodolites, etc. The principal walls of a building, or the position of piles are indicated by pins, lines and profiles. The process calls for great accuracy and on large and complex works a specialised setting out engineer may carry out this part of the work. The architect is responsible for the accuracy of the drawings and for providing sufficient information to enable setting out to be completed JCT 98, clause 7. However, he is not responsible for the accuracy of the setting out itself. That is the contractor's responsibility. It is good practice for the architect to provide special drawings showing only the outline of the building on the site and such dimensions as are necessary for setting out.

Unfortunately, it is common for drawings to be deficient in this respect, necessitating the architect visiting the site and assisting the contractor to set out, if only by approving what has been done. The architect should avoid

[598][1958] 2 QB 9, 26 (CA).
[599]*Dole Dried Fruit* v. *Trustin Kerwood* [1990] 2 Lloyds Rep 309, CA.
[600]SI 1986/952.
[601]*VHE Construction plc* v. *RBSTB Trust Co Ltd* [2000] BLR 187.

giving approval to the contractor's setting out by preparing properly dimensioned drawings, otherwise he runs the risk of accepting liability for any inaccuracies.

Settlement (1) In construction terms, it is the movement of a building in response to alterations in the bearing capacity of the ground.

(2) In law, it is an arrangement of property in such a way as to create a trust. It is often done by will or by a deed.

(3) An agreement by parties in dispute to compromise or otherwise put an end to their differences before any court or arbitration hearing takes place. It is always wise for litigants to settle if possible rather than run the risk and expense of court proceedings. It is also prudent to embody the terms of the settlement in contract form.

SFA/99 The abbreviated term given to the Standard Form of Agreement for the Appointment of an Architect, produced by the Royal Institute of British Architects in association with the Royal Incorporation of Architects in Scotland (RIAS), Royal Society of Ulster Architects (RSUA), Royal Society of Architects in Wales (RSAW) and the Association of Consultant Architects (ACA) and published by RIBA Publications. The form provides an agreement for use between employer and architect where the architect is to provide services during both the pre-construction and construction stages on traditional fully designed building projects of all sizes. With incorporation of amendments published by the RIAS the form is also usable for use in Scotland, under Scots Law.

The Form comprises:
— Articles
— Appendix to the Conditions
— Schedule 1 — briefly describing the project
— Schedule 2 — briefly stating, by category, the general services and project specific services activities and services the architect will provide. The full extent of those services is then more fully described and detailed in the 'Services Supplement' which also forms an integral part of the SFA/99
— Schedule 3 — setting out the agreed fee rates and other charges and expenses
— Schedule 4 — whereby the employer provides the names and addresses of other professionals (such as project manager, planning supervisor, structural engineer or surveyor) he proposes engaging directly in connection with the project and/or identifies those parts of the project — such as drainage or the like — which will not be designed responsibility of the architect
— Conditions of Engagement
— Attestation provisions.

The Form also includes useful notes for guidance on completion by the parties (which should be removed before the form is executed) and after completion the agreement is then formally executed by the parties signing the form of attestation.

See: **Conditions of Engagement (RIBA).**

Shop drawings Short for 'workshop drawings'. Architect's and/or engineer's drawings are often not suitable for the manufacture of certain building components. Special drawings, termed 'shop drawings', must be produced to enable joinery, steelwork, sheet metalwork etc. to be produced. These drawings are normally the responsibility of the manufacturer although he may, through the contractor, request the architect's approval to shop drawings before manufacture. The architect is under no obligation to give such approval – indeed it may be dangerous for him to do so – provided his own drawings contain all necessary information. It is usual, however, for the architect to examine any shop drawings sent to him and make any comments necessary while expressly reserving his approval.

Shortage of labour and materials Grounds for extension of time under JCT 98, clause 25.4.10.1 and 25.4.10.2, provided that the shortage is of labour or materials that are essential to the proper carrying out of the works and that such shortage arises for reasons beyond the control of the contractor and which he could not reasonably have foreseen at the base date (qv). IFC 98, clause 2.4.10 and 2.4.11 are optional provisions to the same effect.

Shrinkages A term used in some of the standard forms of contract to indicate a type of defect which the contractor is liable to make good during the defects liability period (qv). The normal meaning of 'shrink' is to grow smaller in size. However, it is a characteristic of many materials used in building that the increase or decrease in size depends on physical factors such as moisture or temperature, or chemical factors such as the reaction which takes place when mixing concrete or plaster. The contractor's liability for shrinkages is commonly misunderstood, not least by the contractor himself. He is liable to make good only if his workmanship or materials are not in accordance with the contract or where the shrinkage is caused by the effects of frost occurring prior to practical completion. The difficulties that such provisions for rectifying defects may cause are exemplified by a situation where the contract specifies the use of internal timber having a moisture content of 7%. Shrinkage of skirting and architraves may be found to have taken place at the end of the defects liability period but it may, nevertheless, not be the contractor's liability. The excessive use of central heating may have reduced the moisture content of the timber, and hence the size, to 4%. It is a complicated point. It is easy to see that shrinkage has taken place but not easy to determine the cause. The architect will probably say that the timber brought on to site by the contractor was or was allowed to become of a greater moisture content than specified and that it was drying out of the excess moisture which caused the shrinkage. The problem is not made easier by the fact that it is quite difficult to determine moisture content within 2 or 3% without removing a sample and testing under laboratory conditions. If a great deal of money is at stake, it may be worth the contractor paying to have such a test carried out. If he proves to be correct and the architect had erroneously withheld his certificate on account of the shrinkages, the contractor would have the basis of a sizeable claim at common law if he wished to press it.

Most contractors are extremely generous in making good shrinkages, probably because of the difficulty of proving the point and the retention money outstanding.

The contractor cannot refuse to make good a shrinkage on the ground that 'it is impossible to maintain a low moisture content in timber until the building is occupied'. It is in fact very difficult, but not impossible, and the contractor contracted to do it.

SI An abbreviation for Statutory Instrument (qv).

Signature The name or mark of a person in his own writing, i.e. written by himself or by proxy (qv). The form is not prescribed. It may be the full name, initials or any combination of the two and in some cases a signature may be valid if it is made by a mark properly witnessed, a rubber stamp or made by another with proper authority. The adding of a signature is taken as a sign of agreement. Many people, particularly those in public life, have an 'official' signature in an attempt to differentiate between an 'autograph' and their signatures on legal documents. Such attempts will only be effective, however, insofar as the parties likely to be affected are aware of the difference. For example, a bank may be informed that a particular form of signature must be the only form recognised for the drawing of cheques.

In Scotland the term 'signature' does not include marks, proxy or rubber stamp except in certain cases authorised by statute.

Similar character Used, in the context of provisions in standard form contracts concerning the valuation of variations, to describe work done pursuant to a variation and which in all material respects is like that already described in the contract bills. It need not be identical, but must be of the same type or nature as that described in the bills.
See also: **Similar conditions.**

Similar conditions A longstanding term used in JCT 98, IFC 98 and other standard form contracts to provide a test for determining whether or not there should be a departure from the rates in the bills of quantities when valuing other work done pursuant to a variation instruction. Generally, only where the varied work is not carried out under similar conditions to those described in the contract bills will a departure from the bill rates and prices be justified. In determining whether or not there is similarity, an objective view must be taken. Extrinsic evidence and/or any particular knowledge gained about the site by the contractor during the tender and pre-contract negotiation stages of the project, even if that additional knowledge were in fact used to arrive at the rates and prices given in the contract bills, must in any event be disregarded. Only those conditions that are, or ought reasonably to have been apparent from the contract documents will be deemed to be taken into account in the contract rates and prices.

Hence, the words 'similar conditions' expressly refer back to the works set out in the contract bills[602] so that, in determining whether or not similar conditions apply to the varied work, the question will be whether or not the actual conditions encountered are strictly comparable with the conditions described in or otherwise discernible from the contract. If they are not then a reasonable adjustment to the rates and prices in the contract bills will be made.

Similar A word found in JCT 98, clause 13 and GC/Works/1 (1998), clause 42(5). For a discussion of its meaning see: **Fair valuation.**

Simple contract A contract which is not made under seal (qv) or executed as a deed (qv) but is made or evidenced in writing and possibly signed or made orally or by conduct.
See also: **Contract.**

Single-stage selective tendering See: **Code of Procedure for Single-stage Selective Tendering 1996.**

Site Not always clearly defined in the contract. A definition is however given in clause 1 (1) of GC/Works/1 (1998) which describes 'the Site' as meaning: '. . . the land or place described in the Contract, together with such other land or places as may be allotted or agreed by the parties from time to time, for the purposes of carrying out the Contract'. A clear understanding of the extent of the site is most important since it will impact upon the contractor's rights in connection with possession (qv) and access (qv). Failure by the employer to give possession of 'the site' is a breach of contract. Adequate definition in the contract documents (qv) is therefore essential.
See also: **Examination of site.**

Site conditions In the absence of any specific guarantee or definite representations by the employer or his architect about site conditions, the nature of the ground and related matters, the contractor is not entitled to abandon the contract or claim extra money on discovering the nature of the soil. Equally, under the general law, he has no claim for damages against the employer. The position may be affected by the express terms (qv) of the contract. GC/Works/1 (1998),

---

[602] *Wates Construction (South) Ltd* v. *Bredero Fleet Ltd* (1993) 63 BLR 128.

clause 7 (1) restates and reinforces the common law rule. It places on the contractor the risk that site and allied conditions may turn out more onerous than he expected although this may be subject (as it is in GC/Works/1 (1998) clause 7 (3)–(5)) to provisions relating to unforeseeable ground conditions under which, in specified circumstances, the contractor may be entitled to extra payment.

Optional clause 2.6 of ACA 3 gives the contractor a potential right of claim if he encounters 'adverse ground conditions or artificial obstructions at the Site' as work progresses. On doing so he must notify the architect immediately and must spell out in that notice what action he proposes taking to overcome the effect those conditions will have on the works. The architect must then issue an appropriate instruction. Compliance with this will rank for payment unless the ground conditions etc. could have been reasonably foreseen by a skilled, experienced and qualified contractor. But importantly, the contractor's ability to foresee the conditions is measured at the date the contract was made and not after the works have begun. Under JCT 98 and its derivatives the position is more complex, but where SMM is used the contractor may well have a claim against the employer in respect of certain adverse ground conditions. In *C. Bryant & Son Ltd* v. *Birmingham Hospital Saturday Fund* (1938)[603], Bryant contracted to erect a convalescent home. The contract was in RIBA form with relevant provisions not dissimilar to JCT 98. The bills formed part of the contract and by clause 11, unless expressly stated otherwise, the bills of quantities '. . . shall be deemed to have been prepared' in accordance with the (then current) SMM.

This required that, where practicable, the nature of the soil should be described and that attention should be drawn to any trial holes, and that excavation in rock should be given separately. The bills referred the contractor to the drawings, a block plan and the site, to satisfy himself of the local conditions and the full nature and extent of the operations. The architect knew that there was rock on site, but it was not shown on the plans or referred to in the bills. They contained no separate item for excavation of rock. The High Court held the contractor to be entitled to treat the excavation in rock as an extra and to be paid the extra cost of the excavation plus a fair profit. To that extent the terms of JCT 98, clause 2.2.2.1 similarly provide that 'the contract bills . . . unless otherwise specifically stated therein in respect of any specified item or items . . . are to have been prepared in accordance with the Standard Method of Measurement' (SMM7). Various provisions in SMM7 require information about ground conditions to be given. The main provision is to be found in section D. Among the information which is to be provided will be the ground water level and the date when that level was established, details and location of trial pits and boreholes and live overground and underground services. In many cases a contractor working under JCT 98 will have a claim and may, under clause 2.2.2.2, treat the correction of departures from the SMM7 or any errors or omissions in description or quantity contained in the contract bills as a variation.

Other remedies may be available to the contractor, e.g. if there has been a misrepresentation (qv) about the ground conditions such as, for example,

---

[603][1938] 1 All ER 503.

where the contractor is misled by site information provided by the employer[604]. The employer may try to protect himself by a disclaimer of liability, but the case law establishes that this is not an easy thing to do and the courts seem prone to impose liability if it is possible to do so.

Site manager A term used in ACA 3, clause 5.2, to describe the contractor's full-time representative on site in charge of the works. He must be appointed before work starts on site and the architect's consent on both his appointment and his removal or replacement is necessary. Some of his duties are described in clause 5.3. He is to attend meetings convened by the architect in connection with the site works, must keep complete and accurate records and make these available for inspection by the architect. Like all the contractor's employees under ACA 3 he must be properly skilled, qualified and experienced. His is a key appointment. Under GC/Works/1 (1998), clause 5, the person to be appointed to represent the contractor on site is stipulated to be a 'competent agent' whereas, under the JCT 98 and IFC 98 contracts (clauses 10 and 3.4 respectively), although the precise term 'site manager' is not used, the contractor nevertheless undertakes at all reasonable times to keep upon the works 'a competent person in charge ...'. The expression 'Site manager' is also used in WCD 98, Supplementary Provision S3.
See also: **Person-in-charge.**

Sit-in An expression of industrial dispute in which people occupy some building or place (usually their place of work) until their demands are satisfied or they are forcibly evicted. It is trespass (qv). Although strike (qv) and lockout (qv) are expressly stated in some contracts as ground for extension of time, a 'sit-in' is not included.

Snagging list An expression commonly used on site for any list of defects. In an endeavour to be helpful, a clerk of works (qv) will often go beyond his duties and provide the contractor with a list of work requiring to be completed or rectified before, in the opinion of the clerk of works, the works will be ready for the architect to certify completion. Such a list is, of course, of no contractual effect and binds neither the architect nor the contractor. The contractor is under no obligation to take notice of the 'snagging list', only to fulfil his obligations under the contract.

Nevertheless, contractors often welcome such a list and architects often encourage the clerk of works to prepare such a list before completion. There is a danger that the contractor will be persuaded to do more than is necessary and the architect should not become associated with such a list unless he wishes to be bound by it. The architect should never mention the clerk of works' snagging list in any correspondence. The 'contractor's list' and the 'architect's list' mentioned in ACA 3 (clause 12.1) are expressly empowered by that contract. They are likely to be referred to indiscriminately on site as 'snagging lists'.

---

[604] *Morrison-Knudsen International Co Inc* v. *Commonwealth of Australia* (1972) 13 BLR 114.

The term is also commonly given to the 'schedule of defects' which the architect prepares at the end of the defects liability period (qv) under JCT 98. Although it is too much to expect that the expression will be obliterated from site conversation, the architect should be meticulous in using the correct terms to avoid confusion.

Special damage(s) Damage of a kind which the law will not presume in the claimant's favour, but which must be specifically pleaded and proved at the trial or arbitration hearing, e.g. interest on money (qv) in some cases, loss of profit, medical expenses, etc. It is contrasted with *general damages* which are the damages the law presumes will have resulted for the defendant's act.
See also: **Damages; General damages.**

Specialist A person who concentrates on a particular facet of his trade or profession. Thus a lawyer may specialise in building contract law, an architect may specialise in the restoration of old buildings, etc. In the context of construction contracts, it refers to a person or firm who concentrates on a particular aspect of the construction process, e.g. lift installation, heating, lighting, etc.

In contrast to the position where an individual architect or engineer is merely especially highly qualified in his general profession[605], where an architect, engineer or other professional professes specialism in a particular facet of his broad profession the law will measure the conduct and relevant standard to be expected of that specialist against that to be expected of the ordinary, competent and skilled professional practising that particular speciality and not merely against the conventional standards of practitioners in the wider profession.

Specialty contract A contract executed as a deed.
See: **Contract; Deed.**

Specific performance Where damages (qv) would be inadequate compensation for breach of contract (qv) the contractor may be compelled by means of an injunction (qv) to perform what he has agreed to do by a decree of specific performance. The court will not grant specific performance of an ordinary building contract which would, in effect, require supervision by the court[606]. However, if someone agrees to lease land and erect buildings on it, he may be granted a decree of specific performance provided:
— The building work is defined by the contract.
— The claimant has a substantial interest in the performance of the contract such that damages would be inadequate compensation for the defendant's failure to build.
— The defendant is in possession of the land. Specific performance is a discretionary remedy and is commonly used to compel performance of

---

[605] *Wimpey Construction UK Ltd* v. *D.V. Poole* (1984) 27 BLR 58.
[606] *Hepburn* v. *Leather* (1884) 50 LT 660; *Ryan* v. *Mutual Tontine Westminster Chambers Association* [1893] 1 Ch 116.

contracts for the sale, purchase or lease of land. It will be granted in the case of contracts of personal service.

Specification A document which, together with the drawings, describes in detail the whole of the workmanship and materials to be used in the construction of a building. In contracts which include bills of quantities (qv) as part of the contract documents (qv) the specification is not always a contract document but is merely to assist the contractor and amplify the drawings. Where no bills are included in the contract documents, which is an option expressly provided for under IFC 98, the specification becomes a very important contract document. In this latter case, it should include preliminaries, as for bills of quantities, and preambles as part of the trade descriptions. The specification must describe:
— Quality of materials.
— Quality of workmanship.
— Assembly.
— Location.
The main body of the document is normally divided into elements of construction in much the same sequence as they would be built. Where the specification is to be priced, every detail of the work should be described although not quantified.

The National Building Specification (available from NBS Services, The Old Post Office, St Nicholas Street, Newcastle upon Tyne NE1 1RH) is available only on a subscription service with new material being issued several times a year via both disk and hard copy for insertion into loose-leaf ring binder, thereby keeping it up to date and providing a comprehensive range of standard clauses to simplify the production of both specifications and bills of quantities. See also: **Performance specification.**

Specified perils A term found in the insurance provision of JCT contracts (e.g. JCT 98, clause 22C.1). The specified perils are defined (by clause 1.3) as fire, lightning, explosion, storm, tempest, flood, bursting or overflowing of water tanks, apparatus or pipes, earthquake, aircraft and other aerial devices or articles dropped therefrom, riot and civil commotion, but excluding excepted risks (qv). They are identical to the former 'clause 22 perils' of JCT 80. Where the employer is to insure existing property, the obligation is to insure against loss or damage due to specified perils in contrast to the obligation of employer or contractor to insure new works against loss or damage due to all risks (qv).

Speed reply A system of answering letters particularly favoured by busy executives. In essence, the system works as follows. The answer to correspondence is typed or written on the bottom of the letter to which it refers. The letter is then photocopied and the copy sent to the correspondent. The original is sent to file. Advantages are, as the name suggests, speed, efficiency and saving on expense. Disadvantages are that it must be brief and may not be fully understood by the recipient. The system is said to have originated in South Africa.

SR & O An abbreviation for statutory rules and orders (qv).

Stage payment A general term often used to indicate any payment made during the progress of the work. It is more accurately used for payments made at specific stages of work, e.g. damp proof course level, first floor level, eaves level, etc. This mode of payment is usually confined to relatively small lump sum contracts (qv) without quantities, where a proportion of the total sum is agreed to be paid over in a number of stages and the proportions are fixed so that they do not depend upon any re-measurement of work. Surprisingly, the simpler form of contract, MW 98 (clause 4.2), does not expressly provide for the possibility of an agreed regime of stage payments in lieu of the traditional method of measurement and valuation for interim certificates. However, such a provision is made within the more complicated JCT 98 (clause 30.2) where the detailed provisions for interim valuation is made subject to, and may be overridden by, any alternative agreement between the parties as to stage payments. A similar facility is also provided under IFC 98 (clause 4.2). It should be noted that an agreement as to stage payments must be made by the parties to the contract. Unless expressly empowered to do so, neither the architect nor the quantity surveyor should take it upon himself to make any such ad hoc agreement.

If the parties do propose adopting a process of stage payments in lieu of the traditional scheme for interim valuation and certification, a number of difficulties could arise and careful consideration must therefore be given to any such proposed agreement before the contract is made. That agreement must also be set out clearly and in detail within the contract if later disputes and differences over matters such as the following are to be avoided:

— Whether interim payment will also be made for unincorporated materials on site.
— Whether and if so how, agreed stage payments will be adjusted if and to the extent that the contractor's proposed sequence of works (and thus the anticipated stages of completion) are forcibly altered, delayed or disrupted.
— Whether and if so how retention will be dealt with.
— Whether and if so how and when the agreed value of any stage payment may be adjusted upwards or downwards on account of any additions, omissions and/or variations to the work in any particular stage; or on account of any ascertained loss and expense or any other matter normally entitling the contractor to interim additional payments.
— How and when notices satisfying the Housing Grants, Construction Regeneration Act 1996 will be given in respect of any stage payment or adjusted stage payment.

— The effect on release of any stage payment where relatively inconsequential elements of the work stage are, in the architect's opinion, incomplete or defective.

The BPF System (qv) provides for stage payments for consultants (qv) in lump sums depending upon the stage reached in the design and development process. The BPF System also provides, in effect, for stage payments to contractors except that the contractor determines the stages and the amount payable in the Schedule of Activities (qv).

GC/Works/1 (1998) provides in clause 48 for monthly advances to be paid in accordance with stage payment charts which are charts, tables or graphs included with the invitation to tender and specifying the periods and amounts of the advance payments made to the contractor during the performance of the works.

See also: **Interim certificates; Interim payment.**

Standard forms of contract A printed form of contract containing standard conditions which are applicable (or can be made applicable by the use of alternatives) to a wide range of building projects. They are generally preferable to specially drafted contracts because they are intended to be comprehensive and avoid most of the pitfalls which surround contractual relations in the building industry. Examples of standard forms are:
— The JCT series of contracts (qv).
— The ACA contract (qv).
— GC/Works/1 (qv).
— NEC (qv).

Standard Method of Measurement (SMM) A document which, in relation to building works, is published by the Royal Institution of Chartered Surveyors and the Construction Confederation. Its purpose is to assist all those connected with the construction industry by standardising and rationalising procedures for the preparation of bills of quantities (qv) for building construction. It lays down rules governing the extent to which items should be separately identified or quantified in the bills of quantities or shall be deemed to be included, or separately referred to in the description of another item. The current edition is number 7, but the document is revised at regular intervals to effect improvements and take account of developments in the industry. JCT 98 expressly requires (at clause 2.2.2.1) that the contract bills (qv), unless otherwise specifically stated therein, are to have been prepared in accordance with the SMM. Other standard methods of measurement have been developed and are appropriate for use in connection with civil engineering, roads and bridges and highways works etc.

Standing offer Where tenders (qv) are invited for the carrying out of work or the supply of goods or services over a period of time at irregular intervals, the tenderer may make a standing offer. Whether or not he does so depends on the terms of his offer (qv) and the acceptance (qv). If the tender is to the effect that the contractor will supply, e.g. 'bricks, if and when required between 1 January and 31 December 1989' this is a standing offer.

It is an offer to supply such quantities as may be required. A standing offer may be withdrawn before it is accepted by placing a specific order. Once an order for a specified quantity is placed, the contractor must supply the goods ordered; the order is the acceptance[607].

Standing orders Rules of procedure which apply in Parliament, local and public authority organisations, etc. Local authority standing orders may lay down rules which must be observed in the making of contracts etc., e.g. as to when a performance bond (qv) is required or when a contract must be entered into as a deed. In that regard it will be noted that whereas Private Editions of JCT 98 incorporate standard attestation clauses for the execution of the contract, Local Authority Editions do not[608].

Standing orders are, however, internal procedures and so, for example, a local authority cannot therefore rid itself of an onerous burden assumed under a contract by pleading that the contract is void because it was entered into contrary to standing orders. This situation must be distinguished from that where an authority has entered into a contract *ultra vires* (qv).

Contractors dealing with local authorities are not affixed with notice of standing orders and so are protected if standing orders have not been complied with. The provision, however, does not validate an otherwise invalid contract, e.g. if in fact the local authority never consented to contract at all[609].
See also: **Local authority; Ultra vires.**

Stare decisis Literally, to stand by things decided. It refers to the binding force of judicial precedent (qv) and is the basis of all legal argument and decision of the common law in England and other countries. In certain circumstances the judge is *bound* to stand by the decided cases, although judges often exercise considerable ingenuity in seeking to avoid the application of precedents which they dislike.

Statement of case The new generic reference to formal legal documents[610] which were formerly called pleadings. They comprise claim form (qv); particulars of claim (qv); defence (qv); defence and counterclaim (qv); reply; Part 20 claim form; particulars of Part 20 claim (qv); request for further information and clarification; and further information and clarification.

The function of these documents is to set out the basis of each party's case, identifying the facts relied upon in the allegations made based on those facts.

---

[607] *Percival Ltd* v. *LCC Asylum & Mental Deficiency Committee* (1918) 16 LBR 367.
[608] The Local Government Act 1972, s. 135, allows local authorities to contract in any way authorised by standing orders.
[609] *North-West Leicestershire District Council* v. *East Midlands Housing Association Ltd* [1981] 3 All ER 364.
[610] CPR Part 16.

They should inform the other party and the court of the case intended to be advanced and what issues are in dispute.

Statement of claim A document in which the plaintiff (now claimant) in litigation or the claimant in arbitration sets out all the material facts and law which he relies on as forming the basis of the case and a statement of the remedy or remedies sought. In litigation this is now called particulars of claim. Although the trend is quickly moving away from doing so, the statement is still usually expressed in relatively formal language and it must always be most carefully drafted. It is served on the defendant (the respondent in arbitration) and on the court or arbitrator and it should be sufficiently particularised and precise that it will enable the court (or arbitrator) and the opposing party to ascertain precisely the issues between the parties. It should serve, too, to establish at the outset any common ground that there might be.
See also: **Statement of case; Particulars of claim.**

Statute An Act of Parliament (qv).

Statute-barred Sometimes actions cannot be brought successfully because of lapse of time even though the cause of action may otherwise be sound. Such actions are said to be 'statute-barred' because of the time limits which are imposed by the Limitation Act 1980 in England and Wales and by the Prescription and Limitation (Scotland) Act 1973, as amended.
See also: **Limitation of actions.**

Statutory demand A written demand made on a company for payment of an overdue debt, pursuant to s. 123 (1) (a) of the Insolvency Act 1986 and Insolvency Rules 1986 rr.4.4−4.6[611]. If unjustifiably ignored by the debtor for more than 21 days it may, in the absence of any other proof, provide evidence of the debtor's insolvency. Although not an essential pre-requisite a statutory demand is often sent as a precursor to service of a formal winding up petition.

Provided the debt concerned exceeds the statutory minimum (presently set at £750) and provided the demand is made and served in the prescribed standard form and manner then, if the company fails to make payment within 21 days and the debt cannot be disputed in good faith the company will be deemed, without further proof, to be insolvent.

Statutory duty, breach of Many statutes impose duties on individuals to do something or not to do something and the statute itself may provide the only remedy (qv). In other cases, e.g. the Factories Act 1961 and related statutes, statute imposes general statutory duties in respect of classes of people, such as employees. Breach of statutory duty in this sense can give rise to a claim for

---

[611]SI 1986/952.

damages in tort (qv) when − as a result of a breach of the statutory duty − a person is injured[612].

It is a question of interpretation whether the statute gives a special remedy or whether it co-exists with an existing common law remedy, e.g. an action for damages for negligence.

In some cases the statutory duty is merely enforceable by sanctions of the criminal law. For example, the Health & Safety at Work etc. Act 1974 imposes general duties on employers, employees and others, but s. 47 of the Act makes it clear that such duties do not generally confer any right of civil action, i.e. if there is a breach of the Act's provisions, the injured person cannot bring a claim for damages for breach of the broken duty.

Statutory duties may be absolute (qv) but this is unusual. Breach of the Building Regulations does not, it seems, give rise to absolute liability[613].

Claims for damages for breach of statutory duty are very common and it is probable that the duty to comply with statutory requirements overrides even an express contractual obligation[614].

Statutory instruments The most important class of subordinate or delegated legislation. For the most part they are regulations made by a Secretary of State, e.g. The Building Regulations 1985, for particular purposes. They have the force of law.

Statutory rules and orders Regulations which were made by the King in Council, Government departments and other authorities. In 1948 they were generally superseded by statutory instruments (qv), but Orders in Council are still made, depending upon the procedure laid down by the relevant enabling statute (qv).

Statutory undertakers Organisations such as water, gas and electricity companies which are authorised by statute (qv) to construct and operate public utility undertakings. They derive their powers from statute, either directly or from previous authorities undertaking the function by virtue of statutory instruments (qv). Although their powers are extensive, they are not absolute and constraints are placed upon the exercise of their powers. Failure to observe these constraints can lead to complaints being laid before the appropriate minister or to an action for damages (qv) or an injunction being pursued in the courts.

Statutory undertakers may be involved in a building contract either in performance of their statutory obligations or as contractors or sub-contractors. When performing their statutory obligations they are not liable in contract[615],

---

[612]*Quinn* v. *J. W. Green* (Painters) Ltd [1965] 3 All ER 785.
[613]*Perry* v. *Tendring District Council* (1985) 3 Con LR 74; *Kimbell* v. *Hart District Council* (1987) 9 Con LR 118.
[614]*Street* v. *Sibbabridge Ltd* (1980) unreported.
[615]*Clegg Parkinson & Co* v. *Earby Gas Co* [1896] 1 QB 56; *Willmore* v. *S. E. Electricity Board* [1957] 2 Lloyds Rep 375.

although they may be liable in tort (qv). Most standard building contracts draw a distinction between statutory undertakers performing their statutory duties as such and those cases where they are acting as contractors. JCT 98, clause 6.3 draws that distinction, which is also relevant in claims (qv) situations.

Delay to the works caused by a statutory undertaker carrying out or failing to carry out work in pursuance of its statutory obligations in relation to the works is a ground which entitles the contractor to an extension of time under JCT 98, clause 25.4.11. Likewise under IFC 98, clause 2.4.13. However, neither contract affords the contractor any entitlement to be reimbursed his loss and expense arising from any such delay or disruption so caused. This is a 'neutral event' (qv). Like JCT 98 and IFC 98, ACA 3 similarly draws a clear distinction between the activities of statutory undertakers performing works in pursuance of their statutory obligations as opposed to merely working directly under contract for the employer.

Stay of proceedings The courts have very wide powers concerning whether or not to put a stop, temporary or permanent, to proceedings brought before them, as part of their inherent jurisdiction. Specific powers in that regard are also conferred on them by statute (qv) in many cases.

The discretion is, however, severely restricted where parties have agreed – in terms such as those in JCT 98, Article 7A and clause 41B – to refer their disputes and differences to arbitration rather than litigation. In that case, where a party seeks, instead, to litigate the dispute, then except in the most exceptional circumstances (i.e. where the arbitration agreement is held to be null and void, inoperative or incapable of being performed or where both parties have taken formal steps in the litigation), the courts must stay the legal proceedings[616].

See also: **Arbitration.**

Strict liability Liability irrespective of fault. It arises under the rule in *Rylands* v. *Fletcher* (1868)[617]. Negligence need not be proved where things likely to cause damage are kept on property. The rule is:

> 'A person who for his own purposes brings on to his land and collects and keeps there anything likely to do mischief if it escapes, must keep it at his peril and, if he does not do so, is *prima facie* answerable for all the damage which is the natural consequence of its escape.'

The following points should be noted:
— The rule only applies to a 'non-natural use of land', e.g. blasting operations, demolition operations, water in a reservoir. It does not apply to things naturally on land or to the use of water etc. for ordinary domestic purposes.
— There must be an *escape* from the land.

---

[616]Section 9 Arbitration Act 1996.
[617](1866) LR 1 Exch 265.

— Liability is strict but not absolute (qv) but it arises independently of either negligence (qv) or nuisance (qv).
— Various defences are available: e.g. Act of God (qv); that the damage was caused by the plaintiff's own act or default; that the escape was due to a third party, statutory authority, etc.

Strike A simultaneous withdrawal of labour by the whole or a part of an employer's workforce (employer used in the general sense). In many forms of building contract, it is expressly stated as a ground for extension of time, e.g. JCT 98, clause 25.4.4; IFC 98, clause 2.4.4.

ACA 3 does not refer to strikes, but it is thought that in certain circumstances an extension could be given under the head of *force majeure* (qv).

Sub-contract A contract made between a main contractor and another contractor for part of the work which the main contractor has already contracted to carry out as part of his contract with the employer. Such other contractor is referred to as a sub-contractor.

See also: **Assignment and sub-letting; Domestic sub-contractor; Named sub-contractors; Nominated sub-contractors.**

Sub-contractor A person or firm to whom part of the main contract works are sub-let.

See also: **Assignment and sub-letting.**

Subject to contract In general, the use of the phrase 'subject to contract' indicates an intention not to be bound. There is no enforceable obligation until the contract (usually a formal document) is made. This is commonly the case in contracts for the sale of land.

However, where the parties are agreed on the terms of the contract and acceptance is made subject to the execution of a formal document, it is a question of construction (qv) and interpretation (qv) for the courts to decide whether or not there is a concluded contract[618].

See also: **Acceptance; Conditional contract; Contract; Offer.**

Subrogation The substitution of one person or thing for another. Someone who discharges a liability on another's behalf is, in general terms, put in the place of that other person for the purpose of obtaining relief against any other person who is liable. The most important practical example arises in the field of insurance where an insurer who compensates a policy holder for loss is entitled to stand in the policy holder's shoes and recover from the person who caused the loss.

---

[618] *Branca* v. *Cobarro* [1947] 2 All ER 101.

Substantial completion In an ordinary lump sum contract (qv) provided the contractor has *substantially* performed his work, he will be entitled to recover the contract price, less a deduction in respect of defects[619]. The nature of any defects (or *de minimis* (qv) works) to be carried out or completed must be taken into account as well as the proportion between the cost of rectifying them and the contract price[620].

Substantial completion means complete in all major particulars and should be contrasted with 'practical completion' (qv).

The form of contract which comes nearest to requiring substantial completion is GC/Works/1 (1998) which, at clause 34, requires that by the completion date the contractor shall '... deliver up the Site and the Works in all respects to the satisfaction of the PM'. It seems, therefore, that the PM may accept a lesser state of completion than that envisaged in the JCT forms. However, substantial completion implies that only very minor items (*de minimis*) works will be outstanding.

See also: **Performance.**

Substantially To a considerable degree, not trivial.

Substitute sub-contractor ACA 3, clause 9.6 makes provision for the situation where the contractor is unable, for any reason beyond his control, to enter into a sub-contract with a named sub-contractor (qv) or supplier. The contractor is given the duty to select another person to carry out the work or supply the materials which must be of equivalent standard and quality. The contractor is not entitled to any damage, loss and/or expense or any extension of time in complying with the provisions of this clause. JCT 98, clauses 35.18.1.1 and 35.18.1.2 use this term to refer to the sub-contractor renominated by the architect where the nominated sub-contractor wrongfully refuses to rectify defects in the sub-contract works which appear after final payment to the nominated sub-contractor but before the issue of the final certificate.

Substituted contract A substituted contract arises where there is a novation (qv) and a new contract is substituted for the old. If the substituted contract incorporates or refers to the original one, the two will generally be read together[621], but the liabilities of the parties are always a question of interpretation.

Suitability for purpose Under s. 14 of the Sale of Goods Act 1979 there is an implied condition that goods are reasonably fit for the purpose required in circumstances where the buyer is relying on the seller's skill and judgment, as is normally the case. Under the Unfair Contract Terms Act 1977 (qv) this – and other terms implied by the Act – can be excluded only to a limited extent. As regards building work generally, it is now settled law that, in the absence of

---

[619] *Hoenig* v. *Isaacs* [1952] 2 All ER 176.
[620] *Bolton* v. *Mahadeva* [1972] 2 All ER 1322.
[621] *A. Vigers Sons & Co Ltd* v. *Swindell* [1939] 3 All ER 590.

some express term removing the liability, 'the builder will do his work in a good and workmanlike manner; that he will supply good and proper materials; and that (the completed structure) will be reasonably fit for the purpose required'[622].

Where contractors and sub-contractors undertake to design (qv) the whole or part of a structure, in the absence of a clear contractual indication to the contrary it is implied that they undertake to design a structure which is reasonably suitable for the purpose made known to them[623]. This obligation is to be equated with the statutory obligation of a seller of goods. Hence, many design and build contracts (qv) (e.g. WCD 98, clause 2.5) modify this liability so that it severely restricts that otherwise onerous obligation. Such a clause is, in effect, an exemption clause (qv).

See also: **Supply of Goods and Services Act 1982.**

Summary judgment The procedure by which the court can decide all or any part of an issue without recourse to a full trial where it is satisfied that the claimant has no real prospect of success or the defendant, likewise, has no real prospect of defending the claim against him.

A claimant may apply for summary judgment under Part 24 of the Civil Procedure Rules (CPR) (qv) after the filing of an acknowledgement of service (qv) or a defence (qv). A defendant may also apply for summary judgment against a claimant. The application may relate to specific issues within the proceedings or relate to all matters in dispute. Application is made on the prescribed form[624] so as to comply with the requirements of CPR Part 23. Any evidence upon which the applying party wishes to rely should be served with the application[625] when it is filed with the court and served on the responding party[626]. The responding party must file any written evidence upon which he intends to rely seven days before the hearing (qv) date[627]. Any evidence in reply must be served three days before the hearing[628].

Provided the court is satisfied that at a full trial the defendant would have no real prospect of defending the claim or in the claimant's case the claim would have no real prospect of success such that the issue did not warrant a full trial, then summary judgment will be given, usually together with the associated costs of the application. Where the application is resisted, the parties will require to be represented at a brief court hearing to show whether there is a triable issue. The court has very wide powers in relation to such applications and may make an absolute or conditional order for the continuation of all or part of the claim, in the latter case imposing such conditions as it deems appropriate such as, for

---

[622]See, for example, *Hancock* v. *B. W. Brazier (Anerley) Ltd* [1966] 2 All ER 901.
[623]*Independent Broadcasting Authority* v. *EMI Electronics Ltd & BICC Construction* (1980) 14 BLR 1.
[624]N. 244.
[625]Evidence in support of such an application may be given on the N244 or the party may rely on his statement of case (qv) or file any necessary witness statement(s) which themselves must comply with CPR Part 22.
[626]Paragraph 9.3 of the Practice Direction to CPR Part 23.
[627]CPR Rule 24.5 (1).
[628]CPR Rule 24.5 (2).

example, requiring some payment to be made into court by way of security. In dealing with such applications, it is not the function of the court, however, to try the dispute. Its function is merely to decide whether and, if so, to what extent the defence (or claim) is other than frivolous.

It is a useful and generally quick way of obtaining judgment for the price of goods or services supplied. The usual defence is that the goods or services were defective. It is also a method of obtaining payment under certificates (qv), especially where the architect has not served any notice(s) of set-off (qv)[629].

Summons A document used in court procedure requiring a person to attend court for a particular reason, e.g. to obtain documents, orders, or to testify as a witness, etc.

See also: **Witness summons.**

Superintending officer A term used in GC/Works/1 Edition 2 to indicate the person who supervised the work, and there abbreviated to SO. He was very roughly in the same position as the architect under the JCT or ACA forms of contract. Clause 1 (2) stated that he was to be designated in the Abstract of Particulars (qv) and indeed he may well have been an architect or an engineer. His duties in relation to the contract were set out within the body of the contract. GC/Works/1 Edition 3 introduced the term project manager (PM) (qv) to describe the person carrying out the function of the superintending officer and the title of PM has been retained in GC/Works/1 1998.

Supervising officer A term formerly used in the JCT forms Local Authorities Editions, and IFC 84. Its purpose was to enable an official in the local authority, who may not be an architect, to act in that capacity in relation to the contract. The title 'architect' is legally protected (see: **Architects Registration Council**) and the appropriate chief officer may be a chartered engineer or member of the Chartered Institute of Building. The term 'supervising officer' no longer appears in the Local Authorities Editions of JCT 98. Nor does it appear in IFC 98. In both cases it has been replaced by a reference to the 'Contract Administrator' which is a term similarly used in MW 98 where, given the likely scale and simplicity of the works, the employer may not wish to engage the services of a registered architect.

Private Editions of JCT 98 make no such equivalent provisions, leaving the appointment under the standard form solely that of an architect.

The term 'supervising officer' is still used in the ACA 3 form of agreement where, as an alternative to the appointment of an architect, the employer may instead (under Alternative 2 of Recital E) appoint a supervising officer to act in that capacity. Where that alternative is used, all references in the agreement to 'Architect' are to be taken as a reference to the 'Supervising Officer'.

---

[629]The need to resort to court action in this case has been greatly reduced by a contractor's statutory or express right to suspend work for non-payment. See: **Suspension; Housing Grants, Construction and Regeneration Act 1996.**

Supervision of works Supervision implies constant inspection and direction. In building contracts, that duty lies principally with the contractor who will normally carry out this duty through his site agent, foreman, etc. It is not the responsibility of the architect under the terms of any of the standard forms.

Nor is constant inspection and direction expected of the architect under the provisions of the Standard Form of Agreement for the Appointment of an Architect (SFA/99) or Conditions of Engagement for the Appointment of an Architect (CE/99) (if used) where it is expressly stated that, if frequent or constant inspection is required, a clerk of works or other resident inspector will be appointed by the employer. It should be noted, however, that by JCT 98, clause 12 supervision is not one of the duties of the clerk of works. Clause 1.5 of JCT 98 puts it beyond doubt that neither architect nor clerk of works is responsible for supervising the work that the contractor is to carry out and complete.

See also: **Inspector.**

Supervisor A person who directs or oversees the works. Under the BPF System (qv) the supervisor is the firm or person responsible for monitoring that the works are built in accordance with the contract documents (qv). Under that system he may be an architect, engineer, building surveyor, clerk of works (qv), etc. The supervisor's main responsibility is to monitor the contractor's design, construction, commissioning and maintenance of the project, ensuring that the workmanship and materials are up to contract standard.

Supplier A person or firm undertaking the supply of goods or materials to a contract. The supplier's contract is with the main contractor.

See also: **Nominated suppliers.**

Supply chain partners A term used in prime contracting (qv). It refers to the long-term supply (of goods and/or services) relationships which are said to be essential to the success of prime contracting. The long-term relationships are intended to improve the value of what the supply team deliver over a period of time and not simply referrable to one project. The aim is to create a situation in which the supply team gains increased share of the market. Prime contracting is intended to allow profit and overhead recovery margins to increase in conformity with the improvement to the underlying benefit to the client.

See also: **Continuous improvement; Right first time; Supply clusters.**

Supply clusters A term used in connection with prime contracting (qv). A cluster designs and delivers a particular part of the project, for example, mechanical and electrical services. There are no standard clusters. Each one is formed as and when required in connection with a specific project. It is likely that a cluster will involve designers, sub-contractors, materials suppliers and component suppliers working together with the aim of delivering the best value. Each cluster has a leader who must be a long-term supply partner of the prime contractor.

See also: **Continuous improvement; Right first time; Supply chain partners.**

Supply of Goods and Services Act 1982 Broadly speaking, this Act introduces statutory implied terms (qv) in contracts for the supply of goods and services which do not fall within the ambit of the Sales of Goods Act 1979. It applies, *inter alia*, to contracts for work and materials, hire, exchange or barter, as well as services. Contracts for the sale of goods and hire purchase are covered by other legislation.

Part I of the Act deals with the supply of goods, and its provisions affect building contracts, e.g. as regards materials supplied in the execution of the work. Sections 1 to 5 cover 'transfer of goods' and extend to contracts for work and materials.

Hire of goods is covered by sections 6 to 10 and these provisions are important in the case of plant hired in by contractors.

Part II of the Act deals with the supply of services – which includes professional services. Where the supplier is acting in the course of business there is an implied term that he will carry out the service with reasonable care and skill – an obligation which is already implied at common law. Under s. 15, where no price is fixed for the service, there is an implied term that a reasonable charge will be paid (see: **Quantum meruit**).

Various exemption orders have been made excluding particular categories from the effect of Part II of the Act, e.g. arbitrators. The Act seems unlikely to have any great impact in the field of building contracts and, to a large extent, it merely gives statutory effect to obligations that were already implied by the general law.

See also: **Unfair Contract Terms Act 1977.**

Support, right of An easement (qv) whereby the owner of one house has the right to have it supported by the adjoining house belonging to his neighbour. However, even where a right of support exists, the adjoining owner against whom the right is claimed ('the servient owner') is under no obligation to maintain his property in such a state of repair so that it gives support to the adjoining owner's property[630]. Where a right of support exists, the adjoining owner must provide equivalent support if the original support is removed. Without such a right or privilege there is no liability on an adjoining owner if he demolishes his property, although there might well be a claim in negligence (qv) if the demolition was undertaken in such a way that damage occurred to the neighbour's property. There is no natural right of support. The general rules may be affected and modified in the case of party walls (see: **Party Wall Act 1996**).

Even if there is a right of support, there is no right to weatherproofing and the right to have one's house protected against the weather cannot exist as an easement. Some of the effects of this rule are circumvented by sections 29, 29A, 29B and 29C of the Public Health Act 1961, under whose provisions a local authority may serve a notice on any person who has begun or who intends to

---

[630]*Bond* v. *Nottingham Corporation* [1940] 2 All ER 12.

begin a demolition etc. requiring him (among other things) to:
— Shore up any building adjacent to the building to which the notice relates.
— Weatherproof any surfaces of an adjacent building which are exposed by the demolition.
— Repair and make good any damage to an adjacent building caused by the demolition or by the negligent act or omission of the person engaged in it.

The recipient of such a notice may appeal on the grounds that the adjoining owner ought to pay or contribute towards the expense of weatherproofing the exposed surfaces, and these provisions do not apply where the building to be demolished is less than 1,750 cubic feet.

Surety A person who agrees to be responsible to a third party for the debts or default of another.
See also: **Bonds; Guarantee.**

Survey The careful inspection and recording of something. Thus, a survey of land or buildings may involve taking and recording measurements and making notes about condition. In a wider sense, it will involve inspections and testing and the taking of samples and cores. Geotechnical surveys report on the ground conditions of a site by using boreholes and reference to geological maps.

Suspension The employer has no power to direct suspension of the work under a building contract unless there is an express term in the contract empowering him to do so. Neither the JCT nor ACA standard forms confer any such express power of suspension on the employer or the architect. However, with the exception of MW 98 all the JCT forms empower the architect to 'postpone the execution of any work to be executed under' the contract (JCT 98, clause 23.2; IFC 98, clause 3.15). The exercise of this power of postponement can amount, in effect, to suspension.

Similarly, the contractor has no power to suspend execution of the work at common law and if he does so then this will amount to a breach of contract on his part[631]. Merely 'going-slow' will not amount to suspension giving rise to a breach of contact under JCT terms[632]. But, without lawful excuse or some express right to do so, the contractor cannot simply choose to suspend performance of the works or of his obligations under the contract; his basic obligation is to take possession of the site and to thereafter proceed regularly and diligently (qv) with the works until completion is achieved.

Some acts or omissions by the employer or his architect, interference by *force majeure* (qv), unforeseen events — such as the occurrence of a specified peril (qv) — terrorist activity or other extreme circumstance may inevitably

---

[631]*Canterbury Pipelines Ltd* v. *Christchurch Drainage Board* (1979) 16 BLR 76.
[632]*J. M. Hill & Sons Ltd* v. *London Borough of Camden* (1980) 18 BLR 31.

cause the works to be suspended. In that case it will generally be a question of fact and degree whether or not such a suspension is of sufficient magnitude to allow the innocent party in the case of default, or the parties generally in the case of a neutral event (qv), to treat the contract as frustrated (qv) and so discharged. Under JCT 98 (clause 28), with regard to certain such events (referred to in clause 28.2.2 as 'suspension events'), this uncertainty is largely removed so that:

— In the case of a 'suspension event' for which the employer or his architect is responsible, and which are listed in clause 28.2.2, the contractor will be entitled to determine his employment under the contract on proper notice if the suspension lasts longer than the period stated in the Appendix — (usually one month or more).

— In the case of suspension caused by one of the neutral events listed in clause 28A.1.1.1 to 28A.1.1.3, either party may determine the contract on proper notice if the suspension lasts longer than the period stated in the Appendix — (usually three months or more);

— In the case of suspension caused by one of the neutral events listed in clause 28A.1.1.4 to 28A.1.1.6, either party may determine the contract on proper notice if the suspension lasts longer than the period stated in the Appendix — (usually one month or more).

Following enactment of the Housing Grants, Construction and Regeneration Act 1996, where the parties are operating under a 'construction contract' (qv) an express statutory right to suspend his obligations under the contract is afforded to the contractor if the employer fails properly to observe his obligations concerning payment. With the exception of ACA 3, that statutory right now also appears as an express term in virtually all standard form contracts, including those drafted by the JCT and Property Advisors to the Civil Estates (PACE) (see JCT 98, clause 1.4; IFC 98, clause 4.4A; WCD 98, clause 30.3.8; MW 98, clause 4.8; GC/Works/1 1998, clause 52). In the case of GC/Works/1 1998, the provisions are rather more comprehensive but for all practical purposes have similar effect to those in JCT contracts. The standard forms of sub-contract intended for use with the JCT main contract forms give the sub-contractor similar express rights to suspend work in the event of the main contractor failing to pay to the sub-contractor amounts properly due.

Under GC/Works/1 1998, clause 40 (2) (g) the project manager has a general power to order 'the suspension or the execution of the works or any other part thereof' and under ACA 3, clause 11.8 the architect has power to 'postpone the dates shown on the Time Schedule for the taking-over of the works', etc.

Notably, JCT 98, clause 34, ACA 3, clause 14, and GC/Works/1 1998, clause 32(3) also oblige the contractor to take measures not to disturb and to preserve any object of archaeological or related interest found during the construction operations. In the case of JCT 98 and GC/Works/1 1998 an immediate obligation to cease work in the area of the find is also imposed if this should be necessary.

See also: **Postponement.**

Taking-over See: **Practical completion.**

Target cost In the BPF System (qv) this term is used to describe the amount which the client (qv) expects to pay for the design and construction of the completed building. The target cost includes all fees, costs of investigations and the forecast tender (qv) price.

Target cost is also a term used to describe a contract in which the contractor is paid his prime cost (qv), but if this exceeds or falls short of an agreed target the difference is shared between the contractor and the employer in pre-agreed proportions.

Taxation of costs An outdated term given to the process of reviewing the amount of reasonable costs recoverable by one party to legal (or arbitral) proceedings against the other where the court (or arbitrator) has ordered that the other should be liable, in principle, for the claiming party's costs of the litigation (or arbitration). The procedure is nothing to do with taxes imposed by the Inland Revenue. The term and a broadly similar process also applied in certain circumstances to a solicitor's costs to his client where it was open to the client to have those costs reviewed by the court.

Under the Civil Procedure Rules, where the costs relate to legal proceedings and they are reviewed by the courts, the process is now referred to as a 'costs assessment' (qv) and takes the form of either:
— Summary assessment – i.e. where the court orders a sum of money to be paid.
— Detailed assessment – i.e. where the amount of the paying party's liability for costs will be decided by a costs judge (qv).

Where the costs arise in arbitration proceedings and the review procedure is to be carried out by the arbitrator under the Arbitration Act 1996, s. 63 of that Act likewise makes the term 'taxation' redundant. Unlike previous legislation, the 1996 Act now draws a clear distinction between the more formal approach taken by the courts in relation to costs assessment and the less rigid procedure to be expected of an arbitrator when assessing the amount of the paying party's liability.

See also: **Costs; Costs assessment.**

Technology and Construction Court (TCC) Previously known as the Official Referees' Court and first established by the Judicature Acts 1873–1875. This name change, which became effective from 1 October 1998, is designed to more closely reflect the type of business for which this court has been and still remains primarily responsible since, from its creation as the Official Referee's Court, by far the greatest majority of its business has been concerned with all aspects of construction disputes and claims. Currently, some 80% of the TCC's workload is construction related with indications that this figure is set to rise.

Beyond the simple name change, other significant changes have been made to the powers, administration and constitution of this court. With the possibility of a growing trend towards a return of litigation in preference to arbitration as a means of dispute resolution and with the increasing caseload the courts now have to contend with resulting from the call on them to review adjudications arising from the introduction of the Housing Grants, Construction and Regeneration Act 1996 (qv), other important changes have also been made aimed at improving efficiency and reducing the time and cost of litigating what often are complex cases involving prolonged examination of documents or accounts, or requiring technical scientific or other investigation.

Among the most radical changes are reviews of the system of case management and the introduction of various pre-action protocols designed, in appropriate cases, to lead parties to positively explore alternative forms of dispute resolution. A centralised diary system has also now been introduced to more evenly distribute the workload of the Technology and Construction Court judges (qv) and a new date setting system has also been introduced, aimed at ensuring that judges who are allocated a case will see it through and 'manage' it during all of the initial stages leading to trial. Where possible and expedient, that judge may also take it through trial. For the first time, a High Court judge, (currently Sir John Dyson, the senior TCC judge), is appointed to sit permanently at the court and to oversee the various reforms brought about by those and the other fundamental changes to administration and procedure of the TCC.

See also: **Technology and Construction Court – Judge of.**

Technology and Construction Court (TCC) – Judge of Specialist judges who deal with claims involving issues or questions that are technically complex. They deal, in particular, with the substantial majority of construction industry business. They are High Court judges in function and are now given equal nominal status as High Court judges, being addressed in court as 'my lord' and not, as previously, as 'your honour'. There are currently seven judges assigned to deal with the business of the TCC in London, one being the senior judge appointed in overall charge of its administration. They are based at the High Court, St Dunstan's House, Fetter Lane, although if the majority of witnesses live at a distance from London or in other special cases they will sit at a location which is convenient to the parties.

Judges appointed to the TCC are not concerned exclusively with the construction industry, but they in effect form a construction industry court because since its formation as the Official Referees' Court in 1876, the industry has been the major user of their services. A case will be assigned to a named TCC judge – the 'assigned judge' (qv) – who then has primary responsibility for and extremely wide powers in relation to the active management of that case to achieve the overriding objectives[633], which are:

— ensuring that the parties are on an equal footing, and
— saving expense, and
— dealing with the case in ways which are proportionate to:
  • the amount of money involved
  • the importance of the case
  • the complexity of the issues
  • the financial position of each party.

The TCC judge will set a timetable for the steps he decides are necessary leading up to an early trial and this will involve active case management by the judge with the aim of:

— Encouraging the parties to co-operate with each other in the conduct of the proceedings.
— Identifying the issues at an early stage.
— Deciding promptly which issues need full investigation and trial and accordingly disposing summarily of the others.
— Deciding the order in which issues are to be resolved.
— Encouraging the parties to use an alternative dispute resolution procedure if the court considers that appropriate and facilitating the use of such procedure.
— Helping the parties to settle the whole or part of the case.
— Fixing timetables or otherwise controlling the progress of the case.
— Considering whether the likely benefits of taking a particular step justify the cost of taking it.
— Dealing with as many aspects of the case as it can on the same occasion.
— Dealing with the case without the parties needing to attend at court.
— Making use of technology.
— Giving directions to ensure that the trial of a case proceeds quickly and efficiently.

The majority of construction industry disputes which proceed to litigation are allocated to the judges of the Technology and Construction Court. It is possible to have disputes resolved by judges with relevant construction dispute experience outside London; however, the highest concentration of specialist judges and lawyers is to be found in London. In addition to specifically marking the claim form to identify it as one being heard in that court, all other documents relating to the case are also specifically marked with the name of the assigned judge. The main types of action dealt with by the TCC judges are:

— Claims by and against architects, engineers, surveyors and other professionals in contract and in tort.
— Claims relating to building, civil engineering and construction generally. These include a great many cases involving the interpretation of the standard form contracts such as those of the JCT.
— Claims by and against local authorities in respect of their statutory duties, especially those relating to the building regulations, public health and building legislation generally.

---

[633]Part 1, Civil Procedure Rules.

— Claims relating to work done and materials supplied or services rendered.

Many of these cases are lengthy and complex and involve highly technical issues as well as difficult points of law. Long cases are often divided into sub-trials. A large number of cases of considerable importance are finally decided by judges assigned to the Technology and Construction Court. All important judgments by judges of the Technology and Construction Courts affecting the building industry are now reported regularly in *Construction Law Reports,* published six times a year by Butterworth & Co (Publishers) Ltd.

Tender An offer (qv) by a contractor, usually in competition, which is accepted without any material qualification by the employer, will form a binding contract. The architect usually invites a number of contractors to tender on a form specially provided for the purpose. The contractors have a stated time in which to prepare their tenders and a date and time by which these must be deposited with the architect. Tenders must be returned in unmarked envelopes. Sometimes a priced bill of quantities (qv) must also be provided in a separate envelope so that it can be returned unopened if the tender is unsuccessful.
See also: **Code of Procedure for Single-stage Selective Tendering 1996; Code of Procedure for Two-stage Selective Tendering 1996; Invitation to tender.**

Term contract Used when services may be required over a period of time at irregular intervals. The chief characteristics of term contracts are that the contractor:
— Undertakes to carry out a particular category of work (e.g. plumbing, general repairs, etc.) within maximum and minimum individual job values.
— Undertakes to do the work for a particular time period.
— Undertakes to do the work within a particular geographical area.
— Undertakes to do the work at a particular rate.

The system finds its most useful application in maintenance work where the general scope of work and the area may be known but the precise jobs which have to be carried out are not known until the need arises. The contractor agrees to a schedule of prices (qv) which are applicable for the duration of the contract. It requires a good deal of experience to decide upon the correct rate for each item of work because some items may be seldom required. The theory is that the contractor will even out his gains and losses over the contract period. Competitive tendering is used to select the successful contractor. Advantages are that contractors gain familiarity with the property, and lower costs can be achieved than by attempting to secure tenders for each job as it arises.

Variations are the *Standard Form of Measured Term Contract 1998,* where the contractor has the opportunity of measuring the work before he tenders; the *Measured Term Contracts Based on Schedules of Rates* (GC/Works/7); the *Specialist Term Contracts* (GC/Works/8), for use where specified maintenance of equipment is required and is capable of being priced on a task by task basis;

Daywork Term Contracts, where the jobs are small and the pricing arrangements are somewhat more complex.

Term of the contract A provision or stipulation in a contract describing some aspect of the agreement. It may be express (written down or explicitly agreed orally), implied (included by the action of common law or statute) or incorporated (see: **Incorporation of documents**). Important terms are generally known as 'conditions' (qv), less important terms as 'warranties' (qv).

Third party Used generally to refer to any person who is not a party to a contract between two or more other parties. Third parties may be brought into a dispute by one of the parties who claims indemnity or joint liability. The term is given more specific meaning and application in the context of the Contracts (Rights Of Third Parties) Act 1999 where a 'third party' within the meaning of the Act will be either:
— someone that the contract expressly provides is such, or
— someone on whom a term of the contract purports to confer a benefit, unless it appears on a true construction of the contract that the contracting parties did not intend him to have the right to enforce it.

Additionally, in a litigation context, this formerly referred to subsidiary proceedings brought by a party for a contribution or indemnity from a third party in respect of a claim being faced by that party. For example, where an employer sued a main contractor because of leaking windows, the main contractor could have issued third party proceedings against the window manufacturer or supplier. This procedure has been replaced under the Civil Procedure Rules (CPR) (qv) by Part 20 proceedings (qv).
See also: **Contracts (Rights of Third Parties) Act 1999; Privity of contract; Part 20 proceedings.**

Time at large Time is said to be 'at large' when there is no specific date for the completion of the contract and in the absence of an express term as to the date for completion (qv) the contractor's common law obligation is then to complete 'within a reasonable time'. What is a reasonable time (qv) is a question of fact depending on all the terms of the contract and the surrounding circumstances. Time is not normally of the essence in building contracts (see: **Essence of the contract**). This is clearly the case where the contract itself provides—as do all the standard form contracts—for extension of time and liquidated damages for delay[634].
Under the normal standard form contracts, time may become at large if:
— The contractor has been delayed by the act or default of the employer or those for whom he is responsible in law and there is no contractual provision to cover the situation, e.g. a clause entitling the architect (on the employer's behalf) to grant an extension of time (qv).

---

[634]*Lamprell* v. *Billericay Union* (1849) 154 ER 850.

— The architect fails properly to grant an extension of time under the contract.

Except in the latter instance, time will seldom become 'at large' under any of the standard form contracts in common use. If it does, then as indicated, the contractor's obligation is to complete within a 'reasonable time' and the employer then forfeits any right to liquidated damages (qv)[635]. The subject is extremely complex, especially where the architect awards extensions of time after completion of the works. It is a question of interpretation whether he is entitled to do so. Some contracts (e.g. JCT 98 and ACA 3) give him specific powers to do so. The general rule, however, is that any extensions of time (qv) must be awarded properly and in accordance with the express contract provisions: failure so to do will result in the completion date (qv) becoming 'at large'[636].

Time Schedule An Appendix (qv) to the ACA 3 contract which offers two alternative schedules, each of which sets out a number of matters which the parties are to insert. Among them are the following.

Alternative 1 – for use where no phased or sectional possession or completion of the works is anticipated:
— Date for possession of the site.
— Date for taking-over of the works.
— Rate of liquidated damages.
— Maintenance period.

Alternative 2 – for use where phased or sectional possession and/or completion of the works is anticipated:
— Dates for possession of each section of the site as identified by description and by reference to an attached plan.
— Date for taking-over of the sections of the site identified by description and an attached plan.
— Maintenance period for each section.
— Rates of liquidated damages relevant for each section.

Title The right to ownership of property or the legal connection between a person and a right. The word is most commonly used in connection with land but applies to all kinds of property.

A title is said to be *original* where the person entitled does not derive his right from any predecessor, e.g. copyright. It is *derivative* where it is derived from someone else, e.g. by gift, purchase, inheritance or judgment of the court.
See also: **Nemo dat quod non habet.**

Tort A civil wrong other than a breach of contract or a breach of trust or other merely equitable obligation and which gives rise to an action for unliquidated damages at common law (Sir John Salmond). Literally the word is French for

---

[635] *Wells* v. *Army & Navy Co-operative Society Ltd* (1902) 86 LT 764.
[636] *Fernbrook Trading Co Ltd* v. *Taggart* [1979] 1 NZLR 556.

'wrong'. The essential point is that it is a breach of a civil duty imposed by the law generally. The most important tort today is negligence (qv), but other torts include nuisance (qv), trespass (qv) and defamation (qv).

Tortfeasor A person who commits a tort (qv).

Trade custom/trade usage See: **Custom.**

Trade discount A discount which is allowed by suppliers to members of the industry. Thus a building contractor will be able to purchase, say, timber at a price considerably below that at which it is available to members of the public. It is not the same as cash discount (qv).

Treasure trove Gold or silver coin, plate, bullion or other valuable items hidden in a house or in the earth or other secret place, the true owner being unknown and undiscoverable. Treasure trove belongs to the Crown. If the property is merely lost or abandoned it is not treasure trove, and the finder acquires a possessory right to it. The finder of treasure trove must report the finding to the coroner for the area, and an inquest will be held to establish whether or not the objects found are treasure trove. If they are, the Crown awards their market value to the finder.

In building contracts, there is usually a specific clause dealing with objects found on site. JCT 98, clause 4.3 provides that as between the contractor and the employer 'all fossils, antiquities and other objects of interest or value' found on the site or during excavation are the property of the employer. GC/Works/1 (1998), clause 32 (3) and ACA 3, clause 14 provide to much the same effect, but none of those clauses can affect the rights of third parties.
See also: **Antiquities; Fossils.**

Trespass A category of the law of tort (qv). There are several types of trespass, but trespass to land is of most concern to the construction industry. If a person enters upon, remains upon or allows anything to come into contact with the land of another, he is committing trespass. For there to be a cause of action, the person bringing the action must be in possession of the land. (Encroaching tree roots are not trespass but nuisance.) Trespass may take place under the land (e.g. foundations), on the surface of the land (e.g. fences and buildings generally) or in the air space for a reasonable height over the land (e.g. erecting cranes, but not aircraft flying over). In order to sue for trespass, there is no necessity to prove damage. Remedies are to take action for damages (if any) and/or an injunction (qv) to prevent continuance. Another remedy which must be exercised with care is forcible eviction if the trespasser has refused to leave peacefully.

A builder is said to have a licence (qv) to be upon the site of the works. He may become a trespasser if he remains on the land or leaves materials there after his work is finished or after his employment has been determined.
See also: **Access to neighbouring land; Occupiers' liability.**

Trust The holding of property by one person for the benefit of another. The property is vested legally in one or more trustees who administer it on behalf of others. The law relating to trusts is set out in a number of Acts of Parliament. Trusts were the creation of equity (qv).
See also: **Fiduciary.**

Trustee in bankruptcy A person who takes charge of all assets of a person who is declared bankrupt, and in whom the bankrupt's property vests. His general functions are specified in s. 305 (2) of the Insolvency Act 1986: to get in, realise and distribute the bankrupt's estate in accordance with the Act. This involves a number of duties:
— To gather in all the bankrupt's discoverable assets.
— To investigate and decide the creditor's claims.
— To distribute the proceeds of the assets according to the statutory order of preference.
See also: **Bankruptcy; Insolvency.**

Turnkey contract The term sometimes used to describe a contract where the contractor is responsible for both design and construction. Alternatively such contracts are called 'Package deal' contracts. They are more often encountered in the industrial field.

The term has no precise legal meaning[637] and its use is best avoided. The alleged advantages of such contracts are project cost, co-ordination and speed. Against this must be set the substantial disadvantage that the client is sometimes deprived of an impartial third-party check. 'Package deal' contracts are most suitable for specialist engineering fields where companies possessing highly developed expertise may offer such proposals as the only access to that expertise.
See also: **Design and build contracts.**

Two-stage selective tendering See: **Code of Procedure for Two-stage Selective Tendering 1996.**

---

[637]*Cable (1956) Ltd* v. *Hutcherson Brothers Pty Ltd* (1969) 43 ALJR 321.

Uberrimae fidei Of the utmost good faith. This expression is applied to a group of contracts where, contrary to the general rule, the party with knowledge of material facts must make full disclosure of those facts. Failure to do so makes the contract voidable (qv). Building contracts are not contracts *uberrimae fidei*. Nor are contracts for sale of goods. The requirement of utmost good faith applies to contracts of guarantee (qv), insurance (qv), partnership (qv) and certain others. If a contract is one of *uberrimae fidei* the party with special knowledge must disclose to the other every fact and circumstance which might influence him in deciding whether to enter into the contract or not.

See also: **Good faith; Misrepresentation.**

Ultra vires Beyond the powers. An act in excess of the authority conferred on a person or body whether by statute or otherwise. The doctrine is largely important in relation to the acts or contracts of local and other public authorities and companies. For example, local authorities may act *ultra vires* if they act in bad faith or exercise their powers for some unauthorised purpose[638]. An architect will act *ultra vires* if he acts outside the terms of his appointment or in excess of the powers conferred upon him by the building contract. The employer is not liable to the contractor for acts of his architect which are not within the scope of the architect's actual or apparent authority, though the architect may be personally liable for breach of warranty of authority (qv) or otherwise. In *Stockport Metropolitan Borough Council* v. *O'Reilly* (1978) the position under JCT terms was aptly summarised:

> 'An architect's *ultra vires* acts do not saddle the employer with liability. The architect is not the employer's agent in that respect. He has no authority to vary the contract. Confronted with such acts, the parties may either acquiesce in which case the contract may be *pro tanto* varied and the acts cannot be complained of, or a party may protest and ignore them. But he cannot saddle the employer with responsibility for them.'[639]

See: **Agency.**

Uncertainty A court may find that a contract (qv) or deed is void because it is unclear about the intentions of the parties. Certainty of terms is an essential requirement if there is to be a valid contract.

See also: **Interpretation of contracts.**

Undertaking This is a promise, usually made in the context of litigation or arbitration between solicitors (e.g. an undertaking to pay photocopying charges) or a promise made to the court, breach of which would amount to a contempt.

---

[638]See, for example, *Hazell* v. *Hammersmith & Fulham London Borough Council* [1992] 2 AC 1 (interest rate swaps); cf. *Kleinwort Benson* v. *Lincoln City Council* [1999] 1 AC 153.
[639][1978] 1 Lloyds Rep 595.

Unfair Contract Terms Act 1977 This statute, which came into force on 1 February 1978, imposes limits on the extent to which 'civil liability for breach of contract, or for negligence or other breach of duty, can be avoided by means of contract terms and otherwise...'. It deals with limitation of liability in contract and in tort. Contrary to common understanding, it does not outlaw 'unfair' contract terms as is often supposed. An important distinction is drawn between those who deal as 'consumers', i.e. private individuals, and those who are in business. The criteria for avoiding liability are more stringent for a businessman dealing with a consumer than a businessman dealing with another businessman. The main provisions are:

— Liability for death or injury caused by negligence can never be excluded by any term in the contract or any notice (for example, displayed on a building site). 'Negligence' includes both the tort of negligence (qv) and situations in contract where one party has a duty to behave with reasonable care and skill: s. 2 (1). Thus, a notice displayed on a building site disclaiming responsibility for injury howsoever caused will be totally ineffective if the injury to a visitor is caused through the contractor's negligence.

— Any other loss or damage due to negligence can only be excluded if it satisfies the Act's requirement of *reasonableness* (see below): s. 2 (2). It should be noted that it is unclear whether the statutory duty of care owed by an occupier under the Occupiers' Liability Act 1984 can be excluded altogether by means of an appropriately worded notice in the case of other entrants, e.g. trespassers[640].

— If one party deals as a consumer or not as a consumer but on the other party's *written standard terms of business* the other party cannot:
  - Exclude or restrict his liability in respect of any breach of contract; or
  - Claim to be entitled to do something substantially different from that which he contracted to do or to do nothing at all, unless he satisfies the reasonableness test.

This is so no matter what terms he includes in the contract: s. 3. This is an extremely important provision since it will affect any contract in the construction industry if one party can be said to be using his own written terms of business. The supply of goods is a common example where suppliers often have printed conditions. It is thought that the only main contract conditions to escape the provisions of the Act are the JCT and ICE forms, because they are negotiated between all sides of industry. Even these forms may fall under the Act if and insofar as they are amended by the employer to suit his special requirements. They would then become his written standard terms of business. For example, if an employer inserted a clause in JCT 98, clause 26, to the effect that he would not be liable for any claim for loss or expense above £10 000, it is unlikely that the court would support him if the contractor could prove that the employer had caused him £20 000 damage. Similarly, if a contractor attempted to show that a term in his standard terms

---

[640]See *Clerk & Lindsell on Torts*, 18th edn at para 10–72 to 10–74, Sweet & Maxwell.

of business allowed him to substitute an inferior material for what he had originally priced (say softwood in place of hardwood), he would be unsuccessful under this Act.

Section 6 is also important in a construction context. No exemption clause can exclude liability in respect of claims brought under s. 12 of the Sale of Goods Act 1979 (as to the title of the seller of the goods) and corresponding provisions in hire purchase contracts. Implied terms as regards description, sample or quality can only be excluded *if reasonable*. In consumer transactions the terms cannot be excluded at all. Section 7 is similar to section 6, but deals with transactions which do not fall under the Sale of Goods Act or hire purchase. Section 8 excludes all attempts to limit or avoid liability for misrepresentation (qv).

Section 10 makes any term in a contract ineffective if it attempts to exclude liability on another contract. Although the point appears not to have been directly tested in the courts, it appears that GC/Works/1 (1998), clause 51, which attempts to give the authority (employer) power to deduct monies owing on the contract from any sums due on any other contract, may be such a term.

The test of 'reasonableness' is important. It has to be applied at the time the contract was made or, in the case of an excluding notice, when the liability arose. Section 11 and Schedule 2 of the Act deal with reasonableness. Schedule 2 lays down the guidelines. The court is only required to have regard 'in particular' to them; they are not intended to be exhaustive. The burden of proof lies on the party who claims that a term is reasonable. The guidelines are:

— The strength of the bargaining positions of the parties relative to each other, taking into account (among other things) alternative means by which the requirement could have been met.

— Whether the customer was induced to agree to the exemption clause or could have made a contract with someone else omitting the term in question.

— Whether the customer knew or ought reasonably to have known of the term.

— Where the exemption clause only operates after non-compliance with a particular condition, whether at the time of the contract it was reasonable to expect that compliance would be practicable.

— Whether the goods were manufactured, processed or adapted to the special order of the customer. Section 11 (4) also provides added guidelines in the case of a party seeking to limit his liability to a specified sum. Regard must be had to the resources he could expect to be available to him to meet the liability and the extent to which he could obtain insurance cover.

The Act does not apply to international transactions or to certain types of contract, e.g. insurance. It does contain provisions, in s. 13, to prevent people evading or contracting out of its requirements. For example, attempts to evade the Act by limiting remedies or restricting rules of evidence or procedure are specifically prevented. But agreements to submit disputes to arbitration (qv) are expressly excluded from this section.

The Act is of great importance to the construction industry, relying as it does upon a mass of contracts, sub-contracts and standard conditions. The Act attempts to make people shoulder and not evade their responsibilities unless it is reasonable to do so. There have been a number of reported cases on the Act including the decision of the House of Lords in *Smith* v. *Eric S Bush*[641] which have amplified the guidelines for satisfying the Act's requirement of reasonableness. In that case, a disclaimer of any duty of care by a valuer engaged by a house mortgage lender was held not to satisfy the test of reasonableness so as to exclude a claim for negligence by house purchasers who had relied on the valuation.

Unincorporated association A group of people not incorporated, under royal charter or statute, and which has no legal existence independent of the members of the association. Common examples are partnerships (qv) and some members' clubs. While partners may sue and be sued in the name of the firm, most other unincorporated associations cannot be so sued. Usually, the best procedure is by way of a representative action when one (or more) of the individuals concerned is authorised to appear on behalf of the group as a whole. A judgment against representative defendants is binding on them all. See also: **Agency; Capacity to contract.**

---

[641][1989] 17 Con LR 1.

Valuation The process by which the quantity surveyor arrives at the value (qv) of the work carried out by the contractor. It normally involves visiting site and checking that the work has been carried out by visual inspection and/or measurement (qv).

Value The meaning of 'value' in the context of interim valuations is sometimes the subject of dispute. JCT 98, clause 30.2.1.1, for example, refers to the 'total value of work properly executed by the Contractor...'. The contractor's view of the matter is that the value is to be found by reference to the bills of quantities (qv) and he is entitled to receive payment for what he has done at bill rates plus a proportion of the preliminaries. This appears entirely fair and reasonable and it is the system most commonly followed in practice. It has been argued, very convincingly, that this system does not represent the value of the work to the employer.

From the employer's viewpoint the value of the work done by the contractor is the value of the whole contract less the cost of completing the work with the aid of another contractor (which would include additional professional fees) if the first contractor went into liquidation immediately after the issue of the certificate. This could result in a minus figure in the early stages of a contract[642]. Contractors argue that the retention fund is designed to take care of that sort of eventuality but the retention fund as provided in most modern contracts is quite insufficient to cover the additional cost involved in the finishing of a contract by another contractor.

Although there are difficulties in operating the latter system, not least the method of evaluating the cost of completion, it does have the merit of assuring the employer of adequate funds if the worst happened. However, in light of the decision of the Court of Appeal in *Townsend* v. *Stone Toms & Partners* (1985)[643] it is suggested that the first view is the better one.

Where the contract is one governed by all of the provisions of Part II of the Housing Grants, Construction and Regeneration Act 1996 (qv), the parties to the construction contract must make adequate provision for valuing the amount of each instalment payment made to the contractor. If they do not, then the value of each such instalment will be calculated according to the mechanism laid down in the Scheme for Construction Contracts (see: **Housing Grants, Construction and Regeneration Act 1996**). All standard form contracts now make relevant valuation provisions satisfying the Act. With the partial exception of MW 98, all variants of the JCT standard form contracts make particularly detailed provision for valuing the amounts due to the contractor both during and on completion of the works.

---

[642]It is unclear whether such an approach would now be possible in light of the payment provisions in the Housing Grants, Construction and Regeneration Act 1996 (qv).
[643](1985) 27 BLR 26.

Value Added Tax (VAT) A tax on purchases, charged by the seller or purveyor and in turn payable by the seller to HM Customs and Excise. Not all goods and services attract the tax and those that do may attract it at differing rates ranging from zero rating to the standard rate which currently stands at $17\frac{1}{2}\%$. Collection and payment of the tax to HM Customs and Excise is, therefore, notoriously difficult in the context of construction works. It may be that works of repair and alteration and new buildings could be standard rated whilst new dwellings and communal residential buildings and some new buildings for charitable use may be zero rated or entirely exempt from tax. Thus, a complicated construction project may involve a combination of exempt, zero and standard rated work. The situation is made yet more complicated by the fact that the contractor will be charging for his work in instalments, as the project proceeds, thus raising the difficult task of correctly apportioning the tax or the appropriate rate of tax to each aspect of the work on a month by month basis as interim certificates and payments fall due. Added to which, the regulations governing VAT are in any event constantly being revised.

For those reasons, contractors almost invariably express their estimate(s) or quotation(s) as VAT exclusive. The resulting contract sum (qv) is then likewise treated as VAT exclusive with special VAT provisions – such as those for use with the JCT standard forms of contract – being used to enable VAT transactions to be treated separately from other contractual payments. It is, therefore, important to check on the position before beginning any building work and before concluding any contract to ensure that the latest such provisions are used.

Where a contractor submits an estimate or quotation without any reference to whether the price is or is not VAT inclusive and where there is no indication that the contractor is registered for VAT, then it appears, following the decision in *Franks & Collingwood* v. *Gates*[644], the courts will hold the price to be deemed VAT inclusive. In that case, the contractor's quotation did not bear a VAT registration number and no mention was made as to whether the fixed and provisional sums set out in the quotation were inclusive or exclusive of VAT. Judge John Newey QC held, therefore, that the contract was for sums which were inclusive of VAT, saying; 'The quotation made no reference to VAT whatsoever. It was intended to be a competitive offer in respect of work to a single house.... It was perfectly proper for the plaintiffs to submit an inclusive offer, and in my view there is no reason why [the employer] should not regard it as such.' However, where the employer is in the construction industry and is aware of the custom of quotations being offered on a VAT exclusive basis, the position may well be viewed differently[645].

Under JCT 98, the position concerning the contractor's entitlement to VAT on his interim and final payments is dealt with by clauses 15.2 and 15.3 and by the Supplemental Provisions (the VAT Agreement) which offer two

---

[644](1983) 1 Con LR 21.
[645]*Tony Cox (Dismantlers) Ltd* v. *Jim 5 Ltd* (1997) 13 Const LJ 209.

alternatives. In summary, they are that:
(1) The contractor must, not later than seven days before the first interim certificate is due, notify the employer of the rate of tax that will become chargeable on the supplies to which that and all future interim and the final certificate will refer and an amount calculated at that rate will be added by the architect to each interim and final certificate then issued by him; or

(2) For the purposes of and prior to the issue of each interim and the final certificate the contractor will provide provisional assessment of each of the amounts on which VAT will be chargeable at zero rate and/or any other rates of VAT and shall specify each rate that is applicable. Unless the employer has reasonable grounds for objecting to that assessment he will then remit to the contractor the amount of tax to which the contractor is due, calculated by reference to the rates specified and provisional assessments given by the contractor, together with the substantive amount shown on the architect's certificate. In those circumstances, exclude reference to VAT.

It should be noted that, notwithstanding that the parties may have agreed under a JCT standard form contract to have any and all disputes or differences referred to arbitration and/or adjudication (qv), any and all disputes relating to the parties' liability for, or the amount of, VAT to be paid to or by the contractor are expressly excluded from those procedures and must be referred to the commissioners.

Value cost contract In this type of contract, the contractor is paid only a fee which fluctuates depending upon the actual cost of work compared with a valuation made on the basis of an agreed schedule of prices (qv). The fee is increased or reduced depending upon the contractor's success or failure in meeting the agreed valuation. The cost of the work is paid directly by the employer. A disadvantage is the complex accounting and measurement procedures required. The value cost contract is useful where a continuous programme of work is involved and time is at a premium.
See also: **Cost reimbursement contract; Management contract; Prime cost.**

Variation of price See: **Fluctuations.**

Variation order An outdated term still commonly used to describe an architect's instruction (See: **Instructions**) requiring alterations, additions or omissions to the quality, quantity or design of the works.

Variations Alterations, additions or omissions in work, materials, working hours, work space, etc.
See also: **Instructions; Variation order.**

Vesting and seizure The majority of building contracts contain clauses dealing with the ownership of materials and/or plant (see: **Vesting clause**). Some contracts also contain an express provision dealing with seizure (sometimes 'vesting and seizure') of materials and plant, usually on determination of the contract or in the case of forfeiture (qv). For example, GC/Works/1 (1998), clause 30 which is headed 'Vesting' is a vesting and seizure clause. Clause 30 (1) transfers ownership of 'the Works and any things brought on the site in connection with the contract which are owned by the contractor or by any company in which the contractor has a controlling interest or which will vest in him under any contract' though risk remains with the contractor.

The object of this provision (and similar clauses in other contracts) is to improve the employer's position in the event of failure by the contractor to complete the contract, especially where that failure is caused by the contractor's insolvency (qv). It transfers the property in both plant and materials to the employer and is effective to defeat claims made by the contractor's trustee in bankruptcy (qv), liquidator (qv), etc. until the contract is completed. However, although clause 30 (1) provides that plant, etc. 'shall become the property of and vest in the Employer', as regards things which will eventually be moved from site, the transfer is only temporary and so property will re-vest in the contractor on completion. Where the plant etc. is owned by third parties, the clause cannot be effective against the third party owner, even if it is the intention that it should be so.

Each such clause must be interpreted strictly and on its wording, against the background of the general law.

See also: **Title; Nemo dat quod non habet.**

Vesting clause A clause in a contract which deals with the transfer of property in goods and materials (qvv), e.g. JCT 98, clause 16. Subject to the important legal rights and liabilities that may be conferred on and against third parties by virtue of the Contracts (Rights of Third Parties) Act 1998 (qv), such a clause will otherwise only be effective as between the parties to the actual contract and cannot affect the rights of third parties, such as suppliers. (See: **Contracts (Rights of Third Parties) Act 1998; Privity of contract.**)

'Vesting clauses are inserted in contracts for the purpose of securing money advanced to the contractor or as security for the due performance of the contract'[646]. The effect of a vesting clause depends on its terms and also on the general law relating to the passing of property (qv). Even if the vesting clause is effective to transfer property in unfixed materials brought on site to the employer, this is qualified by the contractor's right to use the materials for the purpose of the works[647].

See also: **Title; Nemo dat quod non habet**.

Vexatiously With an ulterior motive to oppress or annoy[648]. It is always wrong to take action vexatiously and, in litigation, may cause an action to be

---

[646] *Emden's Building Contracts & Practice*, 8th edn, vol. 1, 1990, p. 336, Butterworths.
[647] *Bennet and White (Calgary) Ltd* v. *Municipal District of Sugar City No 5* [1951] AC 786.
[648] *John Jarvis Ltd* v. *Rockdale Housing Association Ltd* (1986) 10 Con LR 51.

dismissed. JCT 98, clause 27.2.4 is one of many examples in the standard forms which contain an express prohibition on vexatious action.

Vicarious liability The liability of one person for the wrongs done by another. The liability generally arises in tort (qv). The most common examples are the liability of an employer for the actions of his employee and that of a principal for the acts of his agent. There will be no liability, however, if the employee is acting outside the course of his employment or if an agent is acting outside the scope of his authority. In general, the employer is not vicariously liable for the wrongful actions of an independent contractor engaged by him. An employer is, however, liable for the actions of an independent contractor if he is negligent in selecting him, where there is a breach of an absolute statutory duty, and in certain other limited cases, e.g. where the contractor's work involves operations on the highway (qv) and injury is caused.
See also: **Agency; Master.**

Vicarious performance Performance of a contractual obligation by or through another person, e.g. performance of part of a contractual obligation by a sub-contractor (qv). English law draws a distinction between assigning duties (see: **Assignment)** and engaging someone else vicariously to perform them. Vicarious performance is generally permitted except when the nature of the contract calls for personal performance, which is not usually the case with building contracts, although it would be so in a case where the personality of the builder was important.

Vicarious performance is only effective to discharge the contractor's duties if it is perfect. If the vicarious performance falls below the prescribed contractual standard, the original contractor is liable.

This concept is largely important in the context of sub-contracting (qv) and sub-letting. Most standard forms of contract deal with this matter expressly and while such clauses prohibit vicarious performance of the *whole* contract, they permit it in part with the written consent of the architect or the employer.

JCT 98, clause 19 is typical. Clause 19.1 deals with assignment (qv) of the contract, while clause 19.2 deals with the contractor's right of sub-letting. This right can be exercised only with the written consent of the architect.

Vis major Irresistible force whether of nature or act of man. It can be equated with *force majeure* (qv) and covers any overpowering force such as exceptional storms, earthquakes, riotous mobs, armed forces. It is an excuse for damage done or loss of property and is one of the excepted perils in certain insurance policies.
See also: **Act of God.**

Vitiate To make invalid. The word is used in JCT 98, clause 13.2.5 and in IFC 98, clause 3.6 to indicate that the variations stated will not invalidate the contract.

MW 98 states the same thing in clause 3.6 though it uses the word invalidate instead of vitiate.

It should be noted that no action expressly allowed under the terms of a contract can invalidate that contract and the various standard forms appear simply to be stating the common law position. ACA 3 does not include a similar statement.

Void; Voidable Void means of no legal effect, or a nullity. Thus, an illegal contract (qv) is void *ab initio* (from the start) and cannot create any rights or obligations. A contract for an immoral purpose, e.g. to build a brothel, would be void at common law on grounds of public policy. In some cases the innocent party may be entitled to recover money paid or property transferred under a void contract, usually by way of *quantum meruit* (qv)[649].

Voidable, in contrast, means that the transaction is valid until one party exercises the right of rescission (qv), e.g. in the case of fraud (qv) or misrepresentation (qv). For example, a contract of partnership (qv) made by a minor is voidable at his option.

Voucher A document which is evidence of something. Thus JCT 98, clause 13.5.4 requires vouchers specifying the time spent upon the works, workmen's names etc. to be presented to the architect for verification. That type of voucher is commonly known as a 'daywork sheet'. ACA 3, clause 16.1A refers to the documents, vouchers and receipts necessary for computing the total amount due to the contractor.

---

[649] *Craven Ellis* v. *Canons Ltd* [1936] 2 All ER 1066.

Waiver The relinquishment of a right or remedy. It may be express, by a written statement (e.g. letter) to that effect, or implied, by inaction in enforcing a right. Care must be taken to avoid the latter situation which may easily arise on a building contract if the contractor commits a breach for which there is a clear remedy and the employer takes no advantage of the remedy. For example, if the contractor sub-lets part of the work without seeking the architect's consent in accordance with JCT 98, clause 19.2, the architect must immediately take action. If he does nothing, he may be said to have waived his right to object.

A waiver may be given by a planning authority in connection with satisfaction of the requirements of the Building Regulations (qv). Its effect is to remove the requirement to comply with the particular regulation to which it relates. The Secretary of State may also give a general waiver in certain circumstances.

War Open, armed conflict between two or more nations or states, with the object of satisfying a claim. The outbreak of war makes all commerce between British subjects and alien enemies illegal. Any contract with an enemy alien is automatically dissolved by the outbreak of war and even in other cases war may well cause frustration (qv) of the contract.

ACA 3, clause 11.5, Alternative 2 makes war or hostilities (qv) a ground for extension of time. It is also a ground for termination of the contractor's employment under clause 21 (c) if the contractor is prevented or delayed from executing the works for 60 consecutive days. A notice from one party to the other is all that is required.

GC/Works/1 1998 makes no specific provision for war, but under clause 58A, either the employer or contractor may determine the contract in the event that the whole or substantially the whole of the works are suspended for a continuous period (usually 182 days) through some reasonably unforeseeable circumstance outwith the parties' control. That right might well be exercised if war broke out. In any event, as has been indicated, war may result in frustration of the contract. The JCT Agreement for Minor Works MW 98 likewise has no specific provision.

See also: **Force majeure; Frustration; Hostilities**.

Warranty A subsidiary or minor term in a contract, breach of which entitles the other to damages (qv) but not to repudiate the contract. It should be contrasted with a condition (qv) which is a term going to the root of a contract. It is for the court to decide whether a contract term is a warranty or a condition. In *Thomas Feather & Co (Bradford) Ltd* v. *Keighley Corporation* (1953)[650], for example, a clause in a building contract forbidding sub-letting

---

[650](1953) 52 LGR 30.

without the employer's consent was held to be a warranty as opposed to a condition. The court takes account of all the circumstances including the intentions of the parties.

See also: **Collateral contract.**

Warranty of authority, breach of Although the general rule of an agency (qv) is that the agent is not liable personally to the third party, this is subject to an important exception. If the agent exceeds his actual authority and the third party suffers damage as a result, the agent will be liable to the contractor for breach of warranty of authority[651]. The architect's implied authority to bind his principal (the employer) is limited, but clearly if he exceeds his authority he is liable to the contractor in damages[652].

Wayleave A right of way (qv) over, under or through land for such things as a pipeline, an electric transmission line, or for carrying goods across the land. The word is often used as a synonym for an ordinary right of way whether on foot, with vehicles or otherwise. Many statutory authorities may apply to the appropriate minister for a compulsory wayleave over land where the owner refuses his consent. A wayleave is a kind of easement (qv).

Weather It can be a very important influence on the rate of progress of a job. Some contracts are more generous than others in giving the architect power to award extensions of time for delays caused by weather conditions. In the absence of such a provision, bad weather is at the contractor's risk unless it is of such magnitude as to amount to *force majeure* (qv).

See also: **Adverse weather conditions.**

Winding-up The process by which a limited liability company (see: **Limited company**) is brought to an end. The same term is used to describe the operation of putting an end to a partnership (qv). Under the Insolvency Act 1986 there are two broad categories of winding-up:
— Compulsory – by order of the court.
— Voluntary – either a *members'* or a *creditors'* winding-up. A voluntary winding-up may also be effected under the supervision of the court.

The winding-up of a limited company, except for the purposes of amalgamation or reorganisation of the company's structure, is a ground on which the contractor's employment may be determined under most standard forms of building contract, e.g. JCT 98, clause 27.3, ACA 3, clause 20.3, because of its connotations with insolvency (qv).

Without prejudice A phrase used in correspondence or discussions seeking to negotiate a compromise and settle a dispute. Statements made 'without

---

[651] *Yonge* v. *Toynbee* [1910] 1 KB 215.
[652] *Randall* v. *Trimen* (1856) 139 ER 1580.

prejudice' for the purpose of settling a dispute cannot be given in evidence without the consent of both parties. The courts may imply consent if a party, wishing to rely upon the privilege (qv), seeks simultaneously to reveal part of the document which is to their advantage. The basis of the privilege is to be found in an implied agreement arrived at from marking the letter 'without prejudice'[653]. It is important to note that 'without prejudice' statements and discussions will only be privileged if there is a dispute and an attempt to settle or compromise it. Architects and contractors alike must beware of heading letters 'without prejudice' indiscriminately, in the mistaken assumption that it gives them the opportunity to write whatever they wish with impunity. Equally, a letter may be truly 'without prejudice' but not headed as such. Although accurate labelling of correspondence is useful, a court will look at the substance of any correspondence to determine whether it really is 'without prejudice'. In arbitration proceedings a 'without prejudice' offer can never be referred to by either party at any stage of the proceedings, because it is in the public interest that there should be a procedure whereby the parties can discuss their differences freely and frankly and make offers of settlement without fear of being embarrassed by these exchanges if they do not lead to settlement.

Letters written 'without prejudice' which do not result in agreement cannot, therefore, be looked at by the court even on the question of costs, unless both parties consent[654]. There are very limited exceptions to the general rule and it has been held that an offer of settlement, made before trial of an action and containing a 'without prejudice' letter which expressly reserved the right to bring the letter to the notice of the judge on the issue of costs after judgment, is admissible without the consent of the parties. But that will be the case only where what is in issue is something more than a simple money claim in respect of which a payment into court (qv) is appropriate[655]. In *Rush & Tompkins Ltd* v. *Greater London Council* (1988)[656], the House of Lords held that 'without prejudice' privilege is not lost once settlement is reached. Hence, third parties are not entitled to discovery of 'without prejudice' material which might affect their claim.

The phrase 'without prejudice' is also used, in JCT 98, clause 26.6 to mean that the foregoing provisions are not to affect the contractor's common law rights, which are preserved.

See also: **Sealed offer.**

Witness A person who has seen or who can give first-hand evidence of an event or one who gives evidence (qv) in arbitration or litigation of events or facts within his own knowledge. A person who attests to the genuineness of a signature etc. is also described as a witness.

See also: **Attestation; Evidence; Expert witness; Witness summons; Oath.**

---

[653] *Rabin* v. *Mendoza & Co* [1954] 1 All ER 247.
[654] *Computer Machinery Co Ltd* v. *Drescher* [1983] 3 All ER 153.
[655] *Cutts* v. *Head* [1984] 1 All ER 597.
[656] [1988] 3 All ER 18.

Witness summons With the introduction of the Civil Procedure Rules (CPR) (qv) this term now replaces the terms *subpoena ad testificadum* and *subpoena duces tecum*. It is an order made by the courts in a prescribed form, and generally but not necessarily served by the court, requiring a person to attend at court to give evidence or to produce a specified document to the court on the date of the hearing, or such other date as the court directs.

A witness whose attendance is ordered in such a way is entitled to, and must when served with the summons be offered, a sum reasonably sufficient to cover his expenses in attending court and is entitled to a specified rate of compensation for time lost in attending as directed[657]. Failure to obey such an order without proper excuse amounts to a contempt of court.

Work and materials contract Building contracts are contracts for work and materials, which means that they are not subject to the provisions of the Sale of Goods Act 1979. The distinction between contracts for the sale of goods and those for work and materials was formerly more important than it is today and there is a large volume of case law on the topic, much of which is confusing. See also: **Cost reimbursement contracts.**

Working drawings The drawings which the contractor will use to construct the works. They will be accurately dimensioned and together with the specification (qv) or bills of quantities (qv), will contain all the information the contractor requires. In practice, schedules and tables may be included in the term. The SFA/99 and CE/99 Plan of Work substitutes two stages, E: Final proposals and F: Production information, for what used to be called the working drawing stage of the architect's work. See also: **Drawings and details.**

Workmanship Skill in carrying out a task. Building contracts commonly use the word to differentiate between goods and materials and the work done on them to produce the finished building. Thus, JCT 98, clause 2.1 refers to the quality of materials and the standards of workmanship. In the absence of an express term to the contrary, the contractor is under an obligation at common law to carry out his work in a good and workmanlike manner. Express terms of the contract sometimes impose a higher obligation, e.g. ACA 3, clause 5.4.

Workpeople The 1998 edition of JCT fluctuations clauses for use with the JCT 98 Private version of the standard form contract, at clause 38.6.3, uses and defines the term 'workpeople' for the purposes of those price fluctuations as being 'persons whose rates of wages and other emoluments ... are governed by the rules or decisions ... of the Construction Industry Joint Council or some other wage fixing body for trades associations within the building industry'. However, labour-only sub-contractors are not workpeople for the purposes of this provision[658].

---

[657]Civil Procedure Rules Part 34.
[658]*J. Murphy & Sons Ltd* v. *London Borough of Southwark* (1982) 22 BLR 41.

Works The operation on site required to produce a building or structure. Works includes not only the building itself at various stages of construction but also all ancillary works necessary such as scaffolding, site huts, temporary roads etc. even though they may not form part of the finished structure. Most building contracts draw a distinction between 'the Works' and 'work'. Thus, under JCT 98 'the Works' means either the whole of the work contracted for, to the extent shown on, described by or referred to in the contract documents (qv) (article 1), or the site, as in clause 8.6. In contrast, 'work' means 'work carried out under the contract' as in clause 13.1.1.1. Whereas JCT 98 (article 1) and IFC 98 (1st Recital) each, in effect, provide that 'the Works' comprises everything shown or described within the documents forming the contract – including contract drawings, specifications etc., notably – and rather ambiguously – MW 98 seemingly extends the meaning of the term so that, by article 1 it not only covers all work shown, described or referred to in the contract documents but also encompasses 'any changes made to that work in accordance with (the) Agreement'. Although when read literally together with article 2 that then suggests that the contract sum must include for the future costs of any and all alterations, additions and/or omissions to the quality or quantity of the work shown on the contract drawings, specifications etc., there can be little doubt that on a proper construction of the contract that will not, in fact, be how the contract will be interpreted.

Works contractor A contractor engaged by a management contractor to carry out a specific parcel of work as part of a larger project. The term is used in the JCT Standard Form of Management Contract 1998 and in the associated Works Contract documentation: Works Contract/1, /2 and /3.

It is the job of the management contractor to advise the professional team about the best way to break down the total work in the project into suitable work packages (Third Schedule of MC 98, clause 6). Under clause 8.2 the management contractor and the architect must agree in writing on the selection of the works contractors to carry out the work. This agreement is generally achieved by inviting tenders from suitable contractors on Works Contract/1, section 1. The tenders are submitted on Works Contract/1, section 2 and the contractor submitting the accepted tender is required to enter into Articles of Agreement (Works Contract/1, section 3), which incorporate the Works Contract/2 Conditions. The Works Contract Conditions have many similarities in wording to NSC/C and JCT 98.

Works contractors are superficially in the same position as sub-contractors (qv) under a traditional contract, but the management contract contains valuable protections for the management contractor in case of defaults on the part of the works contractors. These protections are principally to be found in clause 3.21 of the management contract.

Writ A largely outdated term referring to an order issued in the name of the Queen requiring the performance of an act. The term most commonly referred to the means of originating proceedings, that is to say the means by which actions in the High Court were commenced (i.e. by way of a *writ of summons* generally referred to merely as 'a writ'). In that context, since April 1999 the process of originating proceedings in the High Court is governed by the Civil Procedure Rules (qv) and from that date onwards only those claim forms (qv) that comply with the Rules will be issued by the court. If an outdated writ of summons is presented to the court by a party wishing to originate proceedings, other than in the most exceptional circumstances it will be returned unissued.

The term does, however, remain in limited use under the Civil Procedure Rules where it still applies to certain other specific types of writ such as, for example:

— Writ of execution, issued for the purposes of enforcing a judgment or order of the court, including the enforcement of any order made in insolvency proceedings in the High Court and for enforcement of an order for payment of costs.

— Writ of *habeus corpus*, issued for the purposes of releasing someone unlawfully imprisoned.

— Writ of possession, issued for enforcement of an order in relation to the possession of land.

— Writ of sequestration where the court seizes assets in contempt proceedings.

Writing Many building contracts require certificates, notices, instructions etc. to be given in writing, e.g. ACA 3, clause 23. 1. This requirement is satisfied by any process which represents the words in visible form and includes hand-writing, typewriting, printing etc., although a particular contract may distinguish writing from printing. In contrast, NEC, clause 13 requires that where, under the contract, any instruction, certificates etc. is to be 'communicated' from one party to the other, that communication must be in a form capable of being read, copied and recorded. If in writing, then the contract specifically calls for it to be written in the language of the contract as specified in the contract data. As a noun, 'writing', means a document produced in permanent form as contrasted with oral communication. Certain contracts are required to be in writing, e.g. assignments of copyright (qv) or to be evidenced in writing, e.g. contracts for the sale of land.

See also: **Electronic data interchange; Management contract; Notices.**

Y

Year A period of 12 calendar months or 365 consecutive days in ordinary years or 366 days in leap years. From 1 January 1753 in England the year has commenced on 1 January. The regnal year commences on the accession of the Sovereign. The income tax year runs from 6 April to 5 April and the Government financial year runs from 1 April to 31 March. The accounting year of limited companies (qv) runs from any date convenient to the company.

BIBLIOGRAPHY

Beale, H. G. *et al.* (1999) *Chitty on Contracts*, 28th edn, and First Supplement (2000). Sweet & Maxwell.

Bernstein, R., Tackaberry, J. & Marriot, A. L. (1998) *Handbook of Arbitration Practice*, 3rd edn. Sweet & Maxwell.

Burns, A. (ed) (1994) *Legal Obligations of the Architect*. Butterworths.

Carnell, N. (2000) *Causation and Delay in Construction Disputes*. Blackwell Science.

Chappell, D. (1996) *Contractual Correspondence for Architects and Project Managers*, 3rd edn. Blackwell Science.

Chappell, D. (2000) *Understanding JCT Standard Building Contracts*, 6th edn. E & F Spon.

Chappell, D. (1998) *Powell-Smith & Sims' Building Contract Claims*, 3rd edn. Blackwell Science.

Chappell, D. & Powell-Smith, V. (1998) *JCT Minor Works Form of Contract*, 2nd edn. Blackwell Science.

Chappell, D. & Powell-Smith, V. (1999) *JCT Intermediate Form of Contract*, 2nd edn. Blackwell Science.

Chappell, D. & Powell-Smith, V. (1999) *JCT Design and Build Contract*, 2nd edn. Blackwell Science.

Chappell, D. & Willis, A. (2000) *The Architect in Practice*, 8th edn. Blackwell Science.

Cornes, D. (1994) *Design Liability in the Construction Industry*, 4th edn. Blackwell Science.

Cox, S. & Hamilton, A. (1998) *Architect's Handbook of Practice Management*, 6th edn. RIBA Publications.

Dugdale, A. M. (2000) *Clerk and Lindsell on Torts*, 18th edn. Sweet & Maxwell.

Eggleston, B. (1998) *Liquidated Damages and Extensions of Time*, 2nd edn. Blackwell Science.

Furmston, M. P. (2000) *Powell-Smith & Furmston's Building Contract Casebook*, 3rd edn. Blackwell Science.

Greenstreet, R., Chappell, D. & Dunn, M. (2002) *Legal and Contractual Procedures for Architects*, 5th edn. Architectural Press.

Harris, B., Plantrose, R. & Tecks, J. (2000) *The Arbitration Act 1996*, 2nd edn. Blackwell Science.

Jackson, R. & Powell, J. (1997) *Professional Negligence*, 4th edn. with Third Cumulative Supplement (2000). Sweet & Maxwell.

Lavers, A. & Chappell, D. (2000) *A Legal Guide to the Professional Liability of Architects*, 3rd edn. A. Lavers.

Lupton, S. (1999) *Guide to JCT 98*. RIBA Publications.

Jones, N. F. (1998) *Set-Off and Adjudication in the Construction Industry*, 2nd edn. Blackwell Science.

MacRoberts (2000) *Scottish Building Contracts*. Blackwell Science.

Marshall, D., Johnston, L. & Chappell, D. (2001) *Which Form of Building Contract?* 2nd edn. E & F N Spon.

Murdoch, J. & Hughes, W. (2000) *Construction Contracts*, 2nd edn. E & F N Spon.

Mustill & Boyd (1989) *Commercial Arbitration*. Butterworths.

Parris, J. (1986) *Effective Retention of Title Clauses*. Collins.

Powell-Smith, V., Sims, J. & Dancaster, C. (2000) *Construction Arbitration*, 2nd edn. Blackwell Science.

Ramsey, V. & Furst, S. (2000) *Keating on Building Contracts*, 7th edn. Sweet & Maxwell.

Redmond, J. (2001) *Adjudication in Construction Contracts*. Blackwell Science.

Reiss, G. (1996) *Project Management Demystified*, 2nd edn. E & F N Spon.

Trickey, G. & Hackett, M. (2001) *The Presentation and Settlement of Contractors' Claims*, 2nd edn. E & F N Spon.

Walker, A. (1996) *Project Management in Construction*, 3rd edn. Blackwell Science.

Wallace, I. N. D. (1995) *Hudson's Building and Engineering Contracts*, 11th edn. Sweet & Maxwell.

TABLE OF CASES

The following abbreviations of reports are used:

AC, App Cas	Law reports, Appeal Cases
All ER	All England Law Reports
ALJR	Australian Law Journal Reports
B & S	Best and Smith's Reports
BCC	British Company Cases
BCL	Building and Construction Law
BLM	Building Law Monthly
BLR	Building Law Reports
Ch, Ch D	Law Reports, Chancery Cases/Division
CILL	Construction Industry Law Letter
CLD	Construction Law Digest
Comm Cas	Commercial Cases
Con LR	Construction Law Reports
Const LJ	Construction Law Journal
CP	Common Pleas
DLR	Dominion Law Reports
EG	Estates Gazette Cases
Ex, Exch, Ex D	Law Reports, Exchequer Cases/Division
H & C	Hurlstone and Coltman's Exchequer Reports
JP	Justice of the Peace & Local Government Review
KB	Law Reports, King's Bench
LGR	Local Government Reports
Lloyds Rep	Lloyds' Reports
LT	Law Times Reports
M & W	Meeson and Welsby's Exchequer Reports
NY	New York Court of Appeal
NZLR	New Zealand Law Reports
PD	Law Reports, Probate Division
QB	Law Reports, Queen's Bench
RVR	Rating and Valuation Reports
SCR	Supreme Court Reports
Sc LR	Scots Law Reports
TLR	Times Law Reports
WLR	Weekly Law Reports
WR	Weekly Reporter

Table of cases